Advances in
ECOLOGICAL RESEARCH

VOLUME 25

Advances in

ECOLOGICAL RESEARCH

Edited by

M. BEGON

Department of Environmental and Evolutionary Biology,
University of Liverpool, UK

A.H. FITTER

Department of Biology, University of York, UK

VOLUME 25

ACADEMIC PRESS

Harcourt Brace & Company, Publishers
London
San Diego   New York   Boston
Sydney   Tokyo   Toronto

ACADEMIC PRESS LTD
24/28 Oval Road
London NW1 7DX

United States Edition published by
ACADEMIC PRESS INC.
San Diego, CA 92101

British Library Cataloguing in Publication Data
Advances in ecological research.
Vol. 25
1. Ecology
I. Begon, Michael
574.5

ISBN 0-12-013925-1

This book is printed on acid-free paper

Typeset by Fakenham Photosetting Ltd,
Fakenham, Norfolk
Printed in Great Britain by T. J. Press (Padstow) Ltd,
Padstow, Cornwall

Contributors to Volume 25

D. ATKINSON, *Population Biology Research Group, Department of Environmental and Evolutionary Biology, University of Liverpool, P.O. Box 147, Liverpool L69 3BX, UK.*

M.G.R. CANNELL, *Institute of Terrestrial Ecology, Edinburgh Research Station, Bush Estate, Penicuik, Midlothian EH26 0QB, UK.*

R.C. DEWAR, *Institute of Terrestrial Ecology, Edinburgh Research Station, Bush Estate, Penicuik, Midlothian EH26 0QB, UK.*

S.L. GUTSELL, *Division of Ecology, Department of Biological Sciences and the Kananaskis Field Stations, University of Calgary, Calgary, Alberta, Canada T2N 1N4.*

J.S. HANAN, *Cooperative Research Centre for Tropical Pest Management, CSIRO Division of Entomology, PB No. 3, Indooroopilly, Q4068, Australia.*

M.J. HUTCHINGS, *School of Biological Sciences, University of Sussex, Falmer, Brighton, Sussex BN1 9QG, UK.*

E.A. JOHNSON, *Division of Ecology, Department of Biological Sciences and the Kananaskis Field Stations, University of Calgary, Calgary, Alberta, Canada T2N 1N4.*

H. de KROON, *Department of Plant Ecology and Evolutionary Biology, University of Utrecht, P.O. Box 800.84, 3508 TB Utrecht, The Netherlands.*

L. MAILLETTE, *Department of Biological Sciences, University of Calgary, 2500 University Drive, Calgary, Alberta, Canada T2N 1N4. Present address: Département de Chimie-Biologie, Université du Québec à Trois-Rivières, CP 500, Trois-Rivières, Québec, Canada G9A 5H7.*

P.M. ROOM, *Cooperative Research Centre for Tropical Pest Management, CSIRO Division of Entomology, PB No. 3, Indooroopilly, Q4068, Australia.*

Preface

It is to be hoped that the bad old days are far behind us when theoretical ecologists, or ecological modellers, field biologists and other empiricists were warring factions, with a relative few spanning the "cultural" divide. Every one of the contributions to this current volume draws on theoretical and empirical work in moving towards a new synthesis.

In the first paper, Atkinson examines data from a wide range of taxa in search of a general rule governing the effects of rearing temperature on the size of ecotherms at any given stage of development. In 83.5 percent of cases, increased temperature led to reduced size, but, as is so often the case, the exceptions were at least as informative as the case conforming to the "general" rule. Past explanations appear to be either erroneous or very limited in their applicability, but a number of new, more general hypotheses suggest exciting opportunities for future work.

For the individual plant, there can be few more fundamental processes than the allocation of carbon to different plant parts. Focusing on trees, Cannell and Dewar take specific functional relationships, in each of which the division of growth between two plant parts can be regarded as being constrained to meet some necessary balance. Many of the ideas have been proposed by modellers, but their evaluation depends on the observations of physiologists. The authors bravely attempt to draw the communities together, but, as they remark, this may result in their having stones thrown at them from both sides.

The following two papers then examine aspects of the disposition of plant parts in space. Room, Maillette and Hanan discuss different ways of modelling the dynamics of plant modules and metamers, remarking that the environmental effects at this intermediate structural level have been neglected relative to internal physiology and the biomass of whole stands. Having reviewed the data to determine what is required, they enter the computer world of "virtual reality" and point to an imminent new era of ecological research into spatial dynamics.

Hutchings and de Kroon then focus more specifically on the functional significance of plant growth with respect to plant foraging. Their (facetious?) references to Triffids, plants that walk, illustrate graphically that an approach that has been familiar in the study of animals is being used here, in a powerful attempt to understand the benefits and disadvantages of different resource-acquiring syndromes, and their ecological and evolutionary consequences.

In the final paper, Johnson and Gutsell reaffirm the now widely-recognized importance of spatial and temporal heterogeneity, and disturbance, in their review of the study of wildfires. They carefully draw together the models of fire frequency, the methods of sampling design, the analysis of data and the empirical work that gives content to the mathematical models. They, like the other authors in this volume, show clearly the mutual interdependence of each approach.

A.H. Fitter
M. Begon

Contents

Temperature and Organism Size—a Biological Law for Ectotherms?

D. ATKINSON

Carbon Allocation in Trees: a Review of Concepts for Modelling

M.G.R. CANNELL and R.C. DEWAR

Module and Metamer Dynamics and Virtual Plants

P.M. ROOM, L. MAILLETTE and J.S. HANAN

Foraging in Plants: the Role of Morphological Plasticity in Resource Acquisition

M.J. HUTCHINGS and H. DE KROON

Fire Frequency Models, Methods and Interpretations

E.A. JOHNSON and S.L. GUTSELL

Temperature and Organism Size—A Biological Law for Ectotherms?

D. ATKINSON

I. SUMMARY

For plants, protists and that vast majority of animals that rely on external sources for body heat (ectotherms), temperature is a particularly important

ADVANCES IN ECOLOGICAL RESEARCH VOL. 25
ISBN 0–12–013925–1

2   D. ATKINSON

and widespread correlate of differences in size between seasons and from one generation to the next. This includes the economically important differences between years in the yields of crops harvested at maturity such as cereals.

Laboratory studies under controlled conditions can isolate the specific effects of temperature and help identify just how similar is the relationship between organism size and (non-extreme) rearing temperature among different species. This chapter reviews the direct effects of rearing temperature on the size of ectotherms at any given stage of development, other environmental conditions being controlled, and within the range of temperatures which allow the organism to reach maturity but are not so stressfully high that an increase causes reduction in the rates of growth, differentiation or both. It thus considers phenotypic plasticity in organism size, and does not examine genetic variation related to temperature differences.

One hundred and nine studies, which included examples of animals, plants, protists and a bacterium, showed a significant effect of rearing temperature on size. In 83·5% of cases, increased temperature led to reduced size at a given stage of development. Only 11·9% of examples showed a size increase, and 4·6% showed a mixture of increases and decreases with temperature. Whilst some of these exceptions to the general size-reduction rule may result from unreported weaknesses in experimental protocol or inappropriate temperature conditions, several detailed studies still appeared to provide genuine exceptions. Because the rule applied to organisms of many different taxa, habitats and lifestyles, general, rather than *ad hoc* explanations were sought.

The explanation for the general rule offered by von Bertalanffy (1960)— that catabolism has a higher temperature coefficient than does anabolism, leading to faster growth and smaller ultimate size at high temperature—is shown strictly to be incorrect, though elaborations on the original hypothesis may be worth testing. No particular support was found for (and a little evidence was found against) the idea that, as size and temperature increased, growth became increasingly constrained due to oxygen shortage in aquatic and water-saturated habitats, or to desiccation in terrestrial habitats.

Of three hypotheses which considered the adaptive value of size at different temperatures, only the advantages for buoyancy in planktonic species could not be eliminated. But because this only applied to a group of aquatic species, other, more general explanations were sought.

Several correlates of temperature that should influence optimal rate of development and size have been identified. In particular, three promising areas require further study: (i) the relationship between temperature and individual growth rate, size at a given developmental stage and mortality. Mortality factors of particular interest are increased predation, accelerated ageing, future risks of drought (for terrestrial species), future risks of oxygen shortage (for aquatic species) and costs of rapid growth at different temperatures; (ii) the use of seasonal cues to adjust relative rates of growth and development; and (iii) the extent to which size is affected by the rate of population growth at different temperatures. A hypothesis related to these latter two problems is based on the advantages of shortening the life cycle when conditions are favourable for population growth or when time available is constrained by season.

Some quantitative effects of temperature on size in natural and managed populations are briefly reviewed.

II. INTRODUCTION

For animals, plants, protists and bacteria, the ecological importance of organism size is clearly demonstrated by its effects on longevity, fecundity, metabolic rates, ability to migrate, competitive, predatory and anti-predator abilities, and ability to withstand starvation and desiccation (e.g. Dingle et al., 1980; Peters, 1983; Calder, 1984; Schmidt-Neilsen, 1984; Reiss, 1989). Its economic value is indicated by the effects of plant size on agricultural crop yields (Monteith, 1981; Warrick et al., 1986).

Temperature has an important effect on many functions relating to an organism's size. Metabolic rates, rates of gaseous exchange, risks of desiccation in terrestrial habitats or oxygen shortage in aquatic habitats are all affected by both temperature and size. This is sometimes because an increase in size without a change in shape decreases the ratio between surface area for exchange of heat, water and gases and the weight or volume of the body that produces or consumes those resources (Schmidt-Neilsen, 1984).

Over 30 years ago, following a series of experiments on a wide range of animals and protists together with a review of previous studies, Carleton Ray (1960) proposed that a general relationship existed between the temperature at which a poikilothermic organism is reared and its subsequent final size. (Body temperatures of poikilotherms vary according to environmental temperature; almost all these species are also ectothermic.) Specifically, he proposed that at increased rearing temperatures, a smaller body size is produced (see also reviews by Bĕlehrádek, 1935; von Bertalanffy, 1960; Precht et al., 1973). Much of the evidence, starting with the work of Standfuss (1895) on Lepidoptera, comes from laboratory studies and field comparisons over small distances.

This relationship appears to mirror that found for endotherms by Bergmann (1847). Bergmann proposed that, intraspecifically, body size decreases with average ambient temperature and hence tends to increase with latitude. However, since the evidence for the relationship in endotherms is based largely on geographical variation, differences may be largely genetic rather than a direct effect of rearing temperature as is observed for ectotherms. Moreover, the original explanation for "Bergmann's Rule", which invokes that large-bodied animals have a superior ability to conserve heat, is inappropriate for most poikilotherms whose body temperatures fluctuate rapidly and markedly as environmental temperatures change (von Bertalanffy, 1960; Calow, 1977). The only widely reported hypothesis for ectotherms is that of von Bertalanffy (1960) who proposed that growth becomes constrained at high temperatures because the rate of anabolism is fundamentally unable to keep up with that of catabolism.

But doubt still remains about whether a general relationship between rearing temperature and ectotherm size even exists. Some authors have produced arguments and data which appear to contradict the relationship, and have cited further counter examples (Calow, 1973; Roff, 1981; Lamb and Gerber, 1985). In one review of the effects of environmental temperature on final size of poikilothermic animals, Laudien (1973, p. 389) states that "individual animal species react in so many different ways that general statements cannot be made".

There is therefore a need to review the evidence critically and assess alternative hypotheses which may account for the effects of raising temperature on ectotherm size. This paper attempts to do this for protists, bacteria, ectothermic animals and plants and then examines the quantitative effect of temperature on size in some natural and managed populations.

III. PARAMETERS OF THE PROBLEM, AND METHODS

A. The Problem Defined

The problem under review here is concerned only with the direct ontogenetic effects of temperature on organism size, and not with genetically distinct strains showing different responses to temperature. Stressfully high or low temperatures or insufficient energy or other resources can obviously produce small size. Yet reviews have suggested that high temperature reduces size even in conditions favourable for growth (Ray, 1960). Also, size should be compared at the same stage of development. Therefore, the problem was defined so as to exclude unfavourable conditions for growth and inappropriate comparisons, thus:

The choice of upper temperature limits, indicated above, should exclude conditions stressful for growth. If organisms did not reach maturity at extreme low or high temperatures, or if daily mortality rate increased as temperature was reduced to very low values, despite reduced metabolic activity, these temperatures were considered stressful and were therefore also excluded from the review.

Not all studies recorded rates of differentiation, and very few recorded rates of growth. But, providing that energy and other resources are unlimiting, both these rates usually increase with temperature within the range

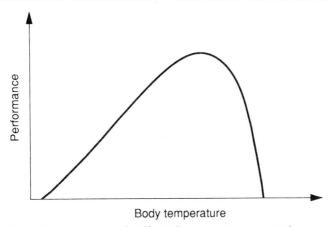

Fig. 1. The hypothetical asymmetric effect of temperature on ectotherm performance (after Huey and Kingsolver, 1989). "Performance" may apply to growth rates, rates of differentiation or other activities. At high temperatures enzyme systems start to break down.

normally encountered by the organism since physiological performance usually increases gradually until a sudden decline near the upper limits when enzyme systems become damaged (Fig. 1; Huey and Kingsolver, 1989). Therefore, the problem can be expressed thus:

> *Under controlled environmental conditions in which energy and other resources are not limiting, at temperatures which allow the organism to reach maturity but not so high that rates of growth, differentiation or both decline—how does rearing temperature affect organism size at a given stage of development?*

Another way to consider the problem is: does an increase in temperature, whilst increasing rates of growth and differentiation, affect the latter more than the former so that body size at a given stage of development is reduced?

B. Scope of Review

The review focused on ectothermic animals, plants—most of which are ectothermic (Gates, 1980)—and protists. One bacterial example was found, so this was also included.

Besides the predominance of English-language references, apparent biases in the review may have arisen because size-at-stage data appear to be recorded particularly commonly in species with easily observed developmental stages and events (e.g. moults), and which cease growing in the adult stage.

Almost every study lacked at least some relevant data (e.g. mortality rates, photoperiod, food quality), so analyses were also performed to investigate the effects of gaps in the data set.

C. Criteria for Inclusion of Studies

1. Size-at-stage Measures

Ideally, the dry or wet weights of an organism at the time of a particular developmental event (e.g. final metamorphosis) would precisely define both the size and developmental stage. However, other correlates of organism weight such as body length, measures of exoskeletal dimensions, insect wing length and, in unicells, cell length or volume, were also accepted in cases where weight measurements were not made. Occasionally, in arthropods, when a paper did not indicate the particular time within a growing developmental stage in which size was measured, it was preferable to use exoskeletal measures rather than weights, since they reflect more accurately the size of the animal at the previous moult (a precise developmental event).

In species in which additional juvenile stages are optionally inserted according to environmental conditions (e.g. Bellinger and Pienkowski, 1987), the size of early stages may correlate particularly poorly with size when at the time of later (and necessary) developmental events such as reproductive maturity. Early stages will also have had less time growing under the experimental temperatures. For these reasons, in studies in which several developmental stages were examined, data from later stages, including adults, were preferentially selected.

In unicells, size at initiation of fission was only available from one study (Adolph, 1929). To increase sample size of protists, cruder measures such as average size over a period of several generations were also included.

2. Experimental Conditions

An experiment was excluded from the review if the amount of energy provided appeared to be limiting in any of the treatments (e.g. if all food was eaten between feeds; if light levels produced slower plant growth than at higher levels), or if treatments had different types of food provided or different photoperiodic regimes. This includes cases in which there was evidence that temperature itself may have altered food quality (e.g. Minkenberg and Helderman, 1990).

Results from animal studies were not included also if the food was both unnatural and caused a reduction in growth rate below that observed on natural food. An example of this is the mayfly, *Leptophlebia intermedia*, which, when reared on unnatural food, also experienced higher larval mortality and produced significantly smaller adults at each temperature than on natural food (Sweeney *et al.*, 1986).

If an excess amount of suitable food is provided for heterotrophs their growth should not be limited by energy supply. Yet for some unicellular algae growth is limited not only by low light intensities, and hence insufficient energy, but also by very high light intensities. Moreover, the high light intensity at which growth retardation starts to occur can increase with increasing temperature, as was found in the planktonic alga *Cryptomonas erosa* (Morgan and Kalff, 1979). Thus it may be impossible to find a single light intensity that will enable some species to grow at each of several temperatures without energy supply limiting growth at at least one of them. Studies which demonstrated that the light intensities which limited growth differed with temperature were not included in the analysis (Morgan and Kalff, 1979; Meeson and Sweeney, 1982).

3. Statistical Significance

Only experiments in which statistical tests had been performed were included in the initial data analysis.

In some studies in which no statistical tests had been performed by authors, it was still possible to perform a rank correlation of mean sizes against temperature using the Hotelling-Pabst statistic (Conover, 1980) or calculate 95% confidence limits, and hence test for significance. If these studies showed significant effects of temperature they were added to the data set for analysis.

Since statistical non-significance ($p > 0.05$) was sometimes associated with very small sample sizes, such results were noted but not considered in detail. Studies without statistical tests, but which showed trends with temperature, were noted in the discussion only.

IV. A BIOLOGICAL LAW FOR ECTOTHERMS?

A. Summary of Data

1. Overall Result

Of 69 experiments in which statistical tests were performed by the authors, only five (7·2%) showed no significant effect of rearing temperature (Ray, 1960; Poston *et al.*, 1977; Palmer, 1984; Pechenik and Lima, 1984; Sims *et al.*, 1984). At least two of these (Ray, 1960 [for *Xiphophorus maculatus*]; Palmer, 1984) had low sample sizes (fewer than nine per temperature).

The subsequent analysis combined the 64 results in which authors found significant effects of temperature, with the 45 others for which the investigation later was able to demonstrate statistical significance. Of these 109 studies (see Appendix), 91 (83·5%) showed a significant reduction in size with

8 D. ATKINSON

an increase in rearing temperature (e.g. Fig. 3), 13 (11·9%) showed an increase (e.g. Fig. 4), and five (4·6%) showed a mixed effect (i.e. combinations of significant increases with significant decreases within the range; e.g. Fig. 5). Thus of the 104 experiments in which a simple significant effect

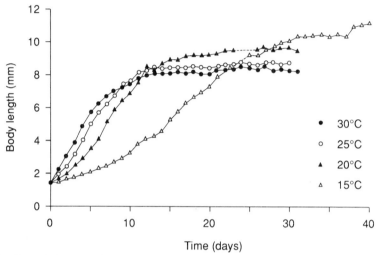

Fig. 2. Growth in mean body length of larvae of the midge *Chaoborus* at different temperatures (after Hanazato and Yasuno, 1989). This shows the typical effect of temperature when resources are not limiting; growth rate increases but final size is reduced with increasing temperature.

(an increase or a reduction) occurred, there is clearly a significantly greater number of reductions in body size than there are increases ($\chi^2 = 58·5$, $p < 0·001$). The proportion of cases giving reductions (83·5%) is similar to the 80% reported by Ray (1960) who applied less strict criteria for accepting studies for his review than were used here.

It is important to note that a decrease in temperature causes the majority of ectotherms to attain a larger size at a given stage of development despite their growing and developing slower. These temperatures which produce large size are often near the bottom of the range normally encountered and appear to be suboptimal (see also Figs. 1 and 2, and Section VI). Thus the widespread reduction in size with increased temperature appears not to be caused by increased stress or a deterioration in enzyme performance with temperature increase since temperatures are apparently well below these stressful levels.

2. Association with Taxonomic Group

A reduction (or increase) in size at a given stage of development caused by increased temperature may be confined only to certain taxa. To test this idea,

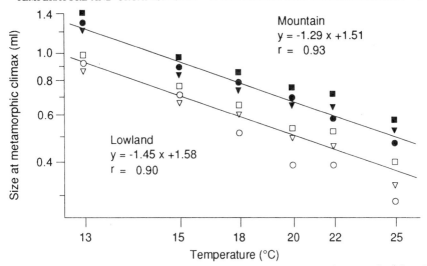

Fig. 3. Temperature dependence of size at metamorphic climax of mountain (closed symbols) and lowland (open symbols) wood frogs, *Rana sylvatica*. Each symbol equals the mean volumes of 24 larvae. Data are presented for frogs obtained from three lowland and three mountain ponds (after Berven, 1982b).

the results were analysed according to the taxonomic groups to which the organisms belonged.

Of the 109 studies, one was a bacterium, seven were of protists, six were of multicellular plants, and 95 were of animals (Fig. 6; Appendix). Of the seven protist examples five were autotrophs and two were heterotrophs. All but one of the protists showed a reduction in size with increased rearing temperature, as did the six plant examples and the bacterium *Pseudomonas*. The only protist exception was the marine diatom *Phaeodactylum tricornutum* which increased in size with increasing temperature (Fawley, 1984). The remaining 12 increases and five cases of mixed effect were found within the animal kingdom.

Eighty of the 95 animal examples were arthropods (Fig. 7). Among animals, only the arthropods showed instances of significant increases in, or significant mixed effects on body size with temperature (Fig. 7).

Effects of temperature on size of multicellular plants, at particular stages of development rather than at a fixed time of harvest, are not frequently cited. Examples are confined to crop plants belonging to the grass and pea families, so the sample is rather limited taxonomically. In addition to the studies of five plant species which followed the general rule (Appendix), Friend *et al.* (1962) found that the weight of wheat plants at anthesis generally decreased with increasing temperature, though statistical tests that might have confirmed the trend were not done. Similarly, whilst no significant tests were performed by

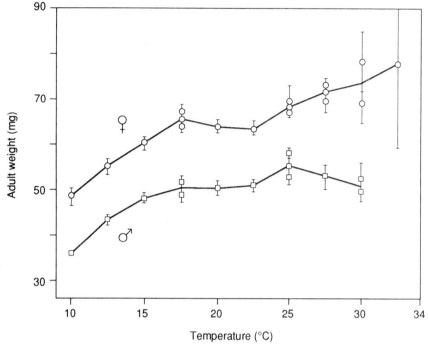

Fig. 4. Average adult weight (± SD) for the beetle *Entomoscelis americana* reared at different temperatures (after Lamb and Gerber, 1985).

Thorne *et al.* (1968), they found that dry weights of ears, shoots and roots at anthesis appeared to decline with increasing temperature, though no obvious effect on plant size was observed at spikelet initiation.

However, studies on maize, whilst most did not satisfy statistical criteria for inclusion in the Appendix, appear to show conflicting results. Consistent with the general rule were leaf weight at a particular developmental stage (determined by leaf number) measured by Brouwer *et al.* (1973) and plant size at anthesis recorded by Hunter *et al.* (1977). But Hardacre and Eagles (1986) and Hardacre and Turnbull (1986) found that plant size at each of three stages (determined by leaf number) was greatest at intermediate temperatures, and Gmelig Meyling (1969) showed that the effects of temperature on plant size at first flowering depended on sowing date.

Whilst all the plants were angiosperms, the animal phyla in which only significant reductions in size were reported, were represented by several classes or subphyla. The chordates included one fish and seven amphibian examples; the aschelminthes included two rotifers and a nematode; and the molluscs included two gastropods and two bivalves.

Two classes of arthropod were represented: the Insecta (67 of the 80 cases),

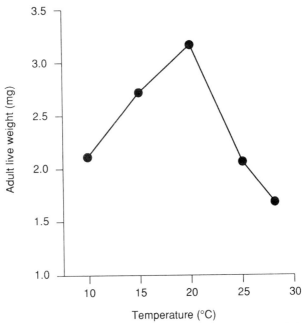

Fig. 5. Effect of rearing temperature on the live weight of apterous adults of the aphid *Acyrthosiphon pisum*. The data come from 15 Australian lines of the aphid (after Lamb and MacKay, 1988).

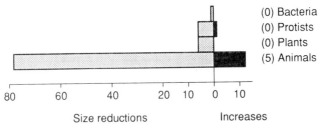

Fig. 6. Distribution of size reductions and increases in response to rearing temperature among four kingdoms (based on data in the Appendix). Numbers in parentheses represent the number of cases in which both size reductions and increases were recorded in the same study.

and the Crustacea (13 cases). Of these, 11 of the 12 cases of increase in size with temperature and all five cases of mixed effect were found among the insects. The only crustacean showing a significant increase in size with temperature was the parasitic copepod, *Salmincola salmoneus* (Johnston and Dykeman, 1987).

Increases in size or mixed effects of temperature were observed in six of the

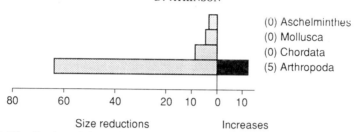

Fig. 7. Distribution of size reductions and increases in response to rearing temperature among four animal phyla (based on data in the Appendix). Numbers in parentheses represent the number of cases in which both size reductions and increases were recorded in the same study.

seven insect orders (Fig. 8). Only in the Orthoptera were no examples of reductions in size found, but here the sample size was only two.

In each of the orders Orthoptera, Coleoptera and Ephemeroptera, more than a third of examples showed an increase in size. These belonged to two families of Orthoptera, three families of Coleoptera and four families of Ephemeroptera (Table 1), and were therefore not confined just to a single atypical family within each order. One family of the Lepidoptera, the Noctuidae, included two instances of a decrease in size, and also one of an increase.

Of the 32 dipteran examples, 30 showed reduced size, a figure which included 13 of the 14 studies of *Drosophila*. The discrepancy between Roff's (1981) claim that in *Drosophila* "phenotypic size increases with temperature" and the findings of the present review is due to his erroneous representation of the literature. Three of the references he cited (Eigenbrodt, 1930; Stanley, 1935; Tantawy and Mallah, 1961) actually found the opposite of what he claimed, and the fourth (McKenzie, 1978) did not examine size.

All simple increases in size except for an isolated diatom example, and all of the mixed effects of temperature were therefore confined to the Insecta. Even so, over three-quarters of insect examples still showed reductions in size with increasing temperature.

3. Association with Sex

If reductions or increases in size are confined mainly to one sex, then explanations related to sex differences at different temperatures may be involved.

In 37 of the 109 examples, statistical tests were performed on males and females separately. Table 2 shows that in 31 of these the effect was the same for both sexes. In no species was a significant increased found in one sex and a significant reduction or mixed effect found in the other. So there is no evidence that reductions or increases are confined mainly to one sex.

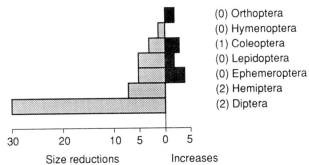

(0) Orthoptera
(0) Hymenoptera
(1) Coleoptera
(0) Lepidoptera
(0) Ephemeroptera
(2) Hemiptera
(2) Diptera

Size reductions   Increases

Fig. 8. Distribution of size reductions and increases in response to rearing temperature among seven insect orders (based on data in the Appendix). Numbers in parentheses represent the number of cases in which both size reductions and increases were recorded in the same study.

Table 1

Effects of temperature increase on size-at-stage in different families of three insect orders

ORDER	No. of cases with size:		
Family	reduction	increase	mixed response
COLEOPTERA			
Chrysomelidae		1	
Dermestidae		1	1
Cerambycidae		1	
Scolytidae	1		
Tenebrionidae	2		
EPHEMEROPTERA			
Ephemerellidae		1	
Caenidae		1	
Oligoneuriidae		1	
Tricorythodidae		1	
Leptophlebidae	1		
Baetidae	1		
Siphlonuridae	3		
ORTHOPTERA			
Acrididae		1	
Gryllidae		1	

4. Association with Cell Size

Laudien (1973) observed that an increased temperature can reduce cell size as well as size of the organism. The relationship between size and temperature found in most species may therefore be explained if the following two premises are true: (i) an increase in temperature causes a reduction in cell

D. ATKINSON

Table 2
Significant effects of temperature increase on size-at-stage on different sexes

Sex	No. of significant size:			
	reductions	increases	mixed responses	Total
M (+ not F)	1	0	0	1
F (+ not M)	3	2	0	5
Both M + F	26	3	2	31
Total	30	5	2	37

M, males; F, females.

size; and (ii) for a given species, individuals at precisely the same stage in development contain the same or similar numbers of cells.

Consistent with the first premise are six of the seven protist examples (Adolph, 1929; Margalef, 1954; Ray, 1960) and the bacterium (Chrzanowski et al., 1988). Other studies of protists, whilst not backed up by statistical tests, are also consistent with the premise (Mučibabić, 1956; James and Read, 1957; James and Padilla, 1959; Johnson and James, 1960; Thormar, 1962; Donnan and John, 1984). Zeuthen (1964) reviewed work on Tetrahymena pyriformis (mainly by Thormar, 1962) showing that cells are smallest at an intermediate temperature which was the optimum for growth and division. Therefore, within the temperature limits considered in the present paper, cell size in protists and the bacterium reduced with increasing temperature.

No obvious effect of temperature was observed in cell size of the prokaryote Salmonella typhimurium (Schaecter et al., 1958). Like some unicellular eukaryotes, some prokaryotes exhibit increased cell sizes at temperatures which are so high that division is inhibited (Dowben and Weidenmuller, 1968).

Support for a reduction in cell size with temperature increase in multicellular animals is provided by the finding that Drosophila reared at lower temperature not only are heavier (Alpatov and Pearl, 1929; Eigenbrodt, 1930; Ray, 1960; Economos et al., 1982), but also have wings that are larger, due almost entirely to increased cell size rather than number (Alpatov, 1930; Robertson, 1959; Masry and Robertson, 1979; Cavicchi et al., 1985). However, Masry and Robertson (1979) also found that when larvae were shifted from 25 °C to 29 °C for a 12-hour period and then returned to complete their development at 25 °C, the reduction in wing size compared with that observed at a constant 25 °C was caused by either reduced cell size or reduced cell number depending on when during larval development the transfer took place. Whilst this does not negate the dual effect of temperature on cell and organism size, it does suggest that the potential exists for high temperature to

lead to the production of a smaller body by mechanisms other than a reduction in cell size.

The complexity of relationship between temperature, cell size and organism size is further illustrated by studies of other taxa. Evidence from multicellular animals and plants suggests that temperature affects the size of actively
dividing cells differently from that of differentiated cells which they subsequently produce. For example, the coelenterate *Cordylophora* was larger
and had larger cells in the growth zone when reared at 10 °C than at 20 °C, but
other cells changed shape without a noticeable size change with temperature
(Kinne, 1958). Cuadrado *et al.* (1989) described an increase in cell size with
temperature increase in root meristem cells of *Allium cepa*, but observed no
effect on cells of the differentiated portions of the root. Consistent with this
difference between actively growing and mature cells is the apparent lack of
effect of temperature on mature, fully differentiated root cells of *Zea mays*
(Erickson, 1959), *Pisum* (Van't Hof and Ying, 1964), *Allium cepa* (Lopez-
Saez *et al.*, 1969) and *Helianthus* (Burholt and Van't Hof, 1971). Despite this,
Burström (1956) reported that the final lengths of root cells of wheat decreased with increased temperature, though statistical tests were not performed. Moreover, Grif and Valovich (1973) are reported, by Francis and
Barlow (1988), to have found no difference in meristematic cell length in
roots of *Secale cereale* between 1 °C and 23 °C.

Overall, the idea that the effect of temperature on organism size is just a
consequence of a universal effect on cell size, and that an explanation should
be sought at the level of cell physiology, is not well supported by the data.

B. Possible Limitations of the Data Set

Non-significant results may be under-represented in the sample because of
the following possibilities: (i) a reluctance of researchers to publish non-
significant results; (ii) a reluctance to test for significant differences between
individual temperatures if no overall trend appears evident; and (iii) a
tendency for authors to overlook non-significant results when citing other
studies that relate to their own statistically significant findings.

Even if a general relationship does exist, a figure of 100% reductions would
hardly be expected given the inadequacies of the available data set. The
following numbers of cases for which no values were available illustrate this:
rates of differentiation, 25 (22·9%); rates of growth 81 (74·3%); and rates of
mortality 80 (73·4%). In the absence of such data temperatures which,
unknown to me, contravened the acceptance criteria (Section III.C) may
have been included in the review. For example, cases in which small size was
associated with high juvenile mortality may have been due to inadmissibly
stressful temperatures (e.g. Tsitsipis, 1980).

Rank correlations are more likely to identify significant increasing or decreasing trends rather than significant mixed size responses (significant increases *and* decreases between temperatures) over a range of several temperatures. Significant mixed size responses will thus be under-represented in the results.

Photoperiods were not stated in 44 cases (40·4%). Thus day length, which may affect the amount of time available for feeding or photosynthesis per day and which can provide information about the time of year, may not have been controlled in some cases. The effects of photoperiod and seasonality are discussed further in Section VI.C.

Animal studies in which the food both was unnatural and was observed to reduce growth rate below that found on natural food were excluded where possible. Yet since many studies did not indicate whether food quality was adequate, some invalid examples may have been included in the review. It is noteworthy that the mayfly *Leptophlebia intermedia* not only suffered higher juvenile mortality and slower growth on the poor diet (Section III.C), but produced larger adults at high rather than low temperatures: this is in contrast to the reduction with increasing temperature found on natural food (Sweeney *et al.*, 1986). Other comparisons show similar effects of unnatural and poor quality food on size (Dixon *et al.*, 1982; Galliard and Golvan, 1957).

Light energy was not demonstrated to be unlimiting in the studies of unicellular algae. Also, the effects of possible genetic adaptation to the temperature regimes were a possibility in studies in which the organism had spent several generations at the experimental temperatures.

Exceptions to the general trend may be caused by the inclusion of inappropriate measures of size-at-stage. Yet the proportions of exceptions found when indirect measures of size were used (e.g. body length, width or length of exoskeletal part) was never much greater, and often less, than when direct size measures (dry weight, live or wet weight) were made. Exceptions were also found over a wide range of developmental stages: reproductive maturity, final juvenile moult, pupa, other juvenile moult, adult, at maximum juvenile weight, average of all stages (in some unicells). Proportions of exceptions were not related to how early in the life cycle the measurements were taken. In addition, although no examples of hatchlings were included in the present analysis, the general reduction in size with increased temperature appears to hold for various fish fry hatching from eggs (reviews by Bělehrádek, 1935; Ray, 1960; Laudien, 1973). However, the size of hatchling locusts (*Schistocerca gregaria*) (as measured by hind femur length; Bernays, 1972) were significantly increased following incubation at high temperature, and effects on alligator hatchling size do not seem to show a simple trend (Deeming and Ferguson, 1988). So there is no evidence that exceptions to the general rule are particularly associated with a particular stage of development. When size and stage are considered together, the 18 exceptions comprised nine different types of size-at-stage measure.

There is thus no evidence that the increases or mixed size response to temperature resulted primarily from the choice of a small number of inappropriate size-at-stage measures.

Overall then, because the data available are so incomplete and their quality so variable, some apparent exceptions to a biological law may result from unreported weaknesses in experimental protocol or inappropriate temperature conditions. Yet a small number of detailed meticulous studies still appear to provide genuine exceptions to such a "law" (e.g. Guppy, 1969; Lamb and Gerber, 1985).

Whether or not the relationship turns out to be a universally applicable "law", rather than just a "rule" that can be demonstrated for the majority of cases (Mayr, 1956), the effect of temperature on size at a given stage of development still requires an explanation. Few have been offered hitherto. These will now be critically evaluated and new explanations will be proposed. Since the relationship is so widespread across different taxa, habitats and lifestyles, general rather than individual *ad hoc* explanations will be sought.

V. EXPLANATIONS INVOKING CONSTRAINTS ON GROWTH

A. A General Reduction in Growth

The present review is restricted to temperatures over which individual growth was observed or expected (see Fig. 2) to increase with temperature. Growth rate was actually observed to increase with temperature in 26% of cases. Despite this, it is still possible that the general trend was largely an artefact due to growth constraints at high temperature in the 74% of cases in which growth was not reported. To test this, the proportion of cases showing reductions in size at increased temperature was examined only among those studies in which growth rates were reported, and hence known to increase with temperature.

Of the 28 cases for which growth rates were known to increase, 22 (79%) showed a significant reduction in size at high temperatures, and six (21%) showed an increase—results not very different from the proportions in the whole data set. There is no evidence, therefore, that the high proportion of reductions in size at high temperature was due primarily to the inclusion of invalid studies in which growth rates were reduced at increased temperatures.

B. Von Bertalanffy's Hypothesis

Von Bertalanffy (1960) suggested that in general terms, catabolic processes were mainly of a chemical nature and would therefore have a high temperature coefficient. Conversely, he argued, anabolic processes ultimately depend on physical processes such as permeation and diffusion, which would likely have a low temperature coefficient. He also argued that anabolism was limited potentially by the rate of intake of substances such as respiratory gases and hence by the size of the areas through which they were absorbed. On the other hand, catabolism was proportional to body weight. Von Bertalanffy produced a growth equation which expresses the rate of growth as the difference between anabolism and catabolism:

$$dw/dt = mw^a - nw^b \qquad (1)$$

in which w is weight, t is time, and m, n, a and b are indices specific to particular combinations of genotype and environment: m, which affects anabolism, is almost constant in response to temperature, but n, which affects catabolism, increases with temperature. The value of a depends on the shape of the organism and specifically on the surface areas through which food or respiratory gases are absorbed. In the simplest case, the surface area of a spherical (say unicellular) organism will increase at a rate of 2/3 the power of body weight. In other species, such as insects, a appears to be nearer to 1 (von Bertalanffy, 1957, 1960). Von Bertalanffy asserted that if catabolism had a higher temperature coefficient (n) than did anabolism then an increase in temperature would increase growth rate and reduce final size. Yet, strictly speaking, the former cannot be true since if a rise in temperature increases catabolism more than anabolism then the rate of growth will decrease. A predicted decrease in final size does, however, follow logically from the above premises. Thus the explanation proposed by von Bertalanffy for a small final size at increased temperature does not strictly apply to the problem under investigation as defined in Section III.A.

Von Bertalanffy's argument can, however, be modified slightly to produce increased growth rates at moderately high temperatures (S. M. Wood, personal communication). If rate of anabolism increases linearly with temperature, and the rate of catabolism increases according to a power function, then growth rates may sometimes increase at first with temperature (Fig. 9). In such a case anabolism may be described algebraically as:

$$m = c+d(T-T_0) \qquad (2)$$

in which m is the same as in equation 1, c and d are constants, T is temperature and T_0 is a threshold growth temperature. Catabolism may be described as:

$$n = k(T-T_0)^i \qquad (3)$$

where k and i are constants. Assuming certain values for the parameters of

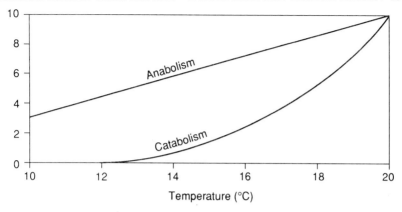

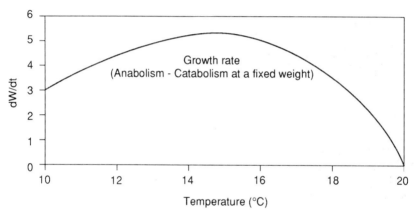

Fig. 9. Hypothetical effect of temperature on anabolism and catabolism (upper graph) based on equations (2) and (3), respectively. Anabolism (m) increases linearly with increasing temperature (T) above at threshold ($10\,°C$) according to the equation $m = 3 + 0·7(T-10)$. Catabolism (n) increases according to the equation $n = 0·01(T-10)^3$. The lower graph shows the combined effect of these curves on growth rate (S.M. Wood, personal communication).

the equations, growth rate, which is the difference between anabolism and catabolism, may at first increase (Fig. 9).

More generally, the highly simplified growth equation (1) described by von Bertalanffy summarizes many different processes. Clearly, there are potentially many other ways in which temperature may constrain rates of growth. Thus whilst the overall effect of temperature increase may be to increase rates of both growth and differentiation, a physical or chemical constraint at high temperatures could still inhibit one activity more than the other.

However, constraints described by von Bertalanffy may be overcome if the surfaces through which metabolites pass, the efficiency of their transport, or both (represented by a in equation (1)) are able to increase greatly at high temperatures. This still requires testing.

In conclusion, the explanation originally given by von Bertalanffy is not supported on theoretical grounds. Alternative explanations are possible, but these require testing.

C. Temperature and Plant Growth

Although plant growth rates are generally increased by increased temperature, this effect is confined mainly to early growth (van Dobben, 1962; Friend *et al.*, 1962; Krol *et al.*, 1984; Grace, 1988) and is due primarily to increased rates of leaf expansion rather than enhanced photosynthetic production per unit area of leaf (Monteith, 1981; Grace, 1988). Over a wide temperature range the rate of plant development increases approximately linearly with temperature (e.g. Roberts and Summerfield, 1987; Ellis *et al.*, 1990). Consequently, increases in plant size at increased temperatures would be expected particularly at early developmental stages when leaf expansion is relatively important. This temperature sensitivity may explain some of the exceptions to the general trend found in early but not late stages of wheat and maize development (Section IV.A.2). However, no unavoidable physiological constraints on growth and development are known which might explain why the sensitivity of plant development rate to temperature prior to maturity is greater on average than is the rate of growth, thereby producing smaller mature plants. Adaptationist explanations (Section VI) which consider the adaptive significance of the relatively greater temperature sensitivity of development need also to be considered.

D. Oxygen Shortage

An oxygen shortage which limits growth at high temperature may be more likely in aquatic environments. The oxygen-carrying capacity of water is reduced as temperature is increased (Wilber, 1964) and may potentially limit growth. But if individual growth rate is still higher, on average, at increased temperature, then it must also be assumed that the oxygen shortage reduces growth rate increasingly as the size of aquatic organisms increases, so growth would slow down and cease at smaller sizes. This constraint would operate if oxygen requirements increase faster than does the rate of oxygen uptake which itself may be limited by the area through which respiratory gases pass.

Even terrestrial species may conceivably experience a shortage of oxygen at increased temperatures. Since cytoplasm and body fluids are aqueous, they will be less capable of holding onto dissolved oxygen, so more effort would have to be directed to obtaining sufficient oxygen from the air. But since the air itself should not be particularly depleted of oxygen at increased temperatures, oxygen shortage should not be as acute as in aquatic environments. The oxygen shortage hypothesis assumes therefore that at increased temperatures, oxygen is in short supply, that (intraspecifically) large organisms are less able to meet their requirements, and that growth is affected more than is differentiation.

Sixty-eight of the 109 studies were on organisms in potentially water-saturated habitats such as ponds, streams and soil (Appendix): 61 of these showed a reduction in size with increased temperature and seven showed an increase. Potentially water-saturated habitats contained slightly but not significantly more of the reductions in body size than did terrestrial habitats ($\chi^2 = 0.73$, $p > 0.20$). Thus, these results do not lend particular support to this oxygen-shortage hypothesis as the primary cause of size reductions. To test for the effects of oxygen shortage at increased temperatures, factorial experiments incorporating different temperatures and different oxygen concentrations (e.g. using an aquarium oxygenator) could be performed.

Again, this explanation assumes that respiratory activities and surface areas are unable to increase enough to compensate for increased oxygen requirements of large organisms at high temperature: and again, this requires testing.

E. Desiccation

In terrestrial habitats, growth may become constrained at high temperature due to desiccation. But if individual growth rate is still higher, on average, at increased temperature, then it must also be assumed that the water shortage reduces growth increasingly as organism size increases, so growth would slow down and cease at smaller sizes. If desiccation is an important part of the explanation for the general effect of temperature on size then terrestrial forms should show the effect more strongly than should aquatic ones. Of the 49 terrestrial examples in the Appendix, 37 (75·5%) showed a reduction in size with increased temperature: this is a lower proportion of the reductions than was found in the total sample. Thus there is no evidence that the main cause of size reductions at high temperature in the whole sample was the detrimental effects of desiccation on growth. Moreover, several studies of terrestrial species indicated that water was unlikely to be limiting (e.g. plants grown in nutrient solution (Brouwer et al., 1973), or said to be grown under conditions with "near optimum moisture content" (Gmelig Meyling, 1969); relative humidities controlled at levels of 80% or more (Muthukrishnan and Pandian, 1983; Wagner et al., 1987)).

In natural populations, whilst warm conditions may promote rapid growth at certain times, they may simultaneously be correlated with *future* drought or oxygen shortage. This possibility will be considered in discussions of adaptive explanations (Section VI.C.1).

VI. ADAPTIVE EXPLANATIONS

A. The Nature of the Adaptation to Temperature

Adaptive explanations, which suggest *why* size should vary with temperature, can complement mechanistic explanations which suggest *how* growth and differentiation are affected by the immediate environment. Variations in life history between organisms can be produced by genetic differences and by environmentally induced differences among individuals of a given genotype. The latter phenomenon, phenotypic plasticity, is thus exhibited by organisms reared at different temperatures. Indeed, among members of a clone, it is the only source of observed variation. The set of phenotypes that a single genotype could develop if exposed to a specified range of environmental conditions is called a Norm of Reaction (Stearns and Koella, 1986). Phenotypic plasticity may allow an organism to utilize information present in prevailing environmental conditions to adjust its resource allocation in an adaptive manner between different activities (Atkinson, 1985; Atkinson and Thompson, 1987) though some of the phenotypic variation may be imposed by immediate environmental constraints. The adaptive nature of phenotypic plasticity, including attempts to model norms of reaction for age and size at maturity, have been the focus of considerable interest in recent years (Stearns and Crandall, 1984; Via and Lande, 1985; Stearns and Koella, 1986; Dodson, 1989; Stearns, 1989; Ludwig and Rowe, 1990; Rowe and Ludwig, 1991; Gomulkiewicz and Kirkpatrick, 1992; Houston and McNamara, 1992; Kawecki and Stearns, 1993).

Within a species large size is often associated with high survivorship and fecundity (see Section II). But the increased sizes at reduced rearing temperatures are associated with low rates of population increase and low individual fitness, at least compared with individuals at increased temperatures which also increase rates of individual growth and differentiation (Cooper, 1965; Orcutt and Porter, 1984; Foran, 1986; Kindlmann and Dixon, 1992). This runs counter to the notion of an optimal thermal regime described by 'Sweeney and Vannote (1978) and Vannote and Sweeney (1980). In their thermal equilibrium hypothesis a thermal regime was considered optimal when an individual's weight and fecundity were maximized. In these terms there is a paradox because conditions seem to be favourable for individual growth and for fitness at high temperature, so why become small at a given developmental stage and suffer its apparently detrimental effects on fecundity and survivorship?

When trying to solve this problem, it is useful to recall that relevant selection pressures may act on different but correlated facets of the life cycle. Thus selection may operate directly not only on size at a given developmental stage, but also on the growth rate required to achieve this, and on the rate (or duration of previous stages) of development (Table 3). Size at one stage of the life cycle may also be closely correlated with sizes at other stages. For example, large offspring or dispersers may be favoured by natural selection but these sizes may also be correlated with size at maturity or, in animals, at the time of the last juvenile moult. Thus Berven (1982a) found that the large adult wood frogs occurring at high altitudes and low temperatures produced large offspring, and he hypothesized that adult size resulted indirectly from selection on offspring size in cold environments.

Possible general explanations for the rule relating temperature to size are provided by life-history theories. Those predictions which include potential likely effects of temperature are summarized in Table 3 and discussed in the following sections.

Other life-history predictions which currently lack a mechanism incorporating an effect of temperature are listed here to encourage the ingenious reader to seek mechanisms. An increase in final adult size is favoured when:

(i)   the relationship between the development time and size becomes more nearly linear rather than convex (when viewed from below) (Sibly and Calow, 1986a);

(ii) the relationship between fecundity and size becomes more nearly linear rather than concave. Thus if an organism, by maturing at a large size can achieve a large increase in fecundity for a small increase in development time, then a large size would be favoured (Sibly and Calow, 1986a);

(iii) size-specific fecundity is reduced (Roff, 1981).

Two other hypotheses can be dismissed fairly promptly. Roff (1981) predicted, using a model which he applied to *Drosophila*, that an increase in rate of development would favour an increased body size. However, this hypothesis is clearly unimportant in the majority of ectothermic organisms since an increased temperature increases the rate of development yet usually reduces size at maturity. Likewise, the idea that in terrestrial habitats the effects of temperature on size are due primarily to the selective advantage of small size in desiccating conditions associated with high temperatures has no empirical support. Desiccating conditions favour large size: small individuals appear to suffer from desiccation more than large ones, at least in cricket frogs (Nevo, 1973), fruit flies (Barker and Barker, 1980), hatchling grasshoppers (Cherrill, 1987), and ants (J.H. Cushman, personal communication). This is probably due to their large surface area in relation to their volumes making them prone to losing a high proportion of their body water. It is this advantage of large size that Schoener and Janzen (1968) suggested could partly explain geo-

Table 3
Conditions favouring changes in size-at-stage and in rate of development: potential effects of temperature

Trait	Conditions favouring increase in trait	Potential effects of temperature	Section of chapter
Size at a given stage of development	If survival, fecundity or both increase with size	Large individual plankton may be better able to: (i) move through cool water which is more viscous than is warm water; (ii) maintain a similar level of buoyancy at low as at high temperature (at high temperature, sinking rate can be increased by the relatively low density of the liquid medium; small size can help reduce sinking rate).	VI.B.1
		Risks of predation will vary with temperature: protection from this may be achieved by being big	VI.B.2
		Increased size may confer greater ability to achieve high body temperature when amount of solar radiation is low. (Conversely, small size in sunny conditions may help avoid overheating)	VI.B.3
Rate of development	High mortality associated with conditions during juvenile period	(i) Predation by ectotherms (but no endotherms) is likely to increase during an unseasonally warm period (ii) Total predation is generally higher in warm seasons (iii) Temperature increases rate of ageing (iv) High temperature may be associated with future drought (terrestrial species), oxygen shortage or habitat loss (aquatic species) (v) Rapid growth at high temperature may incur a cost	VI.C.1
	Rapid increase in population size	Fast individual growth when resources are unlimiting and temperature is high may correlate with population growth	VI.C.2
	Time constraints imposed by season or need for synchrony of developmental event	Slow-growing individuals (at low temperatures) which must reach a particular developmental stage by a particular time, will do so at the expense of size at that stage	VI.C.2 VI.C.3

graphical interspecific variation in insect size. So desiccation stress at high temperature seems unable to account for the general reduction in size-at-stage in terrestrial ectotherms.

B. Temperature and Size

1. Size, Water Viscosity, and Buoyancy

The production of large individuals at low temperatures may be an adaptive response to the high viscosity and hence resistance of liquid media to movement by small organisms at these temperatures (Loosanoff, 1959). No data were found to test this hypothesis precisely. A quantitative analysis of this problem would also be useful, to help assess how important this selection pressure is likely to be. If the hypothesis is true and important there is likely to be selection pressure also to put a high proportion of resources into loco-motory organs and streamlining at low temperatures. It has been speculated that this idea may explain why appendages found on species of dinoflagellates in tropical waters are longer than those on their counterparts in arctic waters (see Walsby and Reynolds, 1980). But Walsby and Reynolds (1980) also reported that no evidence for a direct effect of temperature on appendage length had been found from studies with cloned dinoflagellates in culture.

Another effect of increased size, at least among unicellular plankton, is that rates of sinking are increased (Eppley et al., 1967; Walsby and Reynolds, 1980). Also, since warm water is less dense than cold water, buoyancy will be reduced, although the importance of this needs to be ascertained (Walsby and Reynolds, 1980). Thus at high temperature, plankton may be able to offset some of the effects of reduced buoyancy caused by reduced water density (and also viscosity) by reducing their size. Even if the effect of temperature on water density is negligible, the explanation may be adapted to those unicellular algae which at low temperature cannot tolerate as high a light intensity as they can at high temperature (Morgan and Kalff, 1979; Meeson and Sweeney, 1982) though in other species no effect of temperature on size was evident (Yoder, 1979; Meeson and Sweeney, 1982). In these species, an increased size at reduced temperatures would allow the algae to sink to levels where light intensities are no longer too high. This is consistent with the effects of temperature on size of at least two species (Morgan and Kalff, 1979; Meeson and Sweeney, 1982). Also, no advantages to the pennate diatom *Phaeodactylum tricornutum* are evident from the greater sinking rate associated with large size when subjected to high temperatures (Fawley, 1984), since these gave the fastest rates of carbon fixation and cell division. Finally, since a separate explanation is still needed for terrestrial ectotherms, a more comprehensive explanation will be sought.

2. Size and Predation Risk

Tests of hypotheses which predict how mortality rates affect optimal size (e.g. Williams, 1966; Wilbur and Collins, 1973; Roff, 1981; Stearns and Koella, 1986; Kozlowski and Wiegert, 1986; Ludwig and Rowe, 1990; Rowe and Ludwig, 1991; Kawecki and Stearns, 1993) may be misleading if only data from laboratory studies which exclude natural predation are used.

If risk of predation varies according to environmental temperature, and if protection from predation can be gained by achieving a large size, then (all else being less important) organisms should grow faster in conditions with a high risk of predation. Conversely, if protection from predation is achieved by being small, then an organism under increased risk of predation should decrease growth rates, reproduce at a small size, or both. Chrzanowski et al. (1988) suggested that reduced cell size of planktonic bacteria at increased temperatures may reduce predation pressure which is more acute at warm times of year. This change in size might then affect the size of predators: Pace (1982) and Pace and Orcutt (1981) found that the zooplankton in Lake Oglethorpe was dominated by macroplankton in the cooler seasons, and by microplankton during the warmer seasons. To decide on the most likely selective effects of temperature-mediated predation on body size, the relationships between temperature and predation risk, and between predation risk and size need to be ascertained.

At high temperatures, predation by ectotherms, is likely to increase because of their increased energy demands, although that by endotherms may be unaltered or may even decrease due to a reduction in their energy demands. However, the ectotherm prey also feed, grow and develop faster at high temperatures. So in *physiological time* of the prey, rates of predation of ectotherms by ectotherms may not alter with temperature, but rates of predation by endotherms may decrease. It is worth emphasizing, however, that this statement applies mainly to temperature variations away from seasonal averages. Between seasons, risks of predation are complicated by the following numerical responses of predators: (i) in temperate habitats both endotherms and ectotherms may be feeding offspring as well as themselves in the warm (breeding) season; (ii) at high latitudes, predation in summer is increased due to an influx of summer migrants and emergences from hibernation.

At low temperatures, rates of predation by ectotherms in chronological (daily) time appears generally to be reduced. Culver (1980) found that size at first reproduction in each of seven species of cladoceran declined as temperatures increased between spring and summer, and then sizes increased again in the autumn. He suggested that because small size usually provided protection from fish predation (Zaret, 1980) a rapid temperature change (or some close correlate) was used as a cue for altered predation rates by fish.

But within a season, in ectotherm physiological time the amount of ecto-therm predation may be relatively unaltered by temperature. The percentage juvenile mortality due to ectotherm predation thus may not differ substantially between high and low temperature since increased feeding by ectotherm predators at high temperatures may be balanced by the increased speed of passage through the juvenile period by the ectotherm prey. This argument still needs to be tested rigorously, however. Contrary to the rates of ecto-therm predation, rates of predation by endotherms are likely to increase in both chronological and physiological time in conditions colder than the seasonal average: this is due partly to their increased metabolic requirements at low temperatures, and partly to the slowdown in ectotherm physiological time. So it is suggested that, in physiological time, risks of predation by endotherms are increased at low temperatures.

Examples in which large ectotherms appear to gain protection from ecto-therm predation do exist (e.g. Paine, 1976), but little evidence has been found for this applying to predation by endotherms. Large cockles appear to be protected from predation by small wading birds but not from predation by oystercatchers (Seed and Brown, 1978): whether or not this represents a significant reduction in endotherm predation at large sizes is not known. There is little evidence that overall size of plants is closely related to size-selective grazing, though the size of parts, such as seeds, may be adaptations to size- and shape-specific predators (Janzen, 1971; Dodson, 1989).

Since endotherm predation appears to be relatively unimportant for many ectotherms (Zaret, 1980), this hypothesis is unlikely to be widely applicable.

But an additional argument is that increased size of predatory ectotherms caused by increased endotherm predation may favour an increase in size of their prey, thus allowing these prey to maintain their anti-predator protection. Larger fish fry are preferred less by their ectothermic predators (Miller et al., 1988) but small size provides a refuge for zooplankton experiencing mortality from size-selective predators such as fish and large insects (Zaret, 1980; Dodson, 1989). So even this further explanation is unable to account for the increase in size in planktonic species at low temperatures (e.g. Pechenik, 1984; Lonsdale and Levinton, 1985).

In conclusion, there appears to be little support for the idea that large sizes are produced at low temperatures mainly to provide protection from increased endotherm predation. However, the effect of predation risk on rate of juvenile development, and hence indirectly on size at maturity, still needs to be considered (see Section VI.C).

3. Size and Body Temperature

Two effects of size on body temperature are well recognized. First, for ecto-therms absorbing solar radiation, heat transfer theory predicts that at thermal equilibrium larger organisms will have larger differences between body temperature and air temperature than will smaller organisms (Porter and Gates, 1969; Stevenson, 1985). Experimental work generally supports this theory (Willmer and Unwin, 1981). However, the second effect, in which heating and cooling of large bodies is too slow to follow the daily cycling of the thermal environment, becomes dominant for very large ectotherms. This "inertial homeothermy" (McNab and Auffenberg, 1976) reduces the range of body temperatures that they can experience. Evidence from experiments on reptiles illustrates the relatively slow rates of heating and cooling by large animals (Bartholomew and Tucker, 1964; McNab and Auffenberg, 1976). Large animals may also have a greater potential than small ones to reduce body temperature by evaporative cooling (Willmer, 1991).

Stevenson (1985), using a simple heat balance model which incorporated these two effects, calculated the relationship between body size and predicted maximum daily range of body temperature. The predicted daily body temperature excess above ambient temperature was found to increase at an increasing rate with body size up to about 0·1 kg when the effect then more or less levels off (Fig. 10). At body weights above 1 kg, the temperature excess then reduces rapidly. Thus for ectotherms below 1 kg, relatively large individuals may risk overheating in warm conditions (Willmer, 1991) but may also be better able to take advantage of limited amounts of solar radiation in less sunny conditions. This phenomenon may therefore help explain the smaller sizes at high temperature in species subject to heating from solar radiation. Yet if this is a major explanation, the increasing rate at which the temperature excess increases with body weight implies that temperature should cause the greatest reductions in size-at-stage in larger terrestrial animals weighing up to 0·1 kg.

No support for a greater reduction in live weight among larger species was found: the percentage reduction from the maximum weight over a 5 °C temperature increase showed no trend with species size, except that the range of percentage weight losses was generally greater in the larger species (Fig. 11). No species weighed more than 1 kg, so effects due to inertial homeothermy could not be tested. In conclusion, the hypothesis that either a reduction in size at a given developmental stage at increased temperature is primarily to avoid overheating, or that an increased size is mainly to take advantage of limited amounts of solar radiation in less sunny conditions, has no empirical, albeit indirect support from this review. Moreover, another explanation would still be necessary for fossorial species and aquatic species which do not inhabit surface waters.

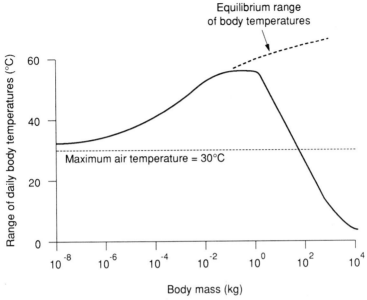

Fig. 10. The maximum daily range of available body temperature as a function of body mass, derived from the heat balance model of Stevenson (1985). For masses smaller than about 0·1 kg, the range of available body temperatures is the same as the range of body temperatures at equilibrium with environmental conditions. For masses greater than 1 kg the range of body temperatures rapidly decreases because of "inertial homeothermy".

C. Temperature and Developmental Rate

1. Mortality at Different Stages and Ages

(a) Juvenile mortality. For an organism with a particular growth rate, increases in size at any given stage can be achieved by slowing development and ultimately delaying maturity. If the benefits of increased size are negligible, the cost of delaying maturity is a delay in the production of offspring, which if the organism has a fixed length of life, will reduce lifetime fecundity. Under such conditions, early maturation and hence smaller size-at-maturity will be favoured. Another condition favouring early maturation is when risks of juvenile mortality are unavoidably high. Organisms should then minimize the time spent in the high-risk stage (Williams, 1966; Wilbur and Collins, 1973; Roff, 1981; Ludwig and Rowe, 1990; Rowe and Ludwig, 1991). Only the life-history model of Stearns and Koella (1986) predicted a delayed maturation in response to an increased risk of juvenile mortality but their model included, for instance, the assumption that increased parental age and size reduces juvenile mortality: this effect, it is suggested here, is not likely to be

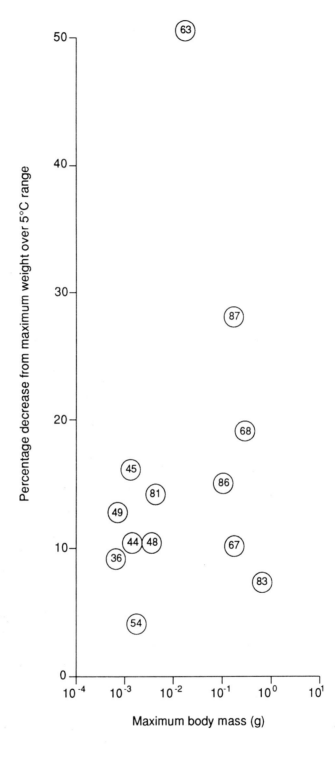

strong in the many ectotherms which do not provide parental care. This assumption also implies that some of the increased juvenile mortality is avoidable.

It may seem at first that when resources are unlimiting and individual growth rate is able to increase with temperature rises, mortality rates are unlikely to increase too. Yet daily risks of predation by other ectotherms will also likely increase. This increased selection pressure to mature early will need to be judged against the pressures to mature at a large size when juvenile growth is rapid (Wilbur and Collins, 1973; Rowe and Ludwig, 1991). The net effects on optimum size may not therefore be easy to predict. Further tests of effects of temperature on the ratio of individual growth to mortality, and more thorough theoretical analysis of the combined effects of individual growth and mortality risk on optimal size at maturity should help solve this problem.

It is important to recall, however, that endotherm predation risk is not likely to be any higher at unseasonally high temperatures since their energy requirements are likely to be less during warm spells. However, when the temperatures experienced are also strongly indicative of season, then in temperate latitudes expected mortality rates in physiological as well as chronological time may well be higher during the warm season due to increased *numbers* of both ectotherm and endotherm predators, as well as increased individual rates of ectotherm predation. Thus the selection pressure to mature early may be more intense during warm seasons than during unseasonal warm spells. Adaptations to season are discussed in detail in Section VI.C.3.

(b) Ageing and other future mortality. Another source of mortality may affect young adults and possibly juveniles, particularly at high, but not stressfully high, temperatures. It has been known for many decades that when resources are unlimiting, ectotherm lifespans are reduced at high temperatures (Loeb and Northrop, 1917; Shaw and Bercaw, 1962; review by Sohal, 1976). When resources are abundant, increased temperature causes rates of metabolism to increase (Sohal, 1986) and thereby brings ageing-related mortality forward in time (Fig. 12). Conditions that bring ageing and mortality risks earlier should generally favour precocious maturation. Indeed, a higher average risk of environmentally induced mortality throughout the whole of the life of an organism should also favour investment in rapid growth and

---

Fig. 11. The percentage decreases in weight below maximum over a 5 °C range, plotted against maximum recorded fresh weight. A consequence of the model of Stevenson (1985) is that temperature may be expected to cause greatest reductions in size in larger terrestrial animals weighing up to 0·1 kg. No such trend is observed. Numbers refer to studies listed in the Appendix.

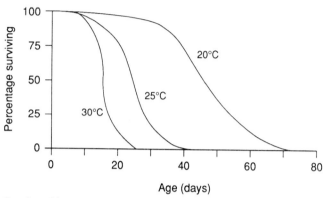

Fig. 12. Survivorship curves of male houseflies at different ambient temperatures (after Sohal, 1986).

reproduction at the expense of maintenance of somatic tissues (Kirkwood and Rose, 1991).

In addition to effects of ageing, future mortality in some terrestrial habitats might be increased following periods of high temperature if water shortage often ensues. In such a case the adaptive response would be to mature early to allow sufficient time for reproduction before the onset of a drought that may kill the organism or significantly reduce reproductive success. Since this correlation between present temperature and future drought is often likely to be very weak (e.g. in high rainfall areas) it at first appears unlikely to explain the majority of size reductions among terrestrial species. Yet even a rare drought can lead to early death and total failure of reproduction and hence local extinction, so genotypes should evolve that accelerate development rate in response to temperature, to allow at least some reproduction in drought years. According to this hypothesis, if organisms are reared under controlled experimental conditions with adequate amounts of water provided, temperature should especially accelerate developmental rate (compared with growth) in those with genotypes adapted to conditions in which a temperature rise correlates strongly with risk and intensity of future drought. This hypothesis remains to be tested.

Risks of future mortality in aquatic habitats may be increased at high temperatures if either oxygen shortage often ensues or the water body dries out. Of course neither of these effects necessarily follows periods of high temperature: exceptions to the general reduction in size at high temperature would therefore be expected under this hypothesis especially in organisms adapted to conditions in whch the correlation with these future mortality factors is very weak or non-existent. This too requires empirical testing.

An important distinction between these adaptationist hypotheses and those invoking constraints on growth due to oxygen shortage (Section V.D) or desiccation (Section V.E) is that in the adaptationist case growth need not be constrained: instead the primary effect of rearing temperature on size-at-maturity is the provision of information about possible future conditions.

For each of these adaptationist hypotheses, the problem referred to in the discussion of juvenile mortality still remains: to what extent is the effect of increased mortality on optimal size balanced by that of increased individual growth at high temperatures?

(c) Cost of rapid growth. Another potential source of mortality at high temperatures is the cost incurred as a result of rapid growth. Sibly *et al.* (1985) and Sibly and Calow (1986a) provide support for the idea that organisms do not always maximize rates of growth but may optimize them. If rapid growth incurs a severe cost at particular temperatures, then selection may favour a reduction below the maximum possible. If the rate of differentiation is unaltered, the result will be a smaller size at a given developmental stage. Yet even an optimum growth rate probably would not eliminate the costs altogether. The benefits from reducing the costs of rapid growth will have to be weighed against the costs of not achieving particular sizes rapidly. Thus an optimum growth rate would be a compromise between the benefits of attaining a particular size in a given time period and the costs incurred directly from rapid growth *per se*. This compromise, between ends and means of achieving particular sizes in given periods of time, would be observed as a trade-off between growth rate and some other trait which affects fitness such as juvenile mortality.

Another slightly different view of optimal (cf. maximal) growth, which links it with ageing, arises if increased temperature increases the rate of molecular damage. Then more resources may need to be diverted from growth to repair (which involves molecular turnover; Calow, 1978) in order to reduce risks of mortality. Thus the costs of increased repair would be a reduction of growth below the maximum possible. If cumulative damage increases mortality risk, there would be selective pressure at high temperature not to slow down development but to bring forward the onset of reproduction. This argument thus assumes a mortality risk which can be reduced by diverting resources away from growth but maximizing the rate of development towards the stage at which offspring are produced.

(d) Adult mortality. For a growing juvenile, one particular set of risks of future mortality may be specifically associated with the adult period. Marked changes in mortality risk at adulthood will be particularly likely in organisms with complex life cycles such as amphibians and those insects that have aquatic larval and terrestrial adult phases (Rowe and Ludwig, 1991).

Roff (1981) and Rowe and Ludwig (1991) predicted that an increased adult mortality would favour delayed maturity and hence large adult size. Organ-

D. ATKINSON

isms which delay maturity would thus grow for a longer time into larger and hence generally more fecund adults.

This prediction appears to run counter to that for ageing-related and other future mortality, discussed earlier, in which higher adult mortality through ageing, desiccation or oxygen shortage favours earlier, not later, maturity. But the apparent discrepancy is resolved when it is realized that ageing-related and other future mortality is influenced by conditions experienced by juveniles as well as by adults, and may not therefore be delayed by spending longer in the juvenile period.

It is argued here that adult mortality (excluding ageing-related and other future mortality) is unlikely to be a major cause of the widespread reduction in size-at-stage of organisms growing at high temperatures. In many rapidly changeable and unpredictable environments, including terrestrial habitats and small water bodies, temperatures experienced during juvenile growth will be poor predictors of conditions that will be experienced by adults, especially when the adult stage is weeks or months away. Therefore the most likely explanations that involve temperature-dependent mortality are the effects of temperature on juvenile mortality and on mortality related to ageing and future risks of desiccation and oxygen shortage.

(e) Mortality evidence. The effects of mortality on size at a given developmental stage at increased temperatures cannot be predicted unless rates of individual growth are also taken into account (Ludwig and Rowe, 1990; Rowe and Ludwig, 1991). In the absence of adequate data to examine this from studies in the Appendix, estimates of mortality in relation to development rate—a correlate of growth rate—were examined. Whilst this measure, which is equivalent to percentage juvenile mortality, is clearly inadequate for rigorous testing of the hypothesis, it will provide at least some idea of whether there is an overwhelmingly high mortality at high temperatures. In 29 cases in which juvenile mortality was measured, 23 showed no simple effect of temperature on percentage juvenile mortality so no overwhelming increase in stage-specific mortality was observed at high temperatures. Four studies did show an increase in mortality with temperature, and each of these showed reduced size-at-stage (Brust, 1967; Leffler, 1972; Economos et al., 1982; Hanazato and Yasuno, 1989). Moreover, whilst one of the two populations in which percentage juvenile mortality increased showed reduced size (Sweeney et al., 1986), the other showed an increase in individual size (Sweeney, 1978). So there is a suggestion that juvenile mortality risk in the absence of natural predators and parasites may affect individual size in some cases, but that no clear relationship is evident in most studies.

A model predicting that a higher mortality rate favours early maturation and smaller size was used by Myers and Runge (1983), together with measurements or estimations of adult size and growth and fecundity schedules, to predict mortality rates in a species of marine copepod. Their model successfully predicted the seasonal trend towards increased mortality rates in summer and a decrease in winter, whether mortality rates were assumed to be constant or dependent on size (i.e. visibility to fish predators increased at large sizes; Section VI.B.2). These data suggest that at least in one natural population, even when individual growth was accounted for, mortality rates in the warm season were apparently sufficiently increased to favour reduced adult size.

2. Temperature and Population Growth

In populations that are increasing in size, precocious reproduction should be favoured by natural selection since offspring born early will form a larger fraction of the total population than if they were born later, and hence will have a higher Darwinian fitness (Lewontin, 1965). Stearns (1976) explained this by analogy with compound interest in banking: "it will always pay to get your money in the bank as soon as possible so that the compounding interest rate will get to work most rapidly." Conversely, delayed reproduction may be favoured in populations that are decreasing in size (Charlesworth, 1980). However, this latter argument assumes that survival till the age of reproduction is the same even if reproduction is delayed, which seems improbable (Sibly and Calow, 1986b).

Within the temperature ranges in which an increase produces increased rates of growth and differentiation, it is suggested that an increased temperature should normally increase the rate of population growth, as found, for instance by Cooper (1965), Orcutt and Porter (1984) and Foran (1986). Since a temperature increase within the range normally experienced generally increases primary and secondary productivity, availability of food may not usually become limiting at high temperatures. If temperatures experienced by juveniles generally correlate positively with future population size, then especially rapid development, facilitating precocious reproduction, would be favoured by natural selection.

But this argument in favour of precocious reproduction, the "compound interest" hypothesis, does not apply to populations in which generation time is fixed (Meats, 1971). These populations lack the demographic pressure to mature especially early at high temperatures since no individual is capable of fitting in additional generations or parts of generations. Any other advantage from starting to reproduce early in the breeding season applies at low as well as high temperatures. In fact, if the length of growing season is limited, a faster growing individual of such a species should be expected to mature at a larger size, albeit still a little earlier, than one growing more slowly, thereby achieving the optimal compromise between large size and sufficient time for copious reproduction (Rowe and Ludwig, 1991).

Another situation in which "compound interest" may be slight or even non-existent has been discussed recently by Houston and McNamara (1992) and Kawecki and Stearns (1993). These authors considered organisms which move freely among habitats or patches rather than remain as a largely isolated population within a habitat. They showed that the optimal life history for an individual in a given habitat or patch depends on the *overall* rate of increase (*r*, the Malthusian parameter) for the whole range of habitats and patches in which the organisms and their offspring occur. Thus for a highly mobile organism occurring in a patch which benefits greatly from increased tempera-ture, the fitness benefits ("compound interest") accruing from early matur-ation may not be so great as would be calculated from just the local popu-lation growth rate in that patch. This would be especially true if there is free migration to and from surrounding patches, and if in these patches popu-lations are at the same time shrinking. Not surprisingly, data are not currently available to test how important and widely applicable these ideas are. How-ever, data on the flexibility of generation times should be easier to obtain.

Of the species listed in the Appendix, most appeared to have the capacity to alter their generation times. Under certain conditions, populations nor-mally with a fixed number of generations a year may insert additional ones (e.g. Pullin, 1986). Therefore, only examples reported to be univoltine with an obligate diapause between generations were accepted as always having a fixed generation time. Based on the information given in the papers, only one definite example was found, the beetle *Entomoscelis americana* (Lamb and Gerber, 1985), in which adult size increased with increased temperature (see Fig. 4), and thus is consistent with the "compound interest" hypothesis.

Since environmental conditions can determine whether or not a population is univoltine (Pullin, 1986), it is possible that in some cases laboratory con-ditions may have indicated (falsely) conditions favouring univoltinism. This is discussed in the next section.

Whilst no generalizations can be made from the solitary example with fixed generation time, the hypothesis does need to be tested further. Recent results from work on a grasshopper with fixed generation time (K. Vosper, unpub-lished) also support this hypothesis.

A potential criticism of the hypothesis is: if increased (more favourable) temperatures account for small size because of advantages obtained from a shorter life cycle during periods of population growth, why aren't individuals that consume high quality food smaller than those that consume poor quality food (cf. Sweeney *et al.*, 1986)? One answer is that malnourished individuals are likely to be physically incapable of growing big because the size at which food intake fails to compensate for the costs of maintaining tissues will be less than for a well-fed individual: this would act against the prediction of the "compound interest" hypothesis. Achieving a large size should not be such a problem for (well-fed) individuals raised at low temperatures (cf. high temperatures) since respiratory costs will be reduced as well as rates of food or energy uptake.

3. Seasonal Adaptation

The effects of temperature on size at a given developmental stage may also depend on the time of year. In some species of temperate latitudes, the timing of developmental events strongly correlate with photoperiod (Masaki, 1978; Tauber et al., 1986; Roberts and Summerfield, 1987; Nylin et al., 1989; Ellis et al., 1990; Scott and Dingle, 1990)—a strong indicator of time of year. Long- and short-day photoperiods have been shown to affect the rate of development of the grasshopper Melanoplus sanguinipes (Scott and Dingle, 1990) and several crop plants (Ellis et al., 1990) in different ways according to the temperature at which they are grown. Thus in populations with little time left before having to reach maturity or some other stage (e.g. before food deteriorates; Palmer, 1984), or in those which synchronize their adult emergence (e.g. some mayflies), faster juvenile growth at high temperatures could produce large rather than small adults since all would undergo the developmental event at similar times (or photoperiods).

Of only 15 species in the five studies reported to have been performed under natural rather than laboratory controlled photoperiods, ten showed a reduction in size-at-stage at high temperatures and five showed an increase. This relatively high proportion of increases (32%, compared with 11·9% in the total sample of 109; see Appendix) provides rather weak evidence that photoperiod may sometimes override the effects of temperature.

Four of the five increases and one of the reductions were shown by mayflies of temperate latitudes (Sweeney, 1978; Vannote and Sweeney, 1980) and the other increase was found in a parasitic crustacean of fish gills (Johnston and Dykeman, 1987). Six of the ten reductions were by chironomid midges, all in the same study (Rempel and Carter, 1987) and three were crop plants (Gmelig Meyling, 1969): in these cases photoperiod presumably did not override the general effects of temperature. However, further support for effects of photoperiod, though not supported by statistical tests, come from maize plants which were grown under high and moderate light levels at different constant temperatures, and sown at two different times of year (Gmelig Meyling, 1969). July-sown plants flowered at a reduced size at increased temperature and thus followed the general rule, but April-sown plants flowered at increased sizes at the higher temperature.

Another seasonal adaptation found in at least some organisms with several generations a year is the possession of specific generations adapted to particular seasons (Carvalho, 1987; Carvalho and Crisp, 1987). Thus some parts of the normal annual temperature range will be stressful to a given generation reared in the laboratory, whilst in nature relatively little stress would be experienced since each generation is adapted to its own particular seasonal temperatures. This may explain why at temperatures below 15 °C which were nevertheless within the annual range experienced by a population of *Daphnia*, under a 16L/8D (summer) photoperiodic regime sizes at death were smaller and mortalities were higher than at higher temperatures (Orcutt and Porter, 1983). Photoperiod may also have been implicated in the production of small individual aphids at unseasonally but not exceptionally low temperatures (8·8–10·1 °C, Müller, 1966; 10–15 °C, Lamb and MacKay, 1988).

In some other studies, too (Vannote and Sweeney, 1980; Rempel and Carter, 1987), effects of temperature at natural photoperiods may have been partly caused by deviations from the ambient temperature to seasonally inappropriate levels.

VII. DEVELOPING NEW WORKING HYPOTHESES

A. Must Resources Be Unlimiting?

It is notable that the general relationship between size and temperature may be produced even when energy or other resources are not absolutely unlimiting: the general biological rule may therefore have wide application in natural populations. This is suggested by both empirical and theoretical studies. Empirical studies on legumes and grasses showed the reduction in size with increasing temperatures at intermediate and low, as well as high, light intensities (Gmelig Meyling, 1969). Theoretical work (Ludwig and Rowe, 1990; Rowe and Ludwig, 1991) suggests that optimal size is strongly influenced by the ratio of individual growth to juvenile mortality rate. Growth rate can be higher at increased temperatures even when resources are not completely unlimiting. Therefore, the biological rule should still apply when the problem outlined in Section III.A is broadened to include organisms which have not experienced unlimited resources, so long as individual growth is still faster at increased temperature.

B. Favoured Hypotheses

Indirect evidence (Sections IV and VI.C.3) suggests that up to five of the 13 temperature increases and two of the mixed responses to temperature may be due to inadmissible temperatures (Müller, 1966; Vanotte and Sweeney, 1980; Tsitsipis, 1980; Lamb and MacKay, 1988). Lack of data may have prevented the identification of other inadmissible studies, so the proportion of significant reductions may be closer to 90%.

The relationship may conceivably be explained by constraints on growth using a modification of von Bertalanffy's hypothesis (Section V.B; Fig. 9).

But no evidence to support this modification is known to the author, and ways of overcoming constraints on growth may also exist.

The adaptationist hypotheses involving water viscosity and buoyancy (Section VI.B.1) have, of course, no explanatory power outside aquatic habitats. The review of mortality data and mechanisms suggests that increased ageing-related mortality, or mortality caused by predation, future desiccation, oxygen shortage or a cost to rapid individual growth are also possible candidates to explain the phenomenon. Further theoretical work is needed, however, to formulate clear testable hypotheses, appropriate to the wide range of life histories found in ectotherms. The "compound interest" hypothesis (Section VI.C.2), if it also incorporates time constraints on development due to season, also appears to be capable of explaining most of the reductions as well as at least one of the increases.

The reason for the very widespread occurrence of reduced size at increased temperature, may be that when the "compound interest" hypothesis does not operate, those involving high mortality may become important and *vice versa*. This is because high mortality generally favours reduced size but is likely to be associated with slow population growth (little or no "compound interest") whereas low mortality is more likely to be associated with increased population growth. However, before detailed and specific predictions can be made, these arguments need to be evaluated using a robust theoretical framework applicable to the vast majority of ectotherms.

In the meantime, empirical studies should identify and attempt to quantify mortality factors both in laboratory and field experiments using factorial experiments in which temperature and availability of predators and either water or oxygen are manipulated, together (where possible) with environmental or physiological information affecting developmental rates. Tests of the "compound interest" hypothesis in which population growth rates in natural populations are compared under different temperature regimes will be complicated by the variety of rates found in patchy environments (Kawecki and Stearns, 1993). However, if this is the primary cause of size reductions, exceptions to the general rule should be found in populations in which the generation time is fixed or when temperature is known not to be correlated with population increase.

VIII. TEMPERATURE AND SIZE IN NATURAL AND MANAGED POPULATIONS

Other environmental influences on organism size, besides temperature, include competition, water balance, photoperiod, and resource abundance and quality (Schoener and Janzen, 1968; Palmer, 1984; Honěk, 1986).

Most of the changes in size between seasons and between years will be ontogenetic responses to environmental change. Genetic variation will only be important between populations or temporally within populations evolving very rapidly (e.g. Carvalho, 1987; Carvalho and Crisp, 1987). For this reason geographical variation in size will not be discussed in this section except where genetic effects have been accounted for and effects of environmental rearing temperature investigated experimentally.

A. Temporal Variation

If temperature variation during the year is more important than that of other environmental variables, then seasonal changes in size should correlate strongly and negatively with temperature experienced during growth. Whilst size of emerging adults often decreases during a period of increasing temperatures (examples in Sweeney and Vannote, 1978; Atkinson and Begon, 1988) this result can also be explained by within-cohort competition (Atkinson and Begon, 1988).

Reductions in size between seasons, correlating with increased temperatures, have been found within populations of: copepods (Bogorov, 1934; Deevey, 1960; McLaren, 1963; Elbourn, 1966; Smith and Lane, 1987); cladocerans (Culver, 1980); amphipods (Nelson, 1980); dipterans (Mer, 1937; Golightly and Lloyd, 1939; Day et al., 1990); and mayflies (Macan, 1957; Elliott, 1967). These seasonal changes were sometimes substantial. For instance, weights of some aquatic insects may be reduced 50% in the summer relative to other seasons (Sweeney and Vannote, 1978).

However, counter-examples are known. First, Fahy (1973) found that the lengths of final-instar *Baetis rhodani* were greater in the winter-emerging than the summer generation in a subterranean isothermic stream as well as in one in which temperature fluctuated. And second, in two studies of chrysomelid beetles, adults were larger after a warm period of larval growth than after a cool one (Palmer, 1984; Sims et al., 1984). Moreover, in the laboratory, temperature was found to have no significant effect on size at adult emergence in these species. Deteriorating food quality best explained the size reduction in the field in one of these (Palmer, 1984).

Despite these exceptions and qualifications, the weight of evidence is consistent with the idea that variation in temperature is usually a major determinant of seasonal changes in size at a given life-cycle stage.

In annual species, temperature may influence differences in size between years. However, adult female hind femur length (a correlate of body weight) of each of two species of grasshopper were not shorter, but significantly (between 8 and 18%) longer in a warm year compared with a cool one (Atkinson and Begon, 1988). But these grasshoppers are univoltine with an obligate diapause only broken after cold winter temperatures have been experienced (Cherrill and Begon, 1989), so no gains in "compound interest" can be achieved from maturing especially early, and the faster growth at high temperatures favours larger size at maturity (Rowe and Ludwig, 1991). Moreover, in at least one of these species, individuals are also larger when they experience higher average daily temperatures (K. Vosper, unpublished).

The effect of temperature on size can have considerable economic importance. The yield of cereals is positively correlated with the size of the plant, and annual yields of wheat (in tonnes ha^{-1}) between 1963 and 1978 in the east Midlands, England, were increased by up to 25% in the year with the coolest growing period compared with that with the warmest (Monteith, 1981). Mean temperature from May to July explained 38% of the annual variation in yield. Monteith described how the rate of development of wheat is also faster in warm years, and went on to argue that the effect on size was not due to dry conditions in such years.

The same response was demonstrated for the whole of England; and a similar effect has been found for barley yields in the east Midlands (Monteith, 1981) and for wheat yields in the United States (Bryson, 1978). In the latter, an average drop of 1 °C below normal during the month of July was shown to increase income from wheat in the United States by $92 M (1978 prices) due to increased yield.

However, if the "compound interest" hypothesis distinguishes between those with and without the capacity for flexible generation time, why aren't these annual crops exceptions to the usual trend? An answer is that under certain conditions barley can grow as a perennial and wheat can be occasionally biennial (Clapham et al., 1987), so they may have the capacity to alter generation time but are simply harvested before this potential is demonstrated. Another explanation is that crop development may be especially sensitive to temperature because of an historical (evolutionary) association between increased temperature and subsequent drought (Section VI.C.1).

B. Altitudinal Variation: A Well Studied Example

A particularly well studied example of altitudinal variation is provided by Berven (1982b) who separated environmental from genetic differences, and investigated temperature effects experimentally. Reciprocal transplantation experiments demonstrated that 73% of the altitudinal variation observed in the size of wood frogs at metamorphic climax could be explained by environmental rather than genetic differences. Results from laboratory experiments suggested that most of this was probably due to temperature differences (see Fig. 3; Berven, 1982b), producing large adults in the cool mountain sites.

IX. CONCLUSIONS

The conclusion of Ray (1960) is reaffirmed and given greater precision: in the vast majority of ectotherms (or poikilotherms) an increase in environmental temperature which increases rates of growth and differentiation reduces size at a given stage of development. If conditions are otherwise controlled, this may well be a near-universal relationship, and most apparent exceptions may result from unreported weaknesses in experimental protocol or inadmissible temperature conditions. It would be useful, therefore, to repeat the experiments on these taxa but taking especial care to measure growth and mortality rates, and ensuring that controlled environmental conditions are maintained. Yet several apparently meticulous studies still appear to contradict the "law" (e.g. Lamb and Gerber, 1985).

When a single relationship applies to many different taxa from different habitats and with different trophic status, a single explanation would normally be sought. Yet authors have often proposed explanations, or used supporting evidence, specific to a given species or taxon.

In its search for one explanation or a very few wide-ranging ones, the present paper has highlighted the following promising areas which require further study:

(i) the effects of temperature on size when amounts of either supplementary water (for terrestrial species) or supplementary oxygen (aquatic species) are varied. This would investigate the effects of potential constraints on growth at high temperatures and could be combined with studies of anabolic and catabolic rates, the surface areas of organs responsible for the uptake of respiratory gases and nutrients from the environment, and the efficiency of their transport to the tissues;

(ii) the extent to which the relationship between temperature and size applies to populations in which generation time is fixed or in which temperature is not positively correlated with rate of population increases;

(iii) the effects of temperature on individual growth rates, size-at-stage and various mortality factors. Mortality at increased temperatures may be related to seasonal increases in predation, to ageing, to future desiccation (terrestrial species) or oxygen shortage (species in aquatic and water-saturated habitats), and to costs of rapid growth;

(iv) the use of seasonal cues to control relative rates of growth and differentiation.

A hypothesis has been proposed (Section VI.C.2) based on the advantages of shortening the life cycle when conditions are favourable for population growth: opposite predictions can be made depending on whether or not there is scope for variation in generation time between individuals in a population.

The effect of temperature experienced during a period of growth appears to be particularly important in explaining seasonal and between-year differences in size, including the economically important differences between years in the yields of crops harvested at maturity.

Knowledge of how and why temperature influences size-at-stage in so many species should help provide a universal biological basis for the production of predictive quantitative models of size-at-stage, especially and perhaps first of all in aquaculture, agriculture, and other closely controlled environments in which energy and nutrients are not usually limiting. But given the considerable and wide-ranging ecological importance of organism size, and the variability in both weather and climate, knowledge of this biological rule, and possible law, should prove valuable to ecologists in many different fields.

ACKNOWLEDGEMENTS

Thanks are due to Betty Howarth, Thomas Lange and Paul Ward for their help in translating some of the papers. Simon Wood pointed out that only slight modifications of von Bertalanffy's hypothesis were required to make it valid, and kindly allowed me to use Fig. 9. Andrew Cherrill, Andrew Cossins, Linda Partridge and Richard Sibly made helpful comments on earlier drafts of this paper.

REFERENCES

Abdullahi, B.A. and Laybourn-Parry, J. (1985). The effect of temperature on size and development in three species of benthic copepod. *Oecologia* **67**, 295–297.
Adolph, E.F. (1929). The regulation of adult body size in the protozoan *Colpoda. J. Exp. Zool.* **53**, 269–311.
Akey, D.H., Potter, H.W. and Jones, R.H. (1978). Effects of rearing temperature and larval density on longevity, size, and fecundity in the biting gnat *Culicoides variipennis. Ann. Entomol. Soc. Am.* **71**, 411–418.
Alpatov, W.W. (1930). Phenotypical variation in body and cell size of *Drosophila melanogaster. Biol. Bull.* **58**, 85–103.
Alpatov, W.W. and Pearl, R. (1929). Experimental studies on the duration of life. XII. Influence of temperature during the larval period and adult life on the duration of life of the imago of *Drosophila melanogaster. Am. Nat.* **63**, 37–63.
Allsopp, P.G. (1981). Development, longevity and fecundity of the false wireworms *Pterohelaeus darlingensis* and *P. alternatus* (Coleoptera: Tenebrionidae) I. Effect of constant temperature. *Aust. J. Zool.* **29**, 605–619.
Atkinson, D. (1985). Information, non-genetic constraints, and the testing of theories of life-history variation. In: *Behavioural Ecology* (Ed. by R.M. Sibly and R.S. Smith), pp. 99–104. Blackwell Scientific Publications, Oxford.
Atkinson, D. and Begon, M. (1988). Adult size variation in two co-occurring grasshopper species in a sand dune habitat. *J. Anim. Ecol.* **57**, 185–200.
Atkinson, D. and Thompson, D.B.A. (1987). Constraint and restraint in breeding birds. *Wader Study Group Bull.* **49**, 18–19.
Baker, J.E. (1983). Temperature regulation of larval size and development in *Attagenus megatoma* (Coleoptera: Dermestidae). *Ann. Entomol. Soc. Am.* **76**, 752–756.
Barker, J.F. and Barker, A. (1980). The relation between body size and resistance to desiccation in two species of *Zaprionus* (Drosophilidae). *Ecol. Entomol.* **5**, 309–314.

Bartholomew, G.A. and Tucker, V.A. (1964). Size, body temperature, thermal conductance, oxygen consumption and heart rate in Australian varanid lizards. *Physiological Zoology* **37**, 341–354.

Bayne, B.L. (1965). Growth and delay of metamorphosis of the larvae of *Mytilus edulis* (L.). *Ophelia* **2**, 1–47.

Beckwith, R.C. (1982). Effects of constant laboratory temperatures on the Douglas-Fir tussock moth (Lepidoptera: Lymantriidae). *Environ. Entomol.* **11**, 1159–1163.

Bělehrádek, J. (1935). *Temperature and Living Matter. Protoplasma Monograph 8.* Borntraeger, Berlin.

Bellinger, R.G. and Pienkowski, R.L. (1987). Developmental polymorphism in the red-legged grasshopper *Melanoplus femurrubrum* (De Geer) (Orthoptera: Acrididae). *Environ. Entomol.* **16**, 120–125.

Bergmann, C. (1847). Ueber die Verhaeltnisse der Waermeoekonomie der Thiere zu ihrer Groesse. *Goett. Stud.* **1**, 595–708.

Bernays, E.A. (1972). Some factors affecting size in first-instar larvae of *Schistocerca gregaria* (Forskal). *Acrida* **1**, 189–195.

Berven, K.A. (1982a). The genetic basis of altitudinal variation in the wood frog *Rana sylvatica*. I. An experimental analysis of life history traits. *Evolution* **36**, 962–983.

Berven, K.A. (1982b). The genetic basis of altitudinal variation in the wood frog *Rana sylvatica*. II. An experimental analysis of larval development. *Oecologia* **52**, 360–369.

Berven, K.A. and Gill, D.E. (1983). Interpreting geographic variation in life-history traits. *Am. Zool.* **23**, 85–97.

Bogorov, B.G. (1934). Seasonal changes in biomass of *Calanus finmarchicus* in the Plymouth area in 1930. *J. Marine Biol. Assoc., UK* **19**, 585–612.

Brouwer, R., Kleinendorst, A and Locher, J. Th. (1973). Growth responses of maize plants to temperature. In: *Plant Responses to Climatic Factors* (Ed. by R.O. Slatyer), pp. 169–174. UNESCO, Paris.

Brust, R.A. (1967). Weight and development time of different stadia of mosquitoes reared at various constant temperatures. *Can. Entomol.* **99**, 986–993.

Bryson, R.A. (1978). The cultural, economic and climate records. In: *Climate Change and Variability* (Ed. by A.B. Pittock), pp. 316–327. Cambridge University Press, Cambridge.

Burges, H.D. and Cammell, M.E. (1964). Effect of temperature and humidity on *Trogoderma anthrenoides* (Sharp) (Coleoptera Dermestidae) and comparisons with related species. *Bull. Entomol. Res.* **55**, 313–325.

Burholt, D.R. and Van't Hof, J. (1971). Quantitative thermal induced changes in growth and cell population kinetics of *Helianthus* roots. *Am. J. Bot.* **58**, 386–393.

Burström, H. (1956). Temperature and root cell elongation. *Physiologia Plantarum* **9**, 682–692.

Calder, W.A. III. (1984). *Size, Function and Life History.* Harvard University Press, Cambridge, Massachussetts.

Calow, P. (1973). On the regulatory nature of individual growth: some observations from freshwater snails. *J. Zool.* **170**, 415–428.

Calow, P. (1977). Ecology, evolution and energetics. *Adv. Ecol. Res.* **10**, 1–62.

Calow, P. (1978). The Ageing Process. In: *Life Cycles*, p. 120. Chapman & Hall, London.

Carvalho, G.R. (1987). The clonal ecology of *Daphnia magna* (Crustacea: Cladocera). II. Thermal differentiation among seasonal clones. *J. Anim. Ecol.* **56**, 469–478.

Carvalho, G.R. and Crisp, D.J. (1987). The clonal ecology of *Daphnia magna* (Crustacea: Cladocera). I. Temporal changes in the clonal structure of a natural population. *J. Anim. Ecol.* **56**, 453–468.

Cavicchi, S., Guerra, D., Giorgi, G. and Pezzoli, C. (1985). Temperature-related divergence in experimental populations of *Drosophila melanogaster*. I. Genetic and developmental basis of wing size and shape variation. *Genetics* **109**, 665–689.

Charlesworth, B. (1980). *Evolution in Age-structured Populations*. Cambridge University Press, Cambridge.

Cherrill, A.J. (1987). *The Development and Survival of the Eggs and Early Instars of the Grasshopper* Chorthippus brunneus *(Thunberg) in North West England.* Unpublished PhD thesis, University of Liverpool.

Cherrill, A.J. and Begon, M. (1989). Timing of life cycles in a seasonal environment: the temperature-dependence of embryogenesis and diapause in a grasshopper (*Chorthippus brunneus* Thunberg). *Oecologia* **78**, 237–241.

Chrzanowski, T.H., Crotty, R.D. & Hubbard, G.J. (1988). Seasonal variation in cell volume of epilimnetic bacteria. *Microbial Ecol.* **16**, 155–163.

Clapham, A.R., Tutin, T.G. and Moore, D.M. (1987). *Flora of the British Isles*, 3rd edn. Cambridge University Press, Cambridge.

Coker, R.E. (1933). Influence of temperature on size of freshwater copepods (*Cyclops*). *Int. Rev. ges. Hydrobiol. Hydrographie* **29**, 406–436.

Conover, W.J. (1980). *Practical Nonparametric Statistics*. John Wiley, New York.

Cooper, W.E. (1965). Dynamics and production of a natural population of a freshwater amphipod, *Hyalella azteca*. *Ecol. Monogr.* **35**, 377–394.

Cuadrado, A., Navarrete, M.H. and Canovas, J.L. (1989). Cell size of proliferating plant cells increases with temperature: implications in the control of cell division. *Exp. Cell Res.* **185**, 277–282.

Culver, D. (1980). Seasonal variation in sizes at birth and at first reproduction in Cladocera. In: *Evolution and Ecology of Zooplankton Communities* (Ed. by W.C. Kerfoot), pp. 358–366. University of New England, Hanover.

David, J. and Clavel, M-F. (1967). Influence de la temperature subie au cours du développement sur divers caractères biométriques des adultes de *Drosophila melanogaster* Meigen. *J. Insect Physiol.* **13**, 717–729.

Day, J.F., Ramsey, A.M. and Shang, J-T. (1990). Environmentally mediated seasonal variation in mosquito body size. *Environ. Entomol.* **19**, 469–473.

Deeming, D.C. & Ferguson, M.W.J. (1988). Environmental regulation of sex determination in reptiles. *Philos. Trans. R. Soc. Lond. B* **322**, 19–39.

Deevey, G.B. (1960). Relative effects of temperature and food on seasonal variations in length of marine copepods in some eastern American and western European waters. *Bull. Bingham Oceanogr. Coll.* **17**, 54–86.

Dingle, H., Blakley, N.R. and Miller, E.R. (1980). Variation in body size and flight performance in milkweed bugs (*Oncopeltus*). *Evolution* **34**, 371–385.

Dixon, A.F.G., Chambers, R.J. and Dharma, T.R. (1982). Factors affecting size in aphids with particular reference to the black bean aphid, *Aphis fabae*. *Entomol. Exp. Appl.* **32**, 123–128.

Dodson, S. (1989). Predator-induced reaction norms. *Bioscience* **39**, 447–452.

Donna, L. and John, P.C.L. (1984). Timer and sizer controls in the cycles of *Chlamydomonas* and *Chlorella*. In: *The Microbial Cell Cycle* (Ed. by P. Nurse and E. Streiblova), pp. 231–251. CRC Press, Boca Raton, Florida.

Dowben, R.M. and Weidenmuller, R. (1968). Adaptation of mesophilic bacteria to growth at elevated temperatures. *Biochim. Biophys. Acta* **158**, 255–261.

Druger, M. (1962). Selection and body size in *Drosophila pseudoobscura* at different temperatures. *Genetics* **47**, 209–222.

Economos, A.C., Lints, C.V. and Lints, F.A. (1982). On the mechanisms of the effects of larval density and temperature on *Drosophila* development. In: *Advances in Genetics, Development and Evolution of Drosophila* (Ed. by S. Lakovaara), pp. 149–164. Plenum Press, New York.

Eigenbrodt, H.J. (1930). The somatic effects of temperature on a homozygous race of *Drosophila*. *Physiol. Zool.* 3, 392–411.

Elbourn, C.A. (1966). The life-cycle of *Cyclops strenuus strenuus* Fischer in a small pond. *J. Anim. Ecol.* 35, 333–347.

Elliott, J.M. (1967). The life histories and drifting of the Plecoptera and Ephemeroptera in a Dartmoor stream. *J. Anim. Ecol.* 36, 344–362.

Ellis, R.H., Hadley, P., Roberts, E.H. and Summerfield, R.J. (1990). Quantitative relations between temperature and crop development and growth. In: *Climatic Change and Plant Genetic Resources* (Ed. by M.T. Jackson, B.V. Ford-Lloyd and M.L. Parry), pp. 85–115. Belhaven Press, London.

Eppley, R.W., Holmes, R.W. and Strickland, J.D.H. (1967). Sinking rates of marine phytoplankton measured in a fluorometer. *J. Exp. Marine Biol. Ecol.* 1, 191–208.

Erickson, R.O. (1959). Integration of plant growth processes. *Am. Nat.* 93, 225–235.

Fahy, E. (1973). Observations on the growth of Ephemeroptera in fluctuating and constant temperature conditions. *Proc. R. Irish Acad., Sect. B* 73, 133–149.

Fawley, N. (1984). Effects of light intensity and temperature interactions on growth characteristics of *Phaeodactylum tricornutum* (Bacillariophyceae). *J. Phycol.* 20, 67–72.

Foran, J.A. (1986). A comparison of life history features of a temperate and a subtropical *Daphnia* species. *Oikos* 46, 185–193.

Francis, D. and Barlow, P.W. (1988). Temperature and the cell cycle. In: *Plants and Temperature* (Ed. by S.P. Long and F.I. Woodward). pp. 181–201. Company of Biologists, Cambridge.

Friend, D.J.C., Helson, V.A. and Fisher, J.E. (1962). The rate of dry weight accumulation in Marquis wheat as affected by temperature and light intensity. *Can. J. Bot.* 40, 939–955.

Galford, J.R. (1974). Some physiological effects of temperature on artificially reared oak borers. *J. Econ. Entomol,* 67, 709–710.

Galliard, H. and Golvan, Y.J. (1957). Influence de certains facteurs nutritionels et hormonaux, a des températures variables, sur la croissance des larves d'*Aedes (S.) aegypti, Aedes (S.) albopictus*, et *Anopheles (M.) stephensi. Ann. Parasitol. Hum. Comp.* 32, 563–579.

Gates, D.M. (1980). *Biophysical Ecology*. Springer, New York.

Gmelig Meyling, H.D. (1969). Effect of light energy in relation to temperature on the rate of development and growth of some agricultural crops. Report 403 of IBS. In: *Annual Report of IBS, 1969*, Wageningen.

Golightly, W.H. and Lloyd, L.I. (1939). Insect size and temperature. *Nature* 144, 155–156.

Gomulkiewicz, R. and Kirkpatrick, M. (1992). Quantitative genetics and the evolution of reaction norms. *Evolution* 46, 390–411.

Grace, J. (1988). Temperature as a determinant of plant productivity. In: *Plants and Temperature* (Ed. by S.P. Long and F.I. Woodward), pp. 91–107. Company of Biologists, Cambridge.

Grif, V.G. and Valovich, E.M. (1973). Effects of positive low temperatures on cell growth and division in seed germination. *Tsitologiya* 15, 1362–1369. (In Russian.)

Guppy, J.C. (1969). Some effects of temperature on the immature stages of the armyworm *Pseudaletia unipuncta* (Lepidoptera: Noctuidae), under controlled conditions. *Can. Entomol.* **101**, 1320–1327.

Hanazato, T. and Yasuno, M. (1989). Effect of temperature in laboratory studies on growth of *Chaoborus flavicans* (Diptera, Chaoboridae). *Archiv. Hydrobiol.* **114**, 497–504.

Hardacre, A.K. and Eagles, H.A. (1986). Comparative temperature response of Corn Belt Dent and Corn Belt Dent × Pool 5 maize hybrids. *Crop Sci.* **26**, 1009–1012.

Hardacre, A.K. and Turnbull, H.L. (1986). The growth and development of maize at five temperatures. *Ann. Bot.* **58**, 779–788.

Harkey, G.A. and Semlitsch, R.D. (1988). Effects of temperature on growth, development and colour polymorphism in the ornate chorus frog *Pseudacris ornata*. *Copeia* (**4**) 1001–1007.

Honěk, A. (1986). Body size and fecundity in natural populations of *Pyrrhocoris apterus* L. (Heteroptera, Pyrrhocorida). *Zool. Jahrb. Abt. Syst.* **113**, 125–140.

Houston, A.I. and McNamara, J.M. (1992). Phenotypic plasticity as a state-dependent life-history decision. *Evol. Ecol.* **6**, 243–253.

Huey, R.B. and Kingsolver, J.G. (1980). Evolution of thermal sensitivity of ectotherm performance. *Trends Ecol. Evol.* **4**, 131–135.

Hunter, R.B., Tollenaar, M. and Breuer, C.M. (1977). Effects of photoperiod and temperature on vegetative and reproductive growth of a maize (*Zea mays*) hybrid. *Can. J. Plant Sci.* **57**, 1127–1133.

Imai, T. (1937a). Influence of temperature on the growth of *Drosophila melanogaster*. *Sci. Rep. Tohoku Imp. Univ.* **11**, 403–417.

Imai, T. (1937b). The larval shell growth of *Lymnaea japonica* Jay. In special reference to the influence on temperature. *Sci. Rep. Tohoku Imp. Univ.* **11**, 419–432.

James, T.W. and Padilla, G.M. (1959). Physiological and size changes in protozoan cells in response to incubation temperatures. In: *Proceedings of the First National Biophysics Conference* (Ed. by H. Quastler and H.J. Morowitz), pp. 694–700. Yale University Press, New Haven, CT.

James, T.W. and Read, C.P. (1957). The effect of incubation temperature on the cell size of *Tetrahymena pyriformis*. *Exp. Cell Res.* **13**, 510–516.

Janzen, D.H. (1971). Seed predation by animals. *Ann. Rev. Ecol. Syst.* **2**, 465–492.

Johns, D.M. (1981). Physiological studies on *Cancer irroratus* larvae. I. Effects of temperature and salinity on survival, development rate and size. *Marine Ecol. Prog. Ser.* **5**, 75–83.

Johnson, B.F. and James, T.W. (1960). Alteration of cellular constituents by incubation temperature. *Exp. Cell Res.* **20**, 66–70.

Johnston, C.E. and Dykeman, D. (1987). Observations on body proportions and egg production in the female parasitic copepod (*Salmincola salmoneus*) from the gills of Atlantic salmon (*Salmo salar*) kelts exposed to different temperatures and photoperiods. *Can. J. Zool.* **65**, 415–419.

Johnston, N.T. and Northcote, T.G. (1989). Life-history variation in *Neomysis mercedis* Holmes (Crustacea, Mysidacea) in the Fraser River estuary, British Columbia. *Can. J. Zool.* **67**, 363–372.

Jones, R.E., Hart, J.R. and Bull, G.D. (1982). Temperature, size and egg production in the cabbage butterfly *Pieris rapae* L. *Aust. J. Zool.* **30**, 223–232.

Kawecki, T.J. and Stearns, S.C. (1993). The evolution of life histories in spatially heterogeneous environments: optimal reaction norms revisited. *Evol. Ecol.* **7**, 155–174.

Kindlmann, P. and Dixon, A.F.G. (1992). Optimum body size: effects of food quality

48 D. ATKINSON

and temperature when reproductive growth rate is restricted, with examples from aphids. *J. Evol. Biol.* **5**, 677–690.

Kinne, O. (1958). Uber die Reaktion erbgleichen Coelenteratengewebes auf verschiedene Asalzgehalts- und Temperaturbedingungen. II. Mitteilung uber den Einflub des Salzgehaltes auf Wachstum und Entwicklung mariner, brackischer und limnischer Organismen. *Zool. Jahrb. Abt. Allgem. Zool. Physiol. Tiere* **67**, 407–486.

Kirkwood, T.B.L. and Rose, M.R. (1991). Evolution of senescence: late survival sacrificed for reproduction. *Philos. Trans. R. Soc. Lond. B* **332**, 15–24.

Kozłowski, J. and Wiegert, R.G. (1986). Optimal allocation of energy to growth and reproduction. *Theor. Pop. Biol.* **29**, 16–37.

Krol, M., Griffith, M. and Huner, N.P.S. (1984). An appropriate physiological control for environmental temperature studies: comparative growth kinetics of winter rye. *Can. J. Bot.* **62**, 1062–1068.

Lamb, R.J. and Gerber, G.H. (1985). Effects of temperature on the development, growth, and survival of larvae and pupae of a north-temperate chysomelid beetle. *Oecologia (Berl.)* **67**, 8–18.

Lamb, R.J. and MacKay, P.A. (1988). Effects of temperature on developmental rate and adult weight of Australian populations of *Acyrthosiphon pisum* (Harris) (Homoptera: Aphididae). *Mem. Entomol. Soc. Can.* **146**, 49–56.

Lamb, R.J., MacKay, P.A. and Gerber, G.H. (1987). Are development and growth of pea aphids, *Acyrthosiphon pisum*, in North America adapted to local temperatures? *Oecologia (Berl.)* **72**, 170–177.

Larsen, K.J., Madden, L.V. and Nault, L.R. (1990). Effect of temperature and host plant on the development of the blackfaced leafhopper. *Entomol. Exp. Appl.* **55**, 285–294.

Laudien, H. (1973). Changing reaction systems. In: *Temperature and Life* (Ed. by H. Precht, J. Christophersen, J. Hensel and W. Larcher), pp. 355–399. Springer-Verlag, Berlin.

Laybourn-Parry, J., Abdullahi, B.A. and Tinson, S.V. (1988). Temperature-dependent energy partitioning in the benthic copepods *Acanthocyclops viridis* and *Macrocyclops albidus. Can. J. Zool.* **66**, 2709–2714.

Lebedeva, L.I. and Gerasimova, T.N. (1985). Peculiarities of *Philodina rosleoa* (Ehrbg.) (Rotatoria Bdelloida). Growth and reproduction under various temperature conditions. *Int. Rev. Ges. Hydrobiol.* **70**, 509–525.

Leffler, C.W. (1972). Some effects of temperature on growth and metabolic rate of juvenile crabs, *Callinectes sapidus* in the laboratory. *Marine Biol.* **14**, 104–110.

Lewontin, R.C. (1965). Selection for colonizing ability. In: *The Genetics of Colonizing Species* (Ed. by H.G. Baker and G.L. Stebbings), pp. 77–91. Academic Press, New York.

Lints, F.A. and Lints, C.V. (1971). Influence of pre-imaginal environment on fecundity and ageing in *Drosophila melanogaster* hybrids, II. Pre-imaginal temperature. *Exp. Gerontol.* **6**, 417–426.

Lock, A.R. and McLaren, I.A. (1970). The effect of varying and constant temperature on the size of a marine copepod. *Limnol. Oceanogr.* **15**, 638–640.

Loeb, J. and Northrop, J.H. (1917). On the influence of food and temperature on the duration of life. *J. Biol. Chem.* **32**, 103–121.

Lonsdale, D.J. and Levinton, J.S. (1985). Latitudinal differentiation in copepod growth: an adaptation to temperature. *Ecology* **66**, 1397–1407.

Loosanoff, V.L. (1959). The size and shape of metamorphosing larvae of *Venus (Mercenaria) mercenaria* grown at different temperatures. *Biol. Bull.* **117**, 308–318.

Lopez-Saez, J.F., Gonzalez-Bernaldez, F., Gonzales-Fernandez, A. and Garcia

Ferrero, G. (1969). Effect of temperature and oxygen tension on root growth, cell cycle and cell elongation. *Protoplasma* **67**, 213–221.

Ludwig, D. and Rowe, L. (1990). Life history strategies for energy gain and predator avoidance under time constraints. *Am. Nat.* **135**, 686–707.

Macan, T.T. (1957). The life histories and migrations of the Ephemeroptera in a stony stream. *Trans. Soc. Br. Entomol.* **12**, 129–156.

Margalef, R. (1954). Modifications induced by different temperatures on the cells of *Scenedesmus obliquus* (Chlorophyceae). *Hydrobiologia* **6**, 83–94.

Marian, M.P. and Pandian, T.J. (1985). Effect of temperature on development, growth and bioenergetics of the bullfrog *Rana tigrina*. *J. Thermal Biol.* **10**, 157–162.

Marks, E.N. (1954). A review of the *Aedes scutellaris* subgroup with a study of variation in *Aedes pseudoscutellaris* (Theo.). *Bull. Br. Mus. (Nat. Hist.) Entomol.* **3**, 349–414.

Masaki, S. (1978). Seasons and latitudinal adaptations in the life cycles of crickets. In: *Evolution of Insect Migration and Diapause* (Ed. by H. Dingle), pp. 72–100. Springer-Verlag, Berlin.

Masry, A.M. and Robertson, F.W. (1979). Cell size and number in the *Drosophila* wing. III. The influence of temperature differences during development. *Egypt. J. Genet. Cytol.* **8**, 71–79.

Mayr, E. (1956). Geographical character gradients and climatic adaptation. *Evolution* **10**, 105–108.

McKenzie, J.A. (1978). The effect of developmental temperatures on population flexibility in *Drosophila melanogaster* and *D. simulans*, *Aust. J. Zool.* **26**, 105–112.

McLaren, I.A. (1963). Effects of temperature on growth of zooplankton and the adaptive value of vertical migration. *J. Fish. Res. Bd Canada* **20**, 685–727.

McNab, B.K. and Auffenberg, W. (1976). The effect of large body size on the temperature regulation of the Komodo dragon. *Varanus komodoensis*. *Comp. Biochem. Physiol.* **55A**, 345–350.

Meats, A. (1971). The relative importance to population increase of mortality, fecundity and the time variables of the reproductive schedule. *Oecologia* **6**, 223–237.

Meeson, B.W. and Sweeney, B.M. (1982). Adaptation of *Ceratium furca* and *Gonyaulax polyedra* (Dinophycae) to different temperatures and irradiances: growth rates and cell volumes. *J. Phycol.* **18**, 241–245.

Mer, G. (1937). Variations saisonnieres des caractères de *Anopheles elutus* en Palestine. II. *Bull. Soc. Pathol. Exot. Paris* **30**, 38–42.

Miller, T.J., Crowder, L.B., Rice, J.A. and Marschall, E.A. (1988). Larval size and recruitment mechanisms in fishes: toward a conceptual framework. *Can. J. Fish. Aquat. Sci.* **45**, 1657–1670.

Miller, W.E. (1977). Weights of *Polia grandis* pupae reared at two constant temperatures (Lepidoptera: Noctuidae). *Great Lakes Entomol.* **10**, 47–49.

Minkenberg, O.P.J.M. and Helderman, C.A.J. (1990). Effects of temperature on the life history of *Liriomyza bryoniae* (Diptera: Agromyzidae) on tomato. *J. Econ. Entomol.* **83**, 117–125.

Monteith, J.L. (1981). Climatic variation and the growth of crops. *J. R. Meterol. Soc.* **107**, 749–774.

Morgan, K.C. and Kalff, J. (1979). Effect of light and temperature interactions on growth of *Cryptomonas erosa* (Chrysophyceae). *J. Phycol.* **15**, 127–134.

Mučibabić, S. (1956). Some aspects of the growth of single and mixed populations of flagellates and ciliates. The effect of temperature on the growth of *Chilomonas paramecium*. *J. Exp. Biol.* **33**, 627–644.

Mullens, B.A. and Rutz, D.A. (1983). Development of immature *Culicoides variipennis* (Diptera: Ceratopogonidae) at constant laboratory temperatures. *Ann. Entomol. Soc. Am.* **76**, 747–751.

Müller, H.J. (1966). Uber die Ursachen der unterschiedlichen Resistenz von *Vicia faba* L. gegenuber der Bohnenblatlaus *Aphis (Doralis) fabae* Scop. IX. Der Einfluss okologischer Faktoren auf das Wachstum von *Aphis fabae* Scop. *Entomol. Exp. Appl.*. **9**, 42–66.

Murdie, G. (1969). Some causes of size variation in the pea aphid, *Acyrthosiphon pisum* Harris. *Trans. R. Entomol. Soc. Lond.* **121**, 423–442.

Muthukrishnan, J. and Pandian, T.J. (1983). Effect of temperature on growth and bioenergetics of a tropical moth. *J. Thermal Biol.* **8**, 361–367.

Myers, R.A. and Runge, J.A. (1983). Predictions of seasonal natural mortality rates in a copepod population using life history theory. *Marine Ecol. Prog. Ser.* **11**, 189–194.

Nayar, J.K. (1968). Biology of *Culex nigripalpus* Theobald (Diptera: Culicdidae) Part 2: Adult characteristics at emergence and adult survival without nourishment. *J. Med. Entomol.* **5**, 203–210.

Nayar, J.K. (1969). Effects of larval and pupal environmental factors on biological status of adults at emergence in *Aedes taeniorhyncus* (Wied.). *Bull. Entomol. Res*, **58**, 811–827.

Nealis, V.G., Jones, R.E. and Wellington, W.G. (1984). Temperature and development in host-parasite relationships. *Oecologia (Berl.)* **61**, 224–229.

Nelson, W.G. (1980). Reproductive patterns of gammaridean amphipods. *Sarsia* **65**, 61–71.

Nevo, E. (1973). Adaptive variation in size of cricket frogs. *Ecology* **54**, 1271–1281.

Nylin, S., Wickman, P-O. and Wicklund, C. (1989). Seasonal plasticity in growth and development of the speckled wood butterfly, *Pararge aegeria* (Satyrinae). *Biol. J. Linn. Soc.* **38**, 155–171.

Orcutt, J.D. and Porter, K.G. (1983). Diel vertical migration by zooplankton: constant and fluctuating temperature effects on life history parameters of *Daphnia. Limnol. Oceanogr.* **27**, 720–730.

Orcutt, J.D. Jr. and Porter, K.G. (1984). The synergistic effects of temperature and food concentration on life history parameters of *Daphnia. Oecologia (Berl.)* **63**, 300–306.

Pace, M. (1982). Planktonic ciliates: their distribution, abundance and relationship to microbial resources in a monomictic lake. *Can. J. Fish. Aquat. Sci.* **39**, 1106–1116.

Pace, M. and Orcutt, J.D. (1981). The relative importance of protozoans, rotifers, and crustaceans in a freshwater zooplankton community. *Limnol. Oceanogr.* **26**, 822–830.

Paine, R.T. (1976). Size-related predation: an observational and experimental approach with the *Mytilus–Pisaster* interaction. *Ecology* **57**, 858–873.

Palanichamy, S., Ponnuchamy, R. and Thangaraj, T. (1982). Effect of temperature on food intake, growth and conversion efficiency of *Eupterote mollifera* (Insecta: Lepidoptera). *Proc. Ind. Acad. Sci. (Anim. Sci.)* **91**, 417–422.

Palmer, J.O. (1984). Environmental determinants of seasonal body size variation in the Milkweed Leaf Beetle *Labidomera clivicollis* (Kirby) (Coleoptera: Chrysomelidae). *Ann. Entomol. Soc. Am.* **77**, 188–192.

Pechenik, J.A. (1984). The relationship between temperature, growth rate and duration of planktonic life for larvae of the gastropod *Crepidula fornicata* (L.). *J. Exp. Marine Biol. Ecol.* **74**, 241–257.

Pechenik, J.A. and Lima, G. (1984). Relationship between growth, differentiation, and length of larval life for individually reared larvae of the marine gastropod, *Crepidula fornicata. Biol. Bull.* **166**, 537–549.

Pechenik, J.A., Eyster, L.S., Widdows, J. and Bayne, B.L. (1990). The influence of food concentration and temperature on growth and morphological differentiation of blue mussel *Mytilus edulis* L. larvae. *J. Exp. Marine Biol. Ecol.* **136**, 47–64.

Peters, R.H. (1983). *The Ecological Implications of Body Size.* Cambridge University Press, Cambridge.

Porter, W.P. and Gates D.M. (1969). Thermodynamic equilibria of animals with environment. *Ecol. Monogr.* **39**, 222–224.

Poston, F.L., Hammond, R.B. and Pedigo, L.P. (1977). Growth and development of the painted lady on soybeans (Lepidoptera: Nymphalidae). *J. Kansas Entomol. Soc.* **50**, 31–36.

Pourriot, R. (1973). Rapports entre la température, la taille des adultes, la longuer des oeufs et la taux de développement embryonnaire chez *Brachionus calyciflorus* PALLAS (Rotifere). *Ann. Hydrobiol.* **4**, 103–115.

Precht, H., Christophersen, J., Hensel, H. and Larcher, W. (1973). *Temperature and Life.* Springer-Verlag, Berlin.

Pullin, A.D. (1986). Effect of photoperiod and temperature on the life-cycle of different populations of the peacock butterfly *Inachis io. Entomol. Exp. Appl.* **41**, 237–242.

Ray, C. (1960). The application of Bergmann's and Allen's rules to the poikilotherms. *J. Morphol.* **106**, 85–108.

Reiss, M.J. (1989). *The Allometry of Growth and Reproduction.* Cambridge University Press, Cambridge.

Rempel, R.S. and Carter, J.C.H. (1987). Temperature influences on adult size, development and reproductive potential of aquatic Diptera. *Can. J. Fish. Aquat. Sci.* **44**, 1743–1752.

Roberts, E.H. and Summerfield, R.J. (1987). Measurement and prediction of flowering in annual crops. In: *Manipulation of Flowering* (Ed. by J.G. Atherton), pp. 17–50. Butterworths, London.

Robertson, F.W. (1959). Studies in quantitative inheritance XII. Cell size and number in relation to genetic and environmental variation of body size in *Drosophila Genetics* **44**, 869–896.

Roe, R.M., Clifford, C.W. and Woodring, J.P. (1985). The effect of temperature on energy distribution during the last-larval stadium of the female house cricket, *Acheta domesticus. J. Insect Physiol.* **31**, 371–1378.

Roff, D.A. (1981). On being the right size. *Am. Nat.* **118**, 405–422.

Rowe, L. and Ludwig, D. (1991). Size and timing of metamorphosis in complex life cycles: time constraints and variation. *Ecology* **72**, 413–427.

Schaecter, M., Maaloe, O. and Kjeldgaard, N.O. (1958). Dependency on medium and temperature of cell size and chemical composition during balanced growth of *Salmonella typhimurium. J. Gen. Microbiol.* **19**, 592–606.

Schmidt-Neilsen, K. (1984). *Scaling: Why is Animal Size so Important?* Cambridge University Press, Cambridge.

Schoener, T.W. and Janzen, D.H. (1968). Notes on environmental determinants of tropical versus temperate insect size patterns. *Am. Nat.* **102**, 207–224.

Scott, S.M. and Dingle, H. (1990). Developmental programmes and adaptive syndromes in insect life cycles. In: *Insect Life Cycles: Genetics, Evolution and Co-ordination*, pp. 69–85. Springer-Verlag, London.

Seed, R. and Brown, R.A. (1978). Growth as a strategy for survival in two marine bivalves, *Cerastoderma edule* and *Modiolus modiolus. J. Anim. Ecol.* **47**, 283–292.

52   D. ATKINSON

Seikai, T., Tanangonan, J.B. and Tanaka, M. (1986). Temperature influence on larval growth and metamorphosis of the Japanese flounder *Palalichthys olivaceus* in the laboratory. *Bull. Jap. Soc. Sci. Fish.* **52**, 977–982.

Shaw, R.F. and Bercaw, B.L. (1962). Temperature and life-span in poikilothermous animals. *Nature* **196**, 454–457.

Sibly, R.M. and Calow, P. (1986a). *Physiological Ecology of Animals: an Evolutionary Approach.* Blackwell Scientific Publications, Oxford.

Sibly, R.M. and Calow, P. (1986b). Why breeding earlier is always worthwhile. *J. Theor. Biol.* **123**, 311–319.

Sibly, R., Calow, P. and Nichols, N. (1985). Are patterns of growth adaptive? *J. Theor. Biol.* **112**, 553–574.

Simonet, D.E. and Pienkowski, R.L. (1980). Temperature effect on development and morphometrics of the Potato Leafhopper. *Environ. Entomol.* **9**, 798–800.

Sims, S.R., Marrone, P.G., Gould, F., Stinner, R.E. and Rabb, R.L. (1984). Ecological determinants of Bean Leaf Beetle, *Cerotoma trifurcata* (Forster) (Coleoptera: Chrysomelidae), size variation in North Carolina. *Environ. Entomol.* **13**, 300–304.

Smith, S.L. and Lane, P.V.Z. (1987). On the life history of *Centropages typicus*: responses to a fall diatom bloom in the New York Bight. *Marine Biol.* **95**, 305–313.

Smith-Gill, S.J. and Berven, K.A. (1979). Predicting amphibian metamorphosis. *Am. Nat.* **113**, 563–585.

Söderström, O. (1988). Effects of temperature and food quality on life-history parameters in *Parameletus chelifer* and *P.minor* (Ephemeroptera): a laboratory study. *Freshwater Biol*, **20**, 295–303.

Sohal, R.S. (1976). Metabolic rate and lifespan. *Interdisc. Topics Gerontol.* **9**, 25–40.

Sohal, R.S. (1986). The rate of living. In: *Insect Aging* (Ed. by K. -G. Collatz and R.S. Sohal), pp. 23–44. Springer-Verlag, Berlin.

Squire, G.R. (1989). Response to temperature in a stand of pearl millet. *J. Exp. Bot.* **40**, 1391–1398.

Standfuss, M. (1895). On causes of variation in the imago stage of butterflies, with suggestions on the establishment of new species. *The Entomol.* **28**, 69–76.

Stanley, W.F. (1935). The effect of temperature upon wing size in *Drosophila*. *J. Exp. Zool.* **69**, 459–495.

Stearns, S.C. (1976). Life-history tactics: a review of the ideas. *Q. Rev. Biol.* **51**, 3–47.

Stearns, S.C. (1989). The evolutionary significance of phenotypic plasticity. *Bioscience* **39**, 436–445.

Stearns, S.C. and Crandall, R.E. (1984). Plasticity for age and size at sexual maturity: a life history response to unavoidable stress. In: *Fish Reproduction* (Ed. by G. Potts and R.J. Wootton), pp. 13–33. Academic Press, London.

Stearns, S.C. and Koella, J.C. (1986). The evolution of phenotypic plasticity in life history traits: predictions of reaction norms for age and size at maturity. *Evolution* **40**, 893–913.

Stevenson, R.D. (1985). Body size and limits to the daily range of body temperature in terrestrial ectotherms. *Am. Nat.* **125**, 102–117.

Sweeney, B.W. (1978). Bioenergetic and developmental response of a mayfly to thermal variation. *Limnol. Oceanogr.* **23**, 461–477.

Sweeney, B.W. and Vannote, R.L. (1978). Size variation and the distribution of hemimetabolous aquatic insects: two thermal equilibrium hypotheses. *Science* **200**, 444–446.

Sweeney, B.W. and Vannote, R.L. (1984). Influence of food quality and temperature on life history characteristics of the parthenogenetic mayfly, *Cloeon triangulifer*. *Freshwater Biol.* **14**, 621–630.

Sweeney, B.W., Vannote, R.L. and Dodds, P.J. (1986). Effects of temperature and food quality on growth and development of a mayfly, *Leptophlebia intermedia*. *Can. J. Fish. Aquat. Sci.* **43**, 12–18.

Tantawy, A.O. and Mallah, G.S. (1961). Studies on natural populations of *Drosophila*. I. Heat resistance and geographical variation in *Drosophila melanogaster* and *D. simulans*. *Evolution* **15**, 1–14.

Tauber, M.J., Tauber, C.A. and Masaki, S. (1986). *Seasonal Adaptations of Insects*. Oxford University Press, Oxford.

Thormar, H. (1962). Cell size of *Tetrahymena pyriformis* incubated at various temperatures. *Exp. Cell Res.* **27**, 585–586.

Thorne, G.N., Ford, M.A. and Watson, D.J. (1968). Growth, development and yield of spring wheat in artificial climates. *Ann. Bot.* **32**, 425–446.

Tsitsipis, J.A. (1980). Effect of constant temperatures on larval and pupal development of Olive Fruit Flies reared on artificial diet. *Environ. Entomol.* **9**, 764–768.

Tsitsipis, J.A. and Mittler, T.E. (1976). Development, growth, reproduction and survival of apterous virginoparae of *Aphis fabae* at different temperatures. *Entomol. Exp. Appl.* **19**, 1–10.

van den Heuvel, M.J. (1963). The effect of rearing temperature on the wing length, thorax length, leg length and ovariole number of the adult mosquito *Aedes aegypti* (L.) *Trans. R. Entomol. Soc. Lond.* **115**, 197–216.

van Dobben, W.H. (1962). Influence of temperature and light conditions on dry matter distribution, development rate and yield in arable crops. *Neth. J. Agric. Sci.* **10**, 377–389.

van Dobben, W.H. (1963). The physiological background of the reaction of peas to sowing time. Report 214 of IBS. In: *Annual Report of IBS, 1963*, Wageningen.

Vannote, R.L. and Sweeney, B.W. (1980). Geographic analysis of thermal equilibria: a conceptual model for evaluating the effect of natural and modified thermal regimes on aquatic insect communities. *Am. Nat.* **115**, 667–695.

Van't Hof, J. and Ying, H.K. (1964). Relationship between the duration of the mitotic cycle, the rate of cell production and the rate of growth of *Pisum* roots at different temperatures. *Cytologia* **29**, 399–406.

Via, S. and Lande, R. (1985). Genotype-environment interaction and the evolution of phenotypic plasticity. *Evolution* **39**, 505–522.

von Bertalanffy, L. (1957). Quantitative laws in metabolism and growth. *Q. Rev. Biol.* **32**, 217–231.

von Bertalanffy, L. (1960). Principles and theory of growth. In: *Fundamental Aspects of Normal and Malignant Growth* (Ed. by W.N. Nowinski), pp. 137–259. Elsevier, Amsterdam.

Wagner, T.L., Fargo, W.S., Flamm, R.O., Coulson, R.N. and Pulley, P.E. (1987). Development and mortality of *Ips calligraphus* (Coleoptera: Scolytidae) at constant temperatures. *Environ. Entomol.* **16**, 484–496.

Walsby, A.E. and Reynolds, C.S. (1980). Sinking and floating. In: *The Physiological Ecology of Phytoplankton* (Ed. by I. Morris), pp. 371–412. Blackwell Scientific Publications, Oxford.

Warrick, R.A. and Gifford, R.M. with Parry, M.L. (1986). CO_2, climatic change and agriculture. In: *The Greenhouse Effect, Climatic Change, and Ecosystems. Scope No. 29* (Ed. by B. Bolin, D. Doos and J. Jager), pp. 393–473. John Wiley and Sons, New York.

Whitman, D.W. (1986). Developmental thermal requirements for the grasshopper *Taeniopoda eques* (Orthoptera: Acrididae). *Ann. Entomol. Soc. Am.* **79**, 711–714.

Wilber, C.G. (1964). Animals in aquatic environments: introduction. In: *Handbook*

54   D. ATKINSON

of Physiology, Section 4: Adaptation to the Environment (Ed. by D.B. Dill), pp. 661–668. American Physiological Society, Washington, DC.

Wilbur, H.M. and Collins, J.P. (1973). Ecological aspects of amphibian metamorphosis. *Science* **182**, 1305–1314.
Williams, G.C. (1966). *Adaptation and Natural Selection*. Princeton University Press, Princeton, NJ.
Willmer, P. (1991). Thermal biology and mate acquisition in ectotherms. *Trends Ecol. Evol.* **6**, 396–399.
Willmer, P.G. and Unwin, D.M. (1981). Field analyses of insect heat budgets: reflectance, size and heating rates. *Oecologia* **50**, 250–255.
Yoder, J.A. (1979). Effect of temperature on light-limited growth and chemical composition of *Skeletonema costatum* (Bacillariophyceae). *J. Phycol.* **15**, 362–370.
Yu, D.S. and Luck, R.F. (1988). Temperature-dependent size and development of California red scale (Homoptera: Diaspididae) and its effect on host availability for the ectoparasitoid *Aphytis melinus* De Bach (Hymenoptera: Aphelinidae). *Environ. Entomol.* **17**, 154–161.
Zaret, T.M. (1980). *Predation and Freshwater Communities*. Yale University Press, New Haven, CT.
Zeuthen, E. (1964). The temperature-induced division synchrony in *Tetrahymena*. In: *Synchrony in Cell Division and Growth* (Ed. by E. Zeuthen), pp. 99–158. Wiley InterScience, New York.

APPENDIX: EFFECTS OF INCREASED TEMPERATURE ON SIZE-AT-STAGE

No.	KINGDOM PHYLUM Class (Subclass/Order) Species	Effect*	Reference
	BACTERIA (MONERA)		
1	*Pseudomonas* sp.	R	Chrzanowski *et al.* (1988)
	PROTISTA		
	CHLOROPHYTA		
2	*Chlamydomonas reinhardi*	R	Ray (1960)
3	*Chlamydomonas* sp. 1	R	Ray (1960)
4	*Chlamydomonas* sp. 2	R	Ray (1960)
5	*Scenedesmus obliquus*	R	Margalef (1954)
	BACILLARIOPHYTA		
6	*Phaeodactylum tricornutum*	I	Fawley (1984)
	CILIOPHORA		
7	*Colpoda* sp.	R	Adolph (1929)
8	*Tetrahymena gelei*	R	Ray (1960)
	PLANTAE		
	SPERMATOPHYTA		
	Liliopsida		
9	*Pennisetum typhoides*	R	Squire (1989)
10	*Triticum aestivum*	R	Gmelig Meyling (1969)
11	*Zea Mays*	R	Brouwer *et al.* (1973)

No.	KINGDOM PHYLUM Class (Subclass/Order) Species	Effect*	Reference
	Magnoliopsida		
12	*Phaseolus vulgaris*	R	Gmelig Meyling (1969)
13	*Pisum sativum*	R	Gmelig Meyling (1969)
14	*P. sativum*	R	van Dobben (1963)
	ANIMALIA		
	ASCHELMINTHES		
	Nematoda		
15	*Cephalobus* sp.	R	Ray (1960)
	Rotifera		
16	*Brachionus calyciflorus*	R	Pourriot (1973)
17	*Philodina roseola*	R	Lebedeva and Gerasimova (1985)
	MOLLUSCA		
	Gastropoda		
18	*Crepidula fornicata*	R	Pechenik (1984)
19	*Lymnaea japonica*	R	Imai (1937b)
	Pelecypoda		
20	*Mytilus edulis*	R	Pechenik *et al.* (1990)
21	*M. edulis*	R	Bayne (1965)
	ARTHROPODA		
	Insecta		
	(Diptera)		
22	*Ablabesmyia mallochi*	R	Rempel and Carter (1987)
23	*Aedes aegypti*	R	van den Heuvel (1963)
24	*Ae. aegypti*	R	Galliard and Golvan (1957)
25	*Ae. nigromaculis*	R	Brust (1967)
26	*Ae pseudoscutellaris*	R	Marks (1954)
27	*Ae. taeniorhyncus*	R	Nayar (1969)
28	*Ae. vexans*	R	Brust (1967)
29	*Chaoborus flavicans*	R	Hanazato and Yasuno (1989)
30	*Conchapelopia aleta*	R	Rempel and Carter (1987)
31	*Culex nigripalpus*	R	Nayar (1968)
32	*Culicoides variipennis*	R	Mullens and Rutz (1983)
33	*C. variipennis*	R	Akey *et al.* (1978)
34	*Culiseta inornata*	R	Brust (1967)
35	*Dacus oleae*	M	Tsitsipis (1980)
36	*Drosophila equinoxalis*	R	Ray (1960)
37	*D. melanogaster*	R	Alpatov and Pearl (1929)
38	*D. melanogaster*	R	Imai (1937a)
39	*D. melanogaster*	R	Tantawy and Mallah (1961)

No.	KINGDOM PHYLUM Class (Subclass/Order) Species	Effect*	Reference
40	*D. melanogaster*	M	David and Clavel (1967)
41	*D. melanogaster*	R	Economos *et al.* (1982)
42	*D. melanogaster*	R	Lints and Lints (1971)
43	*D. melanogaster*	R	Stanley (1935)
44	*D. persimilis*	R	Ray (1960)
45	*D. pseudoobscura*	R	Ray (1960)
46	*D. pseudoobscura*	R	Druger (1962)
47	*D. simulans*	R	Tantawy and Mallah (1961)
48	*Drosophila sp*	R	Eigenbrodt (1930)
49	*D. willistoni*	R	Ray (1960)
50	*Nilotanypus fimbriatus*	T	Rempel and Carter (1987)
51	*Parametriocnemus lundbecki*	R	Rempel and Carter (1987)
52	*Polypedilum aviceps*	R	Rempel and Carter (1987)
53	*Stempellinella brevis*	R	Rempel and Carter (1987)
	(Hemiptera)		
54	*Acyrthosiphon pisum*	R	Murdie (1969)
55	*A. pisum*	R	Lamb *et al.* (1987)
56	*A. pisum*	M	Lamb and MacKay (1988)
57	*Aonidiella aurantii*	R	Yu and Luck (1988)
58	*Aphis fabae*	R	Dixon *et al.* (1982)
59	*A. fabae*	R	Tsitsipis and Mittler (1976)
60	*A. fabae*	M	Müller (1966)
61	*Empoasca fabae*	R	Simonet and Pienkowski (1980)
62	*Graminella nigrifrons*	R	Larsen *et al.* (1990)
	(Coleoptera)		
63	*Attagenus megatoma*	I	Baker (1983)
64	*Enaphalodes rufulus*	I	Galford (1974)
65	*Entomoscelis americana*	I	Lamb and Gerber (1985)
66	*Ips calligraphus*	R	Wagner *et al.* (1987)
67	*Pterohelaeus darlingensis*	R	Allsopp (1981)
68	*P. alternatus*	R	Allsopp (1981)
69	*Trogoderma anthrenoides*	M	Burges and Cammell (1964)
	(Orthoptera)		
70	*Acheta domesticus*	I	Roe *et al.* (1985)
71	*Taeniopoda eques*	I	Whitman (1986)
	(Ephemeroptera)		

No.	KINGDOM PHYLUM Class (Subclass/Order) Species	Effect*	Reference
72	*Ameletus ludens*	R	Vannote and Sweeney (1980)
73	*Caenis simulans*	I	Vannote and Sweeney (1980)
74	*Cloeon triangulifer*	R	Sweeney and Vannote (1984)
75	*Ephemerella funeralis*	I	Vannote and Sweeney (1980)
76	*Isonychia bicolor*	I	Sweeney (1978)
77	*Leptophlebia intermedia*	R	Sweeney *et al.* (1986)
78	*Parameletus minor*	R	Söderström (1988)
79	*P. chelifer*	R	Söderström (1988)
80	*Tricorythodes atratus*	I	Vannote and Sweeney (1980)
	(Hymenoptera)		
81	*Apanteles rubecula*	R	Nealis *et al.* (1984)
	(Lepidoptera)		
82	*Achaea junta*	R	Muthukrishnan and Pandian (1983)
83	*Eupterote mollifera*	R	Palanichamy *et al.* (1982)
84	*Inachis io*	R	Pullin (1986)
85	*Orgyia pseudotsugata*	I	Beckwith (1982)
86	*Pieris rapae*	R	Jones *et al.* (1982)
87	*Polia grandis*	R	Miller (1977)
88	*Pseudaletia unipuncta*	I	Guppy (1969)
	Crustacea		
	(Copepoda)		
89	*Acanthocyclops vernalis*	R	Coker (1933)
90	*A. vernalis*	R	Abdullahi and Laybourn-Parry (1985)
91	*A. viridis*	R	Abdullahi and Laybourn-Parry (1985)
92	*A. viridis*	R	Coker (1933)
93	*A. viridis*	R	Laybourn-Parry *et al.* (1988)
94	*Cyclops serrulatus*	R	Coker (1933)
95	*Macrocyclops albidus*	R	Abdullahi and Laybourn-Parry (1985)
96	*M. albidus*	R	Laybourn-Parry *et al.* (1988)
97	*Pseudocalanus minutus*	R	Lock and McLaren (1970)
98	*Salmincola salmoneus*	I	Johnston and Dykeman (1987)

No.	**KINGDOM** PHYLUM Class (Subclass/Order) Species	Effect*	Reference
	(Decapoda)		
99	*Callinectes sapidus*	R	Leffler (1972)
100	*Cancer irroratus*	R	Johns (1981)
	(Mysidacea)		
101	*Neomysis mercedis*	R	Johnston and Northcote (1989)
	CHORDATA		
	Amphibia		
102	*Pseudoacris ornata*	R	Harkey and Semlitsch (1988)
103	*Rana pipiens*	R	Smith-Gill and Berven (1979)
104	*R. sylvatica*	R	Ray (1960)
105	*R. sylvatica*	R	Berven and Gill (1983)
106	*R. sylvatica*	R	Berven and Gill (1983)
107	*R. sylvatica*	R	Berven and Gill (1983)
108	*T. tigrina*	R	Marian and Pandian (1985)
	Osteichthyes		
109	*Paralichthys olivaceus*	R	Seikai *et al.* (1986)

* R denotes a significant size reduction with increasing temperature; I denotes a significant size increase, and M denotes significant mixed effects (significant increases *and* reductions).

Carbon Allocation in Trees: a Review of
Concepts for Modelling

M.G.R. CANNELL and R.C. DEWAR

ADVANCES IN ECOLOGICAL RESEARCH VOL. 25
ISBN 0–12–013925–1

I. SUMMARY

The plant is divided into five parts which are functionally interdependent (reproductive organs, temporary storage sinks, shoots, roots and woody parts) and concepts for modelling allocation are reviewed concerning the following six functional relationships.

1. Reproductive sinks, which store assimilates in fruits and seeds, are fundamentally different from utilization sinks, which use assimilates for growth. Reproductive sinks generally possess active phloem unloading mechanisms across the apoplast.
2. Temporary storage sinks exist because inevitably there are times when assimilate supply exceeds demand, and trees with storage reserves have a large competitive and survival advantage. These sinks may have a relatively high priority.
3. In the long term, the assimilation of carbon by foliage, and the acquisition of mineral nutrients by fine roots, must be in balance with the utilization of carbon and mineral nutrients in plant growth. This functional balance can be formulated at the whole-plant level without reference to mechanisms, but the formulation is valid only when carbon and nutrient substrates form either negligible or constant fractions of total plant carbon and nutrients. The whole-plant functional balance can be expressed in terms of carbon consumption in growth and carbon produced by assimilation, enabling root:shoot responses to nutrients that behave differently from nitrogen to be described. Transport-resistance models provide a mechanistic explanation of this functional balance in terms of substrate dynamics, and predict root:shoot responses more generally. Two crucial features of these models are that (a) the growth of each plant part is co-limited by local concentrations of carbon and nutrient substrates, and (b) opposing gradients exist in carbon and nutrient substrate concentrations between the foliage and fine roots. The latter condition can be maintained with xylem–phloem cycling of nutrients. A brief discussion is presented of the pathways of nutrient transport, seasonal switching of utilization sinks, storage, and the carbon cost of nutrient acquisition.
4. Allocation to roots can be optimized with respect to water acquisition to meet the water loss associated with CO_2 uptake. If carbon gain is to be maximized, it appears that an increase in mesophyll resistance will decrease carbon allocation to roots, while an increase in stomatal resistance will increase carbon allocation to roots. However, such concepts need to be coupled with processes of nutrient uptake and utilization.
5. The functional interdependence between foliage and the water conducting system is evident in physiological and mensurational studies, and is embodied in the pipe-model hypothesis. This hypothesis can be used to con-

strain assimilate allocation, but is limited by a variable leaf area/sapwood area ratio and unknown rates of sapwood → heartwood transfer.

6. The structural requirements for carbon can be defined without reference to physiological processes using the structural theory for cantilever beams. A transport-matrix method can be used to define the carbon cost of beams of any shape, but, for many branches and trunks, taper giving elastic self-similarity may be assumed. Additionally, allocation along stems can be constrained assuming that the cambium responds to strain (i.e. movement) so as to equalize the distribution of stress.

In conclusion, it is suggested that more work is needed to combine existing concepts, to couple the carbon, nutrient and water cycles, and to measure internal plant substrate levels and the controls of meristem growth.

II. INTRODUCTION

The purpose of this review is to draw together and evaluate the principal concepts that have been put forward to construct or constrain models of carbon allocation in plants, with special reference to trees. Many of the ideas have been proposed by modellers, but their evaluation depends upon observations made by plant physiologists. This review draws on information presented by both research communities and highlights some of the strengths and weaknesses in the arguments. The purpose is to advance the discussion, although the result may be to have rocks thrown from both sides!

Although there is much information on the distribution of dry matter in plants, there is surprisingly little understanding of the mechanisms that govern carbon allocation (Wardlaw, 1990). Research at the process level on carbon allocation, like that on fine root dynamics, has fallen far behind research on processes such as photosynthesis, respiration and leaf growth, and now severely limits our ability to construct process-based models of whole plants (Landsberg *et al.*, 1991). Some of the best data on assimilate transport in plants still dates from the 1920s (Mason and Maskell, 1928). The only whole-plant, complete nitrogen flow budget that we could find was constructed in the 1970s for *Lupinus alba* (Pate, 1980), and most of the research on sink properties seems to have been focused on the fruits, seeds and specialized storage organs that constitute the yield of crop plants (Gifford and Evans, 1981; Thorne, 1985); relatively little work has been done on shoot, root and cambial sinks in the vegetative plant.

Progress in understanding the ground rules governing assimilate allocation in plants may have been hampered by regarding allocation as a single act of partitioning,* in which assimilates are divided among plant parts. A flux

---

* Consequently, the term allocation (= assign or apportion) is probably more appropriate than the term partitioning (= the act of parting or dividing).

arrow labelled "partitioning" or "allocation" exists in many model flow diagrams, on a par with flux arrows labelled photosynthesis, respiration and decomposition. Such flow diagrams normally reveal that the modellers have ignored the diverse and separate processes governing the transport, storage and utilization of assimilates and mineral nutrients and have instead assigned some allocation coefficients or sink priorities. Allocation is the *outcome* of many processes rather than a process in its own right.

For short-lived annual species, it may be adequate and possible to ascribe empirical values to carbon allocation at each stage of plant development. However, this is not satisfactory or possible for perennial species, because important changes will occur in carbon allocation during each year and during the lifetime of the stand. The problem is, indeed, daunting, because carbon allocation ultimately involves all the internal, environmental and genetic processes that regulate plant growth.

One way of making the problem at least tractable is to divide the plant into parts that have clearly different functions, and to identify the phenomena that determine or constrain the relative allocation of assimilates among them. These phenomena should ideally include the physiological processes concerning assimilate transport, meristem growth, respiration and so on, but at our present state of knowledge they are more likely to be hypothetical constructs that provide a surrogate or proxy for the real physiological processes.

To provide a framework for this review, the plant is divided into five parts: (i) reproductive sinks (fruits and seeds); (ii) temporary storage sinks (carbohydrate and nutrient reserves, not specialized sinks such as tubers or storage roots); (iii) foliage; (iv) woody parts (branches, stems and structural roots); and (v) fine roots (Fig. 1). No attempt is made to consider the allocation of assimilates within these five primary types of sink, much less among biochemical compounds; therefore this review does not consider the branching patterns within tree crowns or root systems.

In this review, six separate functional relationships have been identified among these five plant parts (Fig. 1). These are relationships between: 1, reproductive and "utilization" (vegetative) sinks; 2, "utilization" and storage sinks; 3, foliage and fine roots, concerning carbon assimilation and nutrient uptake; 4, foliage and fine roots, concerning carbon assimilation and water acquisition; 5, foliage and woody parts concerning water loss and water conducting capacity; and 6, tree parts providing structural support and parts being supported. In all six relationships, the division of growth between two plant parts (or sink types) can be regarded as being constrained to meet some necessary functional balance or goal. In each case, this balance or goal is stated, followed by an account and discussion of some hypotheses concerning the mechanisms that regulate the balance.

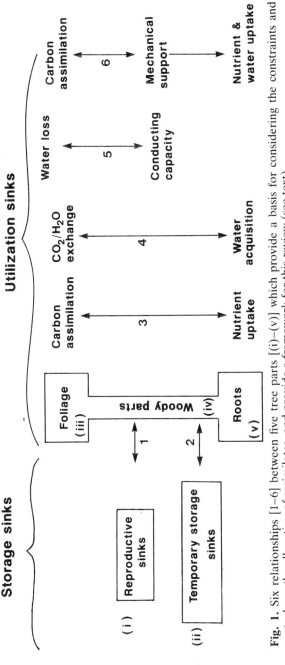

Fig. 1. Six relationships [1–6] between five tree parts [(i)–(v)] which provide a basis for considering the constraints and controls on the allocation of assimilates, and provide a framework for this review (see text).

III. REPRODUCTIVE vs. UTILIZATION SINKS (1 in Fig. 1)

The term "utilization sink" was coined by Ho (1988) to describe the shoot, leaf, root and cambial meristems which use assimilates for growth as opposed to reproductive sinks in which assimilates are transferred to storage pools.

A. The Carbon Cost of Reproduction by Seed

In evolutionary terms, a successful plant genotype must leave at least one descendent. Seeds must contain sufficient reserves of carbohydrate and mineral nutrients to give the embryos a continuation of their maternal support during germination. If carbon is not to be wasted, these reserves must be above a minimal level, and this is ensured by making the seeds and associated structures priority sinks for assimilates. During the period of endosperm filling, the growth rates of seeds (like the bulking rates of potatoes and storage roots) tend to be constant, and, within limits, the final weight per seed is usually much less variable than the numbers of seeds produced per plant (Harper, 1977).

The consequences for other sinks are well documented: metabolite reserves are utilized and the growth rates of the roots, cambium and shoots are all decreased, despite sink-enhanced photosynthesis (see Cannell, 1985). Trees, like other perennial plants, have developed a number of mechanisms to adjust the total number of seeds or fruits that are set (and both the time of year and time in their life cycle when fruiting occurs) to ensure that the assimilate and nutrient resources are adequate to produce full-sized seeds without threatening the survival of the vegetative structure (Harper, 1977; Lloyd, 1980; Cannell, 1985). In this review, we are not concerned with those strategies, but rather with the mechanism by which seeds and fruits operate as powerful sinks.

B. "Strength" of Reproductive vs. Utilization Sinks

The mechanism by which different sinks "attract" assimilates can be explained, without invoking growth regulators (Patrick, 1976), by the way in which carbon is transported over long and short distances within the plant. The following brief account is based on reviews by Peel (1974), Pate (1980), Gifford and Evans (1981), Giaquinta (1983), Thorne (1985), Ho (1988) and Wardlaw (1990).

The consensus view is that long-distance transport of carbon occurs by mass flow in the phloem, along a hydrostatic pressure gradient which is created by a gradient in osmotic pressure between the leaves, where carbon substrates (mainly sucrose, but not exclusively, see Dickson, 1991) are loaded into the phloem and the points where they are unloaded into sinks (Münch, 1930). If the sieve-plate pores are open, little energy is needed to set up and maintain a sufficient gradient to drive mass flow over long distances, and energy need not be applied along the transport pathway itself.

Short distance transport of carbon substrates at the points of loading (principally the leaves) and unloading (the sinks) occurs either through the *symplast*—the mass of living material in the plant interconnected between cells by plasmodesmata—or the *apoplast* (or "free space")—the mass of non-living cell wall material which also forms a continuous network in the plant. The crucial point is that materials can flow relatively unimpeded through either the symplast or apoplast, but if they cross both, they must pass from one to the other through a semipermeable barrier, the plasmalemma membrane. This membrane can then be the site of an energy-requiring proton/C-substrate co-transport process, involving ATPase. This process of active transport enables the substrates to be pumped into or out of the phloem against a concentration gradient.

The simple explanation for the large sink "strength" of fruits and seeds is that, in all species studied, substrates pass through the apoplast from the phloem to the reproductive sink (Fig. 2). Active transport can then occur between the apoplast and the sink, where substrates are used for growth, for the formation of storage products, or are chemically transformed. Active transport keeps the apoplastic substrate concentration much below that in the sieve tubes, and lower than the saturation concentration for sink growth (Fig. 2).

Meanwhile, the sources (leaves) maintain a high concentration of carbon substrates (typically 200–800 mM) in the phloem at the points of loading. In most crop plants of the temperate zone this loading is through the apoplast, involving active transport. However, some woody plants, especially tropical ones, seem to possess less efficient symplastic loading mechanisms which are facilitated by the development of plasmodesmatal connections between the phloem and neighbouring cells (Gamalei, 1991; van Bel and Gamalei, 1992).

The shoot, root and cambial sinks seem to possess only symplastic connections with the phloem, with no active phloem unloading. In these cases, a substrate gradient is maintained between the phloem and sink cells, across plasmodesmatal connections, simply by the utilization of substrates for growth and associated respiration. Consequently, their ability to develop a substrate gradient in the phloem between themselves and the leaves is wholly dependent upon their own growth rate, which itself may depend on the substrate concentration (see below). There is, thus, a fundamental distinction between "utilization sinks"—which use assimilates for growth—and "reproductive sinks"—which store assimilates in fruits, seeds or in specialized storage roots or stems (Table 1). In short, the latter generally possess active phloem unloading mechanisms whereas the former do not (Giaquinta, 1983; Thorne, 1985; Ho, 1988).

66   M.G.R. CANNELL AND R.C. DEWAR

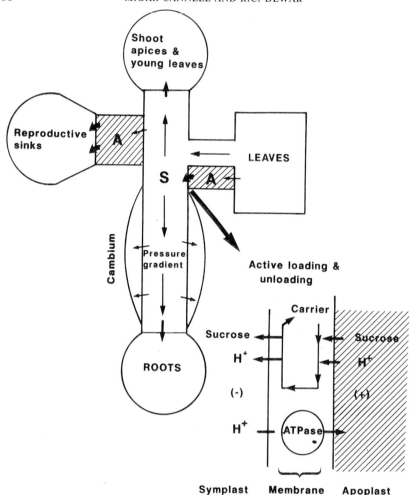

Fig. 2. Simplified schematic representation of the system of carbon assimilate transfer in plants. S = sieve tube/companion cell/vascular parenchyma complex. A = apoplast (cell wall free space) separating the carbon substrate source and the reproductive sinks from the sieve tube complex. The thick curved arrows represent the major points of active transport across the membranes separating the apoplast from the symplast at the points of sucrose loading and unloading, as shown in the inset diagram (from Giaquinta, 1983).

The relative "strength" of different reproductive sinks will depend upon their size (e.g. the number of cells developed in seed tissues) and inherent attributes that determine the rate of utilization, storage or transformation of metabolites. The relative "strength" of different utilization sinks is the subject of discussion later in this review.

Table 1
General attributes of "utilization" and "reproductive" sinks

	Utilization sinks	Reproductive sinks
Plant organs	Vegetative. Shoot and root meristems. Young leaves. Cambium	Reproductive and some storage organs. Seeds, fruits, tubers, storage roots
Phloem unloading mechanism	Symplastic transport through plasmodesmata. Usually irreversible for carbon	Active proton cotransport across plasmalemma membrane from apoplast. Reversible for some storage sinks

IV. TEMPORARY STORAGE vs. UTILIZATION SINKS (2 in Fig. 1)

The main carbohydrate storage tissue in woody plants is the ray parenchyma, which forms a continuous system throughout the branches, stems and structural roots. Parenchyma in the roots seems to be an especially important site of storage (Gholz and Cropper, 1991). Foliage can also contain a temporary store of starch. The parenchyma is a temporary and reversible storage sink, unlike specialized storage sinks, and forms one continuous system, interconnected by plasmodesmata to the plant's symplastic system (Sauter and Kloth, 1986). This temporary storage sink is presumably loaded and unloaded through the symplast, but it may possess some of the sink properties of specialized storage sinks (see below).

Trees also store and retranslocate mobile nutrients such as N, P, K, S, Co and Zn. Again, storage occurs in the ray parenchyma, but in evergreen trees the main storage organs for nitrogen are the leaves (Nambiar and Fife, 1991; Millard and Proe, 1992). In mature trees, over half of the nutrients for new growth may be met by retranslocation (Miller, 1986). According to Nambiar and Fife (1991) nutrient retranslocation is a normal characteristic of trees, which occurs regardless of nutrient supply—it is not a response to nutrient storage; nutrients are remobilized not only from senescing leaves, but also move in and out of young leaves in response to changing demands by growing shoots; but the fine roots, which contain similar nitrogen concentrations to leaves, give up little of their nitrogen before they die.

A. Functional Relationships

Temporary carbohydrate storage sinks play an important role in tree growth simply because inevitably there are times when (i) assimilate supply exceeds the demand of utilization sinks, and (ii) assimilate demand by the utilization sinks exceeds supply. Let us consider those two situations.

In general, small decreases in tissue water potential, like decreases in temperature, reduce rates of cell expansion and leaf growth (sink activity) more than rates of photosynthesis (Hsiao, 1973). Similarly, McDonald et al. (1986) showed that an instantaneous decrease in the rate of nitrogen supply to birch seedlings decreased rates of leaf growth more than rates of photosynthesis. Thus, whenever trees experience modest deficiencies in supplies of water or nutrients they are likely to produce excess assimilates which can be stored. Luxmoore (1991) pointed out that, if the reverse were true, i.e. plants responded to water and nutrient stress by decreasing their rates of photosynthesis more than their rates of growth, they would very soon deplete all reserves and so lack all the advantages of possessing a storage reserve. Stress-tolerant trees may maintain a large storage reserve all the time because they are inherently slow-growing (Grime and Hunt, 1975).

In addition to carbohydrate storage occurring during periods of stress, storage also occurs when sinks are switched off during the phenological cycle. Thus, in non-deciduous temperate-zone trees, much of the carbohydrate storage occurs in early spring, before budburst, and in the autumn, when photosynthetic rates can be quite high (Kozlowski, 1992).

The functional role of carbohydrate reserves is to supplement current assimilates when sink demands are unusually large. One of the crucial periods when reserves of both carbohydrates and mineral nutrients confer a high competitive and survival value is the period of foliage expansion in spring, especially in deciduous trees, when rapid and timely refoliation is important to suppress competitors and to maximize seasonal light interception (e.g. McLaughlin et al., 1980; Cannell, 1989). Reserves are also mobilized during periods when there are brief but crucial large demands for assimilates, such as periods of seed filling or periods of repair following biotic or abiotic damage. Without reserves, the growth of roots and shoots would be severely curtailed during such periods.

The functional role of nutrient reserves is similar to that of carbohydrate reserves, except that mobile nutrients seem to be more readily moved around the plant. Nitrogen appears to be moved day by day in order to maximize carbon gain among shoots within the crown, as well as between leaves and within leaves (Friend, 1991; Nambiar and Fife, 1991).

B. Modelling Storage

Carbohydrate storage sinks are normally modelled as passive reservoirs, which are filled only when assimilate supply exceeds the demands of the utilization sinks. However, it should be noted that the demand of utilization sinks for carbohydrates can be limited by nutrient as well as carbohydrate supply (both are needed for growth) whereas the storage sink capacity may not be limited by nutrient supply. Weinstein *et al.* (1991) allocated excess carbon to storage pools in proportion to the non-structural carbohydrate deficit of each storage tissue. Only the needle storage pool was given a higher priority for carbon assimilates than the utilization sinks. In their model, storage carbohydrate was used when demand exceeded current supply, drawing first on the storage pools that were most saturated with carbohydrate.

C. Discussion

Because the existence of storage reserves of carbohydrate is so important at times when there is an unusually large demand for assimilates, it may not be correct to treat the storage sinks as passive reservoirs. They are not an optional extra. Trees seem to maintain moderate levels of reserve carbohydrates most of the time (Priestley, 1962) which are not used in root, cambial or shoot growth, except at times of very large demand. It is quite normal for new root growth to depend upon a supply of current assimilates while root storage reserves remain unused (Ritchie and Dunlop, 1980). Conversely, when storage sinks are refilled they may well take priority over utilization sinks, suggesting that there is some active unloading of carbon substrates into the sinks. It is certainly true that storage sinks are refilled at the same time as the growth of utilization sinks (Weinstein *et al.*, 1991). For instance, Marshall and Waring (1985) found that fine root growth in Douglas fir was accompanied by a build-up of starch in the fine roots. In short, temporary storage sinks may have quite a high priority, and be capable to some extent of active substrate unloading from the phloem, or at least of acting like strong sinks.

V. CARBON ASSIMILATION vs. NUTRIENT UPTAKE (3 in Fig. 1)

A. Root–Shoot Functional Balance concerning Carbon and Mineral Nutrients

If we take an average over a period when there is no net change in storage reserves, then the assimilation of carbon by foliage, and the acquisition of mineral nutrients by fine roots, must be in balance with the utilization of carbon and mineral nutrients in plant growth (plus associated losses by exudation, leaching etc.). There should, in such circumstances, be a functional balance between the size and activity of the carbon fixation system (shoots) and the size and activity of the nutrient acquisition system (fine roots). This aspect of plant growth has been reviewed and evaluated many times since it was first proposed by Brouwer (1962) and Davidson (1969), and readers are referred to the evidence presented by Ledig (1983), Cannell (1985), Wilson (1988) and Kurz (1989) and the arguments given by Charles-Edwards (1982) and Thornley and Johnson (1990).

Suffice it to say that there is overwhelming evidence that the response of plants to a shortage of carbon, brought about by reduced irradiance or a loss of leaves, is generally to increase assimilate allocation to leaves. Conversely, root pruning or a poor nitrogen supply generally increases assimilate allocation to roots. This functional balance was described by Davidson (1969) as simply a direct proportionality between the supply rates of nitrogen and carbon.

$$\sigma_n W_r \propto \sigma_c W_s \qquad (1)$$

where W_r is root mass, W_s is shoot mass, σ_n is the rate of nitrogen uptake per unit root mass and σ_c is the rate of carbon fixation per unit shoot mass. It follows from equation (1) that:

$$\frac{W_s}{W_r} \propto \frac{\sigma_n}{\sigma_c} \qquad (2)$$

which states that the shoot:root ratio increases with decrease in σ_c or increase in σ_n. However, the observed change in % N content with N supply is inconsistent with equation (2), as we discuss below. Figure 3 presents observations on *Betula verrucosa* which are qualitatively consistent with equation (2). Other evidence has been assembled by Cannell (1985) for trees and by Wilson (1988) for plants in general.

Two important caveats need to be added to this general view of the root–shoot functional balance. First, not all nutrients behave the same way as nitrogen. More recent work by Ingestad and co-workers has shown that reduced rates of supply of S, P and to some extent Fe, increase assimilate allocation to roots in a similar manner to equation (1), but reduced rates of supply of Mg, Mn and to some extent K, behave in the opposite way, as shown in Fig. 4 (Ericsson, 1990). An explanation is given below.

The second caveat is that different species grown in the same environment develop different root:shoot ratios (Grime *et al.*, 1988), indicating that there are inherent differences in the masses of root and shoot, or of fine root and foliage, required to maintain the functional balance. Inherent differences may be related to the structure or function of the roots or shoots, to differences in the fractional content of carbon and nutrients in structural dry matter (see equation (3) below), or to some other inherent sink strength factor. Such differences between species may explain why Nadelhoffer *et al.* (1985) found that broadleaved species growing on fertile sites had more fine roots than conifers growing on infertile sites—an observation that runs counter to most others where trees of the same species have been compared on different sites

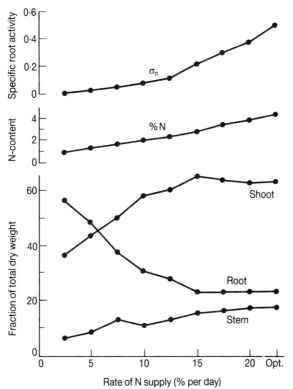

Fig. 3. Effects of supplying nitrogen to *Betula verrucosa* seedlings at different constant rates (using the Ingestad misting system) for 80 days, maintaining constant N concentrations in the tissues at a given rate of supply. (Ingestad, 1979; Ingestad and Lund, 1979). N content is in unit of % dry weight. Specific root activity is in g N per g root dry weight per second $\times\ 10^6$.

(Maggs, 1961; Keyes and Grier, 1981; Alexander and Fairley, 1983; Axelsson and Axelsson, 1986; Kurz, 1989).

B. Hypotheses Put Forward by Modellers

The functional balance between nitrogen and carbon supply and utilization can be represented in models in a non-mechanistic way by assuming that the foliage takes first priority in the use of carbon that they fix, and that the fine roots take first priority in the use of nitrogen that they take up. The model TREGRO of Weinstein *et al.* (1991) deals with carbon and nitrogen allocation in this way, and similar priorities for carbon utilization are given in the CARBON model of Bassow *et al.* (1990), FOREST-BGC (Running and Gower, 1991) and in several crop models (Loomis *et al.*, 1979).

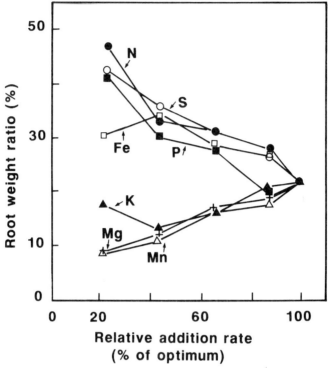

Fig. 4. Effect of varying the relative addition rate of different plant nutrients to *Betula pendula* seedling below the optimum for plant growth (= 100%) on the ratio of root to total plant dry weight. From Ericsson (1990).

It is also possible to derive equation (1) as a logical consequence of the conservation of carbon and nitrogen at the whole-plant level, without reference to internal mechanisms (Thornley, 1972b, his eqn. (43); Charles-Edwards, 1982; Thornley and Johnson, 1990), provided some simplifying assumptions are made. In an Appendix, we outline this derivation of functional balance, which requires that (a) the plant is growing exponentially, (b) non-structural carbon and nitrogen are either negligible or constant fractions of total plant carbon and nitrogen so that the C:N ratio of the plant is constant, and (c) plant maintenance respiration is negligible in comparison to carbon fixation. In the case when substrate concentrations and maintenance respiration are negligible, the expression of functional balance becomes (cf. equation (A6)):

$$\frac{\sigma_c W_s}{\sigma_n W_r} = \frac{f_c}{Y f_n} \tag{3}$$

which provides an interpretation of the constant of proportionality for equation (1) in terms of the fractional contents of carbon and nitrogen in plant dry matter (f_c and f_n), and the efficiency with which carbon substrates (normally sucrose) are converted to new plant dry matter during growth (Y). This may explain some of the species differences referred to above. Equation (3) implies that, in a given environment specified by the ratio of specific activities (σ_c/σ_n), the plant adjusts its shoot:root ratio (W_s/W_r) until the C:N ratio of resource supply ($\sigma_c W_s/\sigma_n W_r$) equals that required for the growth of new plant dry matter ($f_c/Y f_n$). Or, put more simply, C and N are taken up in the same ratio that they are used in growth.

If assumption (b) does not hold at all (as evidenced in Fig. 3 for example), then shoot:root responses to the environment cannot be predicted from the whole-plant carbon and nitrogen budgets alone, because the internal feedbacks between carbon and nitrogen substrates become important (see Appendix). In that case, a mechanistic understanding of substrate dynamics is required, in particular of the relation between utilization sink strength and substrate concentrations, and of the internal transport of substrates between different plant parts. Substrate dynamics are also needed to understand the internal mechanisms by which plants are able to adjust their shoot:root ratios, or to predict the shoot:root behaviour of plants that are not in balanced exponential growth, when, for example, self-shading and competition below ground cannot be ignored (assumption (a)).

Probably the best-known mechanistic treatment of carbon and nitrogen substrate allocation in plants is that developed by Thornley and co-workers (Thornley, 1972a,b; Thornley and Johnson, 1990), recently reviewed by Mäkelä and Sievänen (1987) and considered by Wilson (1988) to account for the majority of observations on the effects of environmental conditions on root:shoot allocation in herbaceous plants.

The assumption on which all the Thornley allocation models is based, is that the relative growth rate of the utilization sinks (roots, shoots and cambium), $\mu = (1/W)\, dW/dt$, is co-limited by the carbon (C) and nitrogen (N) substrate concentrations that they perceive at the point of symplastic unloading from the phloem. This assumption is based on the observation that the reactions that drive the synthesis of proteins from sugars and amino acids in the utilization sinks follow the bisubstrate Michaelis–Menten equation for enzyme kinetics (Thornley and Johnson, 1990, their eqn. (13.7)) for which there is some evidence (Cooper and Thornley, 1976, their Fig. 1; Gifford and Evans, 1981). The simplest version of this assumption is that:

$$\mu = \frac{1}{W}\frac{dW}{dt} \propto C \times N \qquad (4)$$

Thornley used this idea to construct a number of *teleonomic* or goal-seeking models (Fig. 5a). The structural biomass was considered separately from the carbon and nitrogen substrate pools, which were assumed to have the same

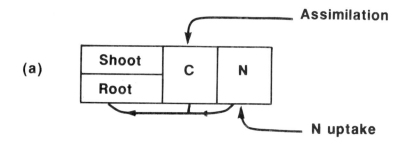

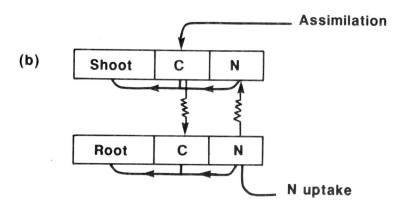

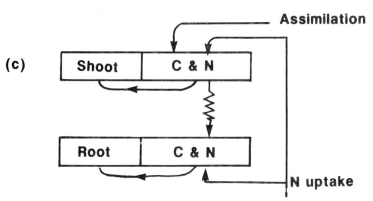

Fig. 5. Basic structure of models in which growth is determined by the product of carbon (C) and nitrogen (N) substrate concentrations. (a) The simple teleonomic model (Reynolds and Thornley, 1982). (b) A shoot–root transport-resistance model (Thornley, 1972a, b). (c) The Münch flow model of combined C and N transport (Dewar, 1993).

concentrations in the shoot and root. Allocation was effected by a coefficient, which was defined as a function of the substrate concentrations, the current shoot and root fractions and the fractional carbon and nitrogen content of shoot and root dry matter. This coefficient could be defined either to constrain the substrate C/N ratio to settle to a fixed value in balanced exponential growth, independent of the external environment (Reynolds and Thornley, 1982), to allow varying C/N ratios (Johnson, 1985), or to maximize plant relative growth rate (Reynolds and Thornley, 1982; Johnson and Thornley, 1987). The latter approach has been extended to include the effect of leaf nitrogen concentration on shoot photosynthesis (Hilbert, 1990; Hilbert and Reynolds, 1991; Hilbert *et al.*, 1991; Gleeson, 1993).

In an earlier *mechanistic* model, Thornley (1972a, b) assumed that there were fluxes of carbon substrates from leaves to other sinks, and of nitrogen substrates from roots to other sinks, driven by concentration gradients divided by resistances to flow (Fig. 5b). As argued by Thornley (1972a), it is physiologically reasonable that transport resistances should decrease with increasing mass of the plant parts being traversed. An analogy with the hydraulic resistance of a pipe (which is proportional to length/cross-sectional area) suggests that transport resistances should be inversely proportional to mass to the power of one-third; however, Thornley (1972a, b) chose a power of one to ensure that his model would attain balanced exponential growth. Observations of gradients in sucrose concentration along the phloem within plants provides good evidence that there are indeed resistances to transport (Mason and Maskell, 1928; Gifford and Evans, 1981) and this view is consistent with the common observation that sinks (such as the roots) that are distant from the leaves, receive least assimilates when, for instance, a reproductive or storage sink is being filled. The assumption concerning nitrogen transport is discussed below.

The Thornley mechanistic model, by imposing resistances to carbon and nitrogen flow, inevitably sets up a gradient of high to low carbon substrate concentration between the shoots and roots and a gradient of high to low nitrogen substrate concentration between the roots and shoots. The growth rates of the shoots and the roots can be made a function of their size, local environment (especially temperature) and an inherent sink strength factor, but the main controlling fact is the product of C and N, as in equation (4). If the nitrogen supply decreases, and there is an equal decrease in nitrogen substrate concentration in the roots and shoots, it is inevitable that the *product* of C and N in the shoots will decrease more than the product of C and N in the roots, because C in the shoots is greater than in the roots. Thus, it is inevitable that a decrease in N will give rise to a smaller decrease in root relative to shoot growth (i.e. an increase in root allocation). The opposite will occur if the carbon substrate supply decreases, because of the gradient in N. The essential ideas are therefore (i) that meristem growth is co-limited by C

and N, and (ii) that gradients in C and N exist in opposite directions in the plant, set up by resistances to transport.

The Thornley transport-resistance approach has recently been used in a forest and generic ecosystem model by Rastetter *et al.* (1991), who assumed that resistances to C and N transport were functions of the length and cross-sectional area of the tissues being traversed. Also, Thornley has constructed a transport-resistance forest model in which the tree is divided into five parts, each with four state variables—C and N substrate pools, meristems and structural tissue (Thornley, 1991; Thornley and Cannell, 1992).

Ågren and Ingestad (1987) suggested that the balance between shoots and roots in plants can be explained as a result of an equilibrium between (i) an internal growth sink for carbon substrate, the strength of which is determined by the total amount of nitrogen in the plant, and (ii) the capacity to supply carbon substrates, set by the amount of shoot. In terms of the present notation they formulated this idea as:

$$\text{Shoot fraction } (f_s) = \frac{P_n N_p}{\sigma_c / f_c} \qquad (5)$$

where P_n is the amount of biomass produced per unit of nitrogen in the biomass (kg dry matter per kg N per unit time), called the nitrogen productivity, and N_p is the plant nitrogen concentration (cf. Ågren and Ingestad, 1987, their eqn. (1)). The numerator in equation (5) effectively describes the consumption of carbon in growth, while the denominator describes the fixation of carbon by photosynthesis.

Ingestad and Ågren (1991) suggested that, by extension, this hypothesis could account for the different root:shoot responses to different mineral nutrients illustrated in Fig. 4. Thus, N, S and P are all major substrates for plant growth and so decreases in their supply will decrease the numerator more than the demoninator, decreasing the shoot fraction. These three elements are the primary ones that are covalently bonded with C, H and O in the organic structures of plants (Clarkson and Hanson, 1980). By contrast, Mg and Mn are not major substrates for growth. Mg is a critical constituent of chlorophyll and is involved in all ATPase reactions, while Mn is required for water-splitting on the oxidant side of photosystem II (among other functions): thus, it is possible that decreasing the supply of Mg and Mn will decrease photosynthetic rates more than the strength of the internal sink, resulting in increased allocation of the shoot (although Mg deficiency will inhibit transport and loading/unloading processes). The responses to K and Fe shown in Fig. 4 can be rationalized in the same way. Clearly, decreased carbon allocation to roots in response to shortages of Mg and Mn will not help to increase their acquisition by the plant and may lead to permanently reduced growth and possibly death.

Ågren and Ingestad (1987) derived equation (5) as a direct consequence of carbon balance at the whole-plant level, independent of any hypothesis of substrate allocation. Their formulation is, in fact, equivalent to a version of the general carbon balance equation (A3) derived in the Appendix, in which carbon substrate concentration (C) and maintenance respiration (R) are assumed to be small. This leads to the relation

$$f_s = \frac{\mu}{(\sigma_c Y / f_c)} \qquad (6)$$

which is identical to equation (5), where the relative growth rate (μ) is expressed in terms of "nitrogen productivity" (P_n) and plant nitrogen concentration (N_p), and where growth respiration (described by Y) is included in the definition of net carbon fixation per unit shoot mass (σ_c). As shown in the Appendix, μ can be eliminated by combining equation (6) with the equivalent expression for the root fraction (f_r) derived from the whole-plant nitrogen balance (equation (A4)), leading to the functional balance expression of equation (3). Therefore, while Ågren and Ingestad (1987) interpret the root:shoot ratio as a balance between nitrogen productivity and photosynthesis, their hypothesis, when combined with the corresponding balance between nitrogen consumption in growth and nitrogen uptake by roots, is formally equivalent to the functional balance between photosynthesis and nitrogen uptake, and is therefore independent of the nitrogen productivity hypothesis. The advantage of Ågren and Ingestad's (1987) approach is that it enables the root:shoot responses to mineral nutrients other than nitrogen to be described in terms of their effect on growth relative to their effect on carbon fixation. However, it should be emphasized that equations (5) and (6) are logical consequences of whole-plant mass balance rather than scientific hypotheses, and offer no explanation of the mechanisms underlying partitioning. The major difference between the Ågren/Ingestad and Thornley approaches is that Thornley assumes that sink strength is controlled by C and N substrate concentrations (equation (4)) while Ågren and Ingestad relate sink strength to the concentration of total (substrate + structural) plant nitrogen (N_p) only.

C. Discussion

Four topics are selected here for discussion: the pathways of nutrient transport, seasonal switching on and off of the utilization sinks, the question of carbon and nutrient storage, and the carbon cost of nutrient acquisition by roots.

1. Nutrient Transport

The simple scheme shown in Fig. 5b for carbon and nitrogen flow ignores the facts that (i) nitrogen moves from the roots to the shoots in the transpiration

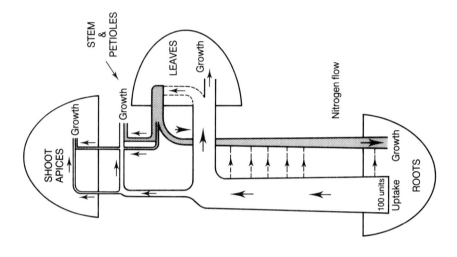

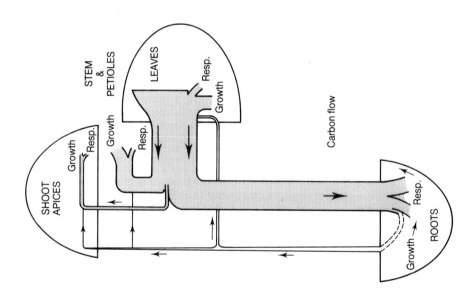

stream, and is therefore not subject to the same kind or magnitude of resistances to flow as carbon moving in the phloem, and (ii) in many plants, much of the nitrogen used in root growth is supplied via the phloem, having passed from roots to leaves and back again, and indeed may cycle back to the shoots (Cooper and Clarkson, 1989). Pate (1980) constructed a complete carbon and nitrogen flow budget for *Lupinus alba* (Fig. 6). Trees, and other species, may not behave in exactly the same way, but *Lupinus*, like most trees, reduces most of its nitrate nitrogen to organic forms in the roots (Peel, 1974; Martin *et al.*, 1981; Dickson, 1991). In *Lupinus*, nitrogen moves in the xylem to those leaves with the greatest transpiration rates, where much of it is transferred to the phloem, using the same transport loading mechanism as used for carbon assimilates. From there, it passes to all sinks, driven by the concentration gradients in carbon substrates—mass flow implies that there can be little bidirectional transport. In *Lupinus*, appreciable quantities of nitrogen are transferred from xylem to phloem in the stems, so that the phloem sap is progressively enriched as it passes to the roots. Consequently, about half of the nitrogen used in root growth in *Lupinus* is received from the leaves, half from the transfer between xylem and phloem, and very little by direct transfer from the sites of uptake to the sites of utilization within the roots themselves.

How can the root:shoot functional balance operate with the *Lupinus* scheme of nitrogen flow? This question was addressed (in words) by Lambers (1983) who proposed that, when the nitrogen supply is low, the shoot is somehow signalled to release a larger fraction of carbon to the root. It could be that, when shoot sinks are less active, there is a build-up of carbon substrates in the shoots, so that more assimilates pass to the roots along a steepened concentration gradient. This explanation seems to be supported by observations on the effects of a step decrease in nitrogen supply to plants (Hirose and Kitajima; 1986; McDonald *et al.*, 1986). However, it is not clear that root sinks would not also become less active. More recently, Dewar (1993) developed a mechanistic model showing that functional balance could operate, using the same principles as in the Thornley models, even though a fraction (and in some circumstances all) of the nitrogen may be passed to the roots via the leaves (Fig. 5c). He pointed out that the essential feature of the Thornley transport-resistance model—that gradients of C and N substrate concentration exist in opposite directions in the plant—could be produced

---

Fig. 6. The flow of carbon and nitrogen in non-nodulated *Lupinus alba* plants, redrawn from Pate (1980). The thickness of the bands are proportional to the mass of carbon or nitrogen being transported, relative to 100 units by weight of nitrogen taken up, which equated to 3200 units by weight of carbon assimilated. The shaded bands are phloem, the unshaded bands are xylem. Almost all carbon substrate is sucrose, and over 90% of the nitrogen is reduced in the roots.

without invoking separate resistances to C and N transport. By assuming that C and N are transported downwards in the phloem, driven by the gradient in C (in accordance with the Münch flow hypothesis) the N substrate concentration in the roots can readily be maintained at a higher level than that in the shoots, even with the *Lupinus* pattern of N transport.

2. Seasonal "Switching" On and Off of Utilization Sinks

So far, it has been assumed that the growth rates of the utilization sinks (roots, shoot and cambial meristems) are entirely a function of their sizes, local environments and internal supplies of substrates. In reality, the sinks of temperature-zone trees (and many other perennial plants) seem to be switched on and off during the year (Lanner, 1976), in a way that cannot be explained by functional interaction between shoots and roots (cf. Borchert, 1973).

The primary switching signals come from the external environment. The clearest such signals are the combinations of photoperiod, chill and warm temperatures which cue the onset of bud dormancy in autumn and budbreak in the spring. These switching mechanisms have the clear "purpose" of preventing shoot growth occurring while there is a risk of subzero temperatures, and the process is amenable to modelling (Cannell and Smith, 1983; Hanninen, 1990). It may be argued that the seasonal changes in cambial and root growth are set in train by the seasonality in bud development and shoot growth. It is well known that the onset of rapid shoot growth in spring is coincident with both a reduction in root growth and the onset of cambial cell division and early-wood production, while the cessation of shoot growth in autumn is roughly coincident with a resumption of root growth and late-wood production. There are, however, two major complications. First, there seems to be *internal* switching between shoot, cambial and root growth which may involve growth regulators as well as internal competition for growth substrates. In particular, it may be argued that shoot extension generates signals which switch on cambial cell division but switch off root growth (Ludlow, 1991). Second, trees with preformed shoots (including most mature temperate conifers) have developed a lag in the process, so that both the rate and duration of shoot extension are determined by the number of leaf primordia formed in the buds the previous late summer and autumn (Cannell and Willet, 1976; Lanner, 1976).

Clearly, any model of seasonal growth will need to deal with the processes of bud set, bud dormancy and budburst, and with a set of internal control mechanisms that seem to follow from the seasonality in shoot growth. However, over a time-scale of years, the functional balance between nutrient absorbing fine roots and carbon-fixing foliage must be maintained. Cannell and Willett (1976) showed that distortions in the root:shoot balance of con-

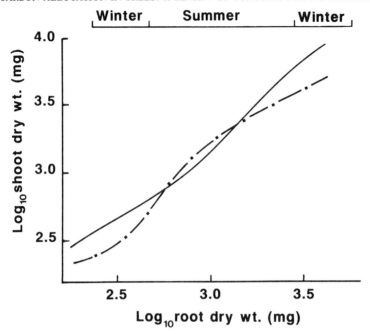

Fig. 7. Oscillations in the shoot:root relationship in *Pinus contorta* ecotypes. The broken line refers to an ecotype with a short duration of shoot growth in summer, which develops a large root:shoot ratio in the autumn which is "corrected" the following spring when shoot growth occurs. The unbroken line refers to an ecotype with a more prolonged period of shoot growth and less seasonal fluctuation in root:shoot ratio (Cannell and Willet, 1976).

ifers that developed each autumn were corrected each spring (Fig. 7). Ludlow (1991) considered how the long-term root:shoot functional balance may be achieved in Sitka spruce with seasonal time lags: for instance, increased carbon fixation in the summer will enable more root growth to occur in late summer, increasing nitrogen uptake, which will increase the rate of leaf primordia formation in the buds, giving larger shoots the following year, and so on. Nevertheless, over a time-scale of decades, the transport-resistance model of Thornley gives a very realistic simulation of forest growth (Thornley, 1991).

3. Carbon and Nutrient Storage

So far, no serious attempt has been made to couple the idea of meristem growth rate being co-limited by the local concentrations of different substrates with the existence in the plant of carbon and nutrient storage pools. As mentioned, there need be no conflict between these two aspects of plant growth if it is recognized that the storage pools, at least of carbon, operate like active sinks rather than as passive reservoirs. It is quite possible that the carbon storage sinks compete with utilization sinks when their storage level is below saturation, and that carbon storage reserves are not mobilized to any extent unless large sink demands lower internal substrate levels below a critical level.

4. Carbon Cost of Nutrient Acquisition by Roots

There is a nutrient requirement for photosynthesis by foliage and a carbon cost of nutrient acquisition by the roots. The change in carbon fixation that results from a change in leaf area and in the nitrogen concentration of the leaves is relatively well researched and is amenable to modelling (Field and Mooney, 1986; Friend, 1991). By contrast, the relationship between (i) fine root mass, surface area, root growth or total carbon consumption by the roots, and (ii) the amount of nutrients returned to the plant, is poorly known, and is, of course, very complex.

Only one major point will be made here, namely, that nutrient uptake is much more likely to be a function of the *growth* of the root-mycorrhizal complex, than of its mass, surface area or ion uptake properties (Clarkson, 1985). The carbon cost of nutrient uptake is probably determined by the cost of acquiring those nutrients which are most expensive to access, notably P (but also Zn and Co) which diffuse slowly in the soil. In order to maintain a flux of P to the plant, the fine roots, and/or the hyphae of mycorrhizal fungi, must continuously extend into undepleted volumes of soil, imposing a continuous carbon cost in fine root and fungal turnover. Models of P uptake in soil show that uptake is highly sensitive to a change in root growth rate (Silverbush and Barber, 1985), and Clarkson (1985) concluded that the carbon costs of root growth and other nutrient capture mechanisms (leakage and exudation) are normally far greater than the carbon cost of ion uptake *per se*. It is noteworthy that the infection of roots by arbuscular mycorrhizal fungi is stimulated by a high carbon substrate concentration in the roots (Hayman, 1974), which is the expected response to N, P and S shortage in the Thornley models. Also, the P return to the plant, per unit of carbon used to develop mycorrhizal hyphae, can sometimes be greater than the return per unit of carbon used to develop roots, although mycorrhizas may become costly when they form a mantle, rhizomorphs and reproductive structures (Jones *et al.*, 1991).

VI. CARBON ASSIMILATION vs. WATER ACQUISITION
(4 in Fig. 1)

A. Root–Shoot Functional Balance concerning Carbon and Water

Carbon dioxide and water vapour share the same diffusive conduits—the stomata—so it is impossible for plants to assimilate carbon without losing water (unless the air is saturated with water vapour). Consequently, changes in stomatal resistance may be regarded as critical in determining the balance between carbon assimilation and the carbon "costs" of transpiration, including allocation of carbon to roots. Natural selection may favour an optimal stomatal behaviour that maximizes daily carbon gain per unit of daily transpiration (Cowan and Farquhar, 1977), which implies an optimal allocation of assimilates between shoots and roots (Givnish, 1986a).

Decreased water potentials in plants increase stomatal resistance, mesophyll resistance, and the activity of enzymes concerned with photosynthesis and growth. As mentioned, the usual effect of water stress is to limit growth more than photosynthesis, in which case the response according to both the Thornley and Ågren/Ingestad hypotheses, will be increased allocation to roots.

However, there are strong interactions among plant traits that determine the balance between carbon fixation and water loss, and it is only by modelling the interactions that the net result can be evaluated.

B. Givnish Hypothesis

Givnish (1986a) proposed that plants modify their stomatal resistance and leaf/root allocation to maximize their net rate of carbon gain. The tradeoffs are shown in Fig. 8. Decreased stomatal resistance can increase whole plant carbon gain by decreasing the resistance of CO_2 diffusion to the plastids, but can decrease carbon gain by decreasing leaf water potential—on the assumption that partially desiccated mesophyll has a reduced photosynthetic capacity. Similarly, increased allocation to leaves can increase carbon gain by increasing the proportion of non-photosynthetic tissues, but can decrease carbon gain by increasing water loss.

Givnish (1986a) developed a model that described the effects of stomatal resistance and root allocation on leaf water potential, and the effects of stomatal resistance and leaf water potential on photosynthesis. His basic assumptions were that photosynthesis is a linear function of the CO_2 concentration in the mesophyll, and that the rate of water absorption by the roots

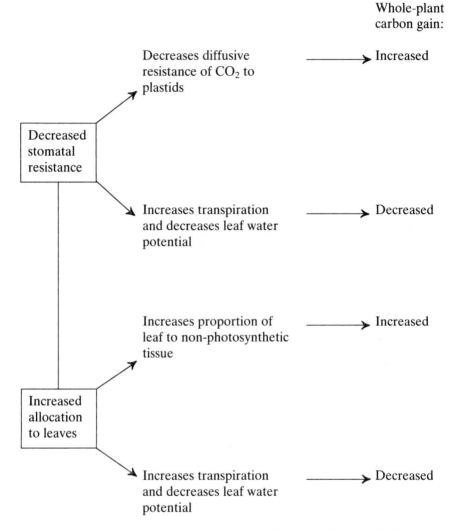

Fig. 8. Qualitative effects of decreased stomatal resistance and increased allocation to levels on whole-plant carbon gain (after Givnish, 1986a, b).

was the product of (i) root hydraulic conductivity, (ii) root biomass and (iii) the difference in water potential between the leaves and the soil. The rate of water absorption must equal the rate of transpiration, so it was possible to determine leaf water potential as a function of stomatal conductance, leaf/root allocation and root hydraulic conductivity. Transpiration and photosynthetic rates can also be expressed as functions of stomatal resistance, and the equations can be combined to find the values of stomatal resistance and shoot:root allocation that maximize whole plant carbon gain.

If a linear relationship is assumed between photosynthesis and the CO_2 concentration in the intercellular air spaces of the mesophyll, then the following relationship holds, regardless of variation in root hydraulic conductivity:

$$\frac{r_m}{r_s} = \frac{f}{1-f} \qquad (7)$$

where r_m is mesophyll resistance, r_s is stomatal resistance, f is the biomass fraction in leaves and $1-f$ is the biomass fraction in roots. Equation (7) results from the assumption that mesophyll resistance only affects CO_2 uptake, whereas stomatal resistance affects both CO_2 uptake and water loss. Thus, if mesophyll resistance is large relative to stomatal resistance, the optimal strategy to maximize carbon gain is to increase leaf fraction. On the other hand, if stomatal resistance is large relative to mesophyll resistance, the optimal strategy is to produce more roots to take up more water and thereby decrease stomatal resistance. It should be added that the assumption that stomatal conductance directly affects the photosynthetic capacity of the chloroplasts is contentious (see Friend, 1991, 1993).

Using data for *Phaseolous vulgaris*, Givnish (1986a) predicted that the fractional allocation to leaves should increase with increase in relative humidity, root hydraulic conductivity, mesophyll resistance and ambient CO_2 concentration, as shown in Fig. 9. In other words, more carbon is allocated to roots when the air is dry, when the resistance to water uptake is large or when the mesophyll resistance to CO_2 transfer is small.

C. Discussion

It is difficult to obtain quantitative solutions to the Givnish model because there are large and complex species differences in the effects of leaf water potential on apparent mesophyll resistance, CO_2 compensation point and effective root hydraulic conductivity. However, his approach successfully couples carbon and water fluxes, largely because both gases are exchanged through the stomata.

Givnish (1986a) points out that the challenge now is to couple the gas exchange properties of plants with the carbon costs of nutrient capture as well as water absorption. The absence of such coupling may partly explain why the Givnish model predicts an increase in allocation to leaves with increase in ambient CO_2 concentration, which runs counter to much of the experimental evidence (Eamus and Jarvis, 1989). One way forward is to optimize short-term carbon gain with respect to stomatal conductance and leaf nitrogen level

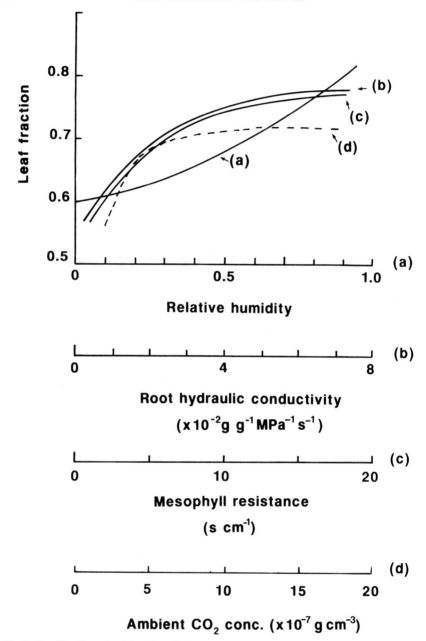

Fig. 9. Predicted optimal fractional allocation to leaves in *Phaseolus vulgaris* to maximize carbon gain, with change in (a) relative humidity, (b) root hydraulic conductivity, (c) mesophyll resistance, and (d) ambient CO_2 concentration, with other values held constant (Givnish, 1986a, b).

and to couple this to carbon and nitrogen allocation to leaves and roots using a mechanistic description of carbon and nitrogen flow in the plant.

VII. STRUCTURAL COSTS OF WATER TRANSPORT
(5 in Fig. 1)

A. Functional Interdependence

There is clearly functional interdependence between the foliage, which fixes carbon and transpires water, and the support structures which provide conduits for the transport of water and mineral nutrients from the roots. In particular, sufficient carbon must be allocated to xylem to meet the transpiration costs of photosynthesis. This functional interdependence implies a strong linkage between foliage and xylem growth.

Whereas root–shoot allocation is greatly altered by light and nutritional conditions, the allocation between foliage and woody parts is more tightly coupled—in the absence of substantial changes in tree height. Thus, in the now classic Swedish experiment on 20-year-old *Pinus sylvestris* (Linder and Axelsson, 1982), the application of nutrients and irrigation increased total dry matter production over 6 years about three-fold, halved the fraction of dry matter allocated to roots, but had almost no effect on the fractional allocation between foliage and woody parts (Cannell, 1985).

However, as trees grow taller, an increasing proportion of the dry weight increment above ground is allocated to the stems, and a decreasing proportion to the foliage (Cannell, 1985; Albrekton and Valinger, 1985; Fig. 10). This is the expected response, because of the increase in length of new xylem to be formed with increase in tree height, and it is also the expected response to the requirements for mechanical support (see below).

The strongest evidence for functional interdependence between foliage and xylem tissues is the now-familiar linear relationship between the sapwood cross-sectional area below the crown and foliage biomass or area (Grier and Waring, 1974; Rogers and Hinckley, 1979; Kaufmann and Troendle, 1981; Whitehead *et al.*, 1984a; Oren *et al.*, 1986), a relationship that apparently extends to the structural roots (Kaipiainen and Hari, 1985; Carlson and Harrington, 1987). This relationship forms the foundation for the pipe-model hypothesis (see below).

The mechanism by which xylem development is kept in balance with leaf growth is presumed to be by the basipetal flow of carbon substrates and growth regulators from new leaves, which stimulate cambial cell division (Zimmerman and Brown, 1971; Coutts 1987). In diffuse-porous trees, cambial activity spreads slowly down the trunk in spring once the foliage starts to grow, adding to the existing functional xylem. In ring-porous trees, cambial

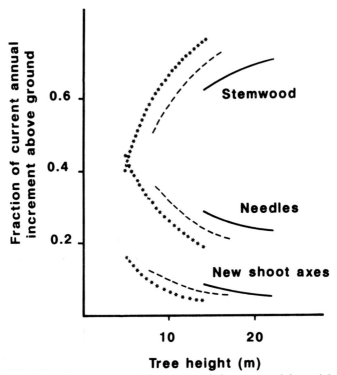

Fig. 10. Allocation of current annual increment of above ground dry weight in trees of *Pinus sylvestris* differing in height and diameter (from Albrekton and Valinger, 1985).

activity must spread downwards much more rapidly to provide new channels for water transport, because many of the previous year's large vessels will have normally developed embolisms (Zimmerman and Brown, 1971). In many trees, the period of cambial cell division is coincident with the period of new foliage growth, and the final sapwood area is largely determined by the number of cells that is formed during the early cell division phase (Denne, 1979).

B. Pipe-model Hypothesis

The pipe-model hypothesis of Shinozaki *et al.* (1964) reasons that each unit of foliage requires a unit pipeline of wood to conduct water from the roots. These pipes also provide mechanical support, and disused pipes (heartwood) provide additional support in large trees. The hypothesis is therefore consistent with the linear relationship between sapwood cross-sectional area and foliage area.

Valentine (1985) used the sapwood area/leaf mass relationship to constrain the distribution of assimilates within a carbon-balance model of stand growth. He also assumed that a constant amount of feeder roots is required to support a unit of foliage. Mäkelä (1986) successfully combined the pipe-model theory with the functional balance between shoots and roots. She also assumed different sapwood/foliage relationships for the stem, branches and roots—thereby allowing some flexibility in the tree's hydraulic architecture. Further division of branches between thick branches and twigs may be necessary to account for the small conductance of twigs (Tyree, 1988). Both the Valentine and Mäkelä models show that, as the trees increase in height, the increasing costs of replacing the sapwood must cause a decrease in both foliage and root growth, leading to a decrease in height increment and/or death.

Ludlow *et al.* (1990) also combined the pipe-model theory with a carbon-balance model, but assumed that the area of *new* sapwood was proportional to the dry weight of *new* foliage. This assumption took into account the observation that new leaf and cambial growth are coupled, but then had to assume that foliage mortality and sapwood mortality were correlated. Ludlow *et al.* (1990) further assumed that, for young conifer plantations, height growth was coupled to foliage growth.

C. Discussion

One of the limitations of the pipe model hypothesis is that the sapwood area–foliage area relationship is empirical and differs greatly among species (Grier and Waring, 1974; Rogers and Hinckley, 1979; Kaufmann and Troendle, 1981; Whitehead *et al.*, 1984a; Oren *et al.*, 1986) among trees that differ in size within species (Espinosa *et al.*, 1987; Keane and Watman, 1987; Long and Smith, 1988), within individual trees (Ewers and Zimmerman, 1984; Zimmerman, 1984; Tyree and Sperry, 1988) and in differing environments (Hinckley and Ceulemans, 1989).

Whitehead *et al.* (1984b) combined equations describing the driving forces for water movement in trees with equations describing the properties of the flow pathway, to suggest that the relationship between leaf area per tree (A_f) to sapwood cross-sectional area (A_s) is linearly proportional to sapwood conductivity, K, and inversely proportional to tree height, h, canopy conductance g_c and the saturation deficit of the air, D_M. That is:

$$\frac{A_f}{A_s} \propto \frac{K}{hg_cD_M} \tag{8}$$

Thus, we might expect a large sapwood area per unit leaf area in trees growing in dry areas, with a large transpirational demand, in tall trees, and in species with xylem properties which have a low hydraulic conductivity. We should also note that there is considerable variation in vessel or tracheid properties within trees, giving a variable hydraulic architecture, and hence a variable optimal sapwood area per unit foliage area throughout the tree.

A second difficulty with the pipe-model hypothesis is the need to assume not only that new sapwood is formed in proportion to new foliage, but that old sapwood ceases to function in proportion to the loss of old leaves. In practice, the rate of sapwood turnover has to be guessed on the assumption that the A_f/A_s ratio remains constant (Mäkelä, 1986). The argument normally advanced is that living sapwood represents a respiratory load on the tree so that maximum net carbon gain is obtained by maintaining the *minimum* sapwood needed to supply water to the foliage (Bamber and Fukazawa, 1985). The transfer of living sapwood to dead heartwood has little effect on the mechanical properties of the wood, but lowers the carbon cost. A further argument is that, whatever their hydraulic architecture, trees seem to operate near the brink of catastrophic xylem dysfunction due to dynamic water stress (Tyree and Sperry, 1988), suggesting that, for any given wood structure, the sapwood area is probably close to the minimum required to meet the maximum rate of water loss from the foliage.

Finally, in a carbon-balance model, variability and uncertainty concerning the A_f/A_s relationship are confounded by variability in the specific gravity of sapwood. The hydraulic property of sapwood is not functionally related to its specific gravity, so the carbon cost of producing a given sapwood area must be estimated using a further empirical relationship.

VIII. STRUCTURAL COSTS OF MECHANICAL SUPPORT
(6 in Fig. 1)

A. Functional Interdependence

The growing literature on plant biomechanics focusses on the carbon costs of providing mechanical support against gravitational and wind forces (Givnish, 1986b). Clearly, some minimum amount of carbon must be invested in supporting structures (stems, branches and petioles), and this investment can be calculated from physical and geometrical properties of the structure. The costs of mechanical support may be viewed as the mass of 'unproductive' tissue required to display foliage, to disperse seeds and to provide anchorage. The carbon cost/gain ratio will depend upon many interacting variables concerning the geometry of the structure, self-shading and respiratory costs.

The theory relating the three-dimensional geometry of structures, their structural properties and their masses is contained in the literature on structural engineering. Using this theory, the mass of any biological structure can be calculated if the geometry, Young's modulus and wind forces are specified. In practice, simplifying hypotheses have been sought to ease the computational difficulties and to introduce some biological meaning. Two notable hypotheses have been advanced for trees: first, that trees can be regarded as assemblages of cantilever beams; and second that the distribution of cambial growth along stems occurs in such a way as to equalize the distribution of stress. Both of these hypotheses can be used to constrain assimilate allocation within the woody structure of trees.

B. Cantilever Beams

The structural theory for cantilever beams can be used to deduce the critical dimensions, shapes and support costs of branches and trunks, if it is possible to define the taper, geometry, specific gravity and Young's modulus of the material. Elementary theory can be used if the end-point deflections are less than about 25% of the branch or trunk lengths, and there is a simple linear taper. For more complex situations, Morgan and Cannell (1987) recommended using a transport matrix method in which the branch or trunk is divided into a large number of short cylindrical segments. A matrix equation can then be written that relates the conditions at the ends of each segment (shear force, bending moment, angle and deflection) to the load on the beam (its own weight plus any side branches, etc.), given the length, diameter and Young's modulus. Matrix equations for successive segments can then be multiplied together, so that the end conditions of the whole branch or trunk can be calculated. Using this method, Cannell et al. (1988) and Cannell and Morgan (1990) were able to specify some of the design features that might minimize the support costs of foliage on trees and hence maximize carbon export from branches to the trunk.

The transport matrix method enables any pattern of taper to be specified. However, McMahon and Kronauer (1976) examined the tapering of branches with respect to the power law:

$$d \propto l^\beta \qquad (9)$$

where d is diameter, l is length to the "virtual origin" where d is zero, and β is the taper exponent. They found that β was approximately $1\cdot5$, which is the condition of elastic self-similarity, that is, when deflection of the tip divided by the overall length is a constant. When $\beta = 1\cdot5$ it implies that the branches are designed with the minimal mass to maintain their shape. This constraint was used by Ford and Ford (1990) and Ford et al. (1990) to calculate assimilate allocation within developing branch networks of trees.

If there were constant stress along branches, as hypothesized for stems (see below) then $\beta = 2\cdot0$. This condition would imply that branches are designed with the minimal mass needed to keep them from breaking under their own weight alone. In reality the combination of gravitational and wind forces applied to foliated branches tends to make the taper exponent closer to $1\cdot5$

than 2·0 (King and Loucks, 1978), whereas the taper exponent for tree trunks is closer to 2·0.

However, the elastic self-similarity model may apply to the stems of *very large* trees, as shown by McMahon and Kronauer (1976) who found a slope of 1·5. between log diameter vs. log height for "big trees" in the US (see also Dean and Long, 1986, *Pinus contorta*; Rich *et al.*, 1986, tropical trees). Clearly, large trees have a large self-weight and, without rapid height growth, they develop thick stems with large margins of safety.

C. Constant Stress Hypothesis

The literature on trees strongly supports the view that cambial growth is stimulated by mechanical forces, although no biochemical or biophysical explanation has been found (Larson, 1965; Wilson and Archer, 1979). Cambial cells may respond to movement, i.e. strain (the fractional change in length) rather than to stress (force/area), and their growth stimulation may be one aspect of the general phenomenon of thigmomorphogenesis (Jaffe, 1973; Hunt and Jaffe, 1980; Telewski and Jaffe, 1986). Studies on the shapes of stems, on stem responses to being displaced from the vertical, and on their responses to being held still by guys, or artificially swayed, all reinforce the view that assimilates passing down the phloem in tree stems are preferentially used in cambial growth in areas of greatest stress (Jacobs, 1954; Zimmerman and Brown, 1971; Coutts, 1987; Mattheck, 1990; Valinger, 1990). That is, cambial growth tends to occur along tree stems to equalize the stress distribution, and the expectation for all but very large trees is that $\beta = 2·0$ in equation (9).

A more direct test of the constant stress hypothesis is contained in the elementary theory for bending beams of circular cross-section, which states that:

$$S = \frac{32M}{\pi d^3} \tag{10}$$

where S is the maximum stress at the outer surface, M is the bending moment, and d is diameter. If S is constant along a tree stem, then d^3 at any position along the stem should be linearly related to distance from that position to a point in the crown where wind forces may be regarded as acting (not generally coincident with the "virtual origin" where $d = 0$). Foresters working over 100 years ago showed that this was indeed a good approximation (Fig. 11), and that the middle part of tree stems (from breast height to crown base) therefore approximated the shape of a cubic paraboloid (Metzger, 1893). Although there are reports for individual trees showing some deviation from constant stress (Milne and Blackburn, 1989; West *et al.*, 1989); the hypothesis seems to

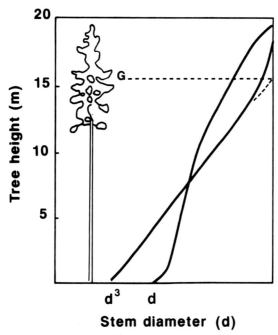

Fig. 11. Stem profile of a 76-year-old tree of *Picea abies*, showing the relationship between stem diameter (*d*) and diameter cubed (*d³*) and height. Note that d^3 forms a straight line, which when extended, meets the centre of gravity (G) of the crown. From Assmann (1970).

be *approximately* true. Indeed, if this were not so, trees would have weak points along their stems where they would be prone to breakage.

The assumption of constant stress can be used to constrain the allocation of assimilates along a stem of given height and thereby determine stem diameter. Most simply, the assumption can be used to predict stem diameter from the relationship:

$$D = c(AH)^{0.33} \qquad (11)$$

where A is the leaf area per tree, H is height to the centre of leaf area, and c is an empirical constant (*ca.* 3·75) (Dean and Long, 1988, 1992). More accurately, the transport matrix method can be used to derive the distribution of diameter down a tree that gives uniform stress, knowing tree height and the distribution of wind force in the crown. Unfortunately, several parameters need to be known, such as the leaf area distribution with height and the drag coefficient; however, variation in average windspeed over the range 2–10 m s⁻¹ has little effect on the stress distribution because of opposing bending and axial forces (Morgan and Cannell, 1993).

D. Discussion

The use of structural mechanical principles to calculate the amount of carbon that must be partitioned to structural tissues has the advantage that it can be done quite independently of the physiological processes involving carbon, water and nutrients. However, some simplifying principles may need to be assumed to avoid having to define a complex three-dimensional geometry. Also, some optimality principle has to be accepted related to wind and gravitational forces, plus a margin of safety. Nevertheless, a growing literature is emerging which deals with adaptation and functional constraints on support structures and their implications for light capture and competitive interactions among plants.

Overall, the structural mechanics approach seems sounder than the hydraulic architecture approach given the variable relationship between hydraulic conductance and carbon cost. Although xylem cell number and size may be determined by hydraulic requirements, the degree of cell wall thickening and overall mass of new xylem tissue is likely to be more closely coupled with mechanical requirements.

We must, however, be wary of oversimplification. Recently, Bertram (1989) reported that the slender twigs at the whole crown periphery of trees are very small in diameter for their length and so do not conform to the elastic self-similarity model. These same twigs have a high hydraulic resistance to water flow and may protect the main branches and stems from excessive cavitation and embolism (Tyree and Sperry, 1988). They also flex in the wind, reducing drag on the whole crown, and, because they are thin, the cambium experiences little strain (relative change in length) and perhaps for this reason the cambium fails to grow as much as in thicker branches subject to the same or less flexing.

IX. CONCLUDING REMARKS

It is clear that several useful concepts have emerged that have, or can be, used to construct or constrain models of carbon allocation among the functionally interdependent parts of trees. Almost all of the concepts use some form of optimality principle: that is, unspecified regulatory mechanisms are assumed to operate such that the tree behaves as if it were seeking some goal—of producing seed, maximizing carbon assimilation, minimizing the carbon cost of supporting tissues, or producing a mechanically sound structure. The immediate challenge for modellers is to combine some of these concepts, particularly to couple the carbon, nutrient (NPK and Mg, Mn) and water-use optimizations and to simulate both the mass and dimensions of tree structures. The challenge for plant physiologists is to extend our knowledge of the processes that govern carbon allocation at the tissue and cellular level. We know surprising little, for instance, about variation in substrate concentrations in trees, the controls on storage sinks, the transport of nutrients, how meristem growth is controlled, and how roots sense the demand of the plant for nutrients. It is no longer sufficient to measure the dry weights of tree parts: mechanistic models demand that we develop methods of measuring the concentrations and fluxes of carbon and nutrient substrates in both the phloem and xylem.

It is noteworthy that it has not been possible in this review to identify any hypotheses concerning growth substances that might usefully be used to build mechanistic models of carbon allocation. There are a few well-reported instances of root–shoot signals that may involve growth substances (Davies and Zhang, 1991), and intuitively one suspects that there may be several rapid-response control-and-command systems in the plant (Thomas, 1985). However, the traditional theories of growth substances being rate-limiting factors that are produced at one place and act at another, no longer seem to hold (Trewavas, 1981, 1991) and new theories are as yet untested and ill-defined.

ACKNOWLEDGEMENTS

We are grateful to John Thornley and Andrew Friend for many helpful suggestions on a draft of this paper.

REFERENCES

Ågren, G.I. and Ingestad, T. (1987). Root/shoot ratios as a balance between nitrogen productivity and photosynthesis. *Plant Cell Environ.* **10**, 579–586.

Albrekton, A. and Valinger, E. (1985). Relations between tree height and diameter, productivity and allocation of growth in Scots pine (*Pinus sylvestris* L.) sample tree material. In: *Crop Physiology of Forest Trees* (Ed. by P.M.A. Tigerstedt, P. Puttonen and V. Koski), pp. 95–105. University of Helsinki, Finland.

Alexander, I.F. and Fairley, R.I. (1983). Effects of N fertilization on populations of fine roots and mycorrhizas in spruce humus. *Plant and Soil* **71**, 49–53.

Assmann, E. (1970). *The Principles of Forest Yield Study*. Pergamon Press, Oxford.

Axelsson, E. and Axelsson, B. (1986). Changes in carbon allocation patterns in spruce and pine trees following irrigation and fertilization. *Tree Physiol.* **2**, 189–204.

Bamber, R.K. and Fukazawa, K. (1985). Sapwood and heartwood: a review. *For. Prod. Abstr.* **8**, 265–278.

Bassow, S.L., Ford, E.D. and Kiester, A.R. (1990). A critique of carbon-based tree growth models. In: *Process Modelling of Forest Growth Responses to Environmental Stress.* (Ed. by R.K. Dixon, R.S. Meldahl, G.A. Ruark and W.G. Warren), pp. 50–57. Timber Press, Portland, Oregon.

Bertram, J.E.A. (1989). Size-dependent differential scaling in branches: the mechanical design of trees revisited. *Trees* **4**, 241–253.

Borchert, R. (1973). Simulation of rhythmic tree growth under constant conditions. *Physiol. Plant.* **29**, 173–182.

Brouwer, R. (1962). Distribution of dry matter in the plant. *Neth. J. Agric. Sci.* **10**, 361–376.

Cannell, M.G.R. (1985). Dry matter partitioning in tree crops. In: *Attributes of Trees as Crop Plants* (Ed. by M.G.R. Cannell and J.E. Jackson), pp. 160–193. Institute of Terrestrial Ecology, UK.

Cannell, M.G.R. (1989) Physiological basis of wood production: a review. *Scand. J. For. Res.* **4**, 459–490.

Cannell, M.G.R. and Morgan, J. (1990). Theoretical study of variables affecting the export of assimilates from branches of *Picea. Tree Physiol.* **6**, 257–266.

Cannell, M.G.R. and Smith, R.I. (1983) Thermal time, chill days and prediction of bud burst in *Picea sitchensis. J. Appl. Ecol.* **20**, 951–963.

Cannell, M.G.R. and Willet, S.C. (1976). Shoot growth phenology, dry matter distribution and root:shoot ratios of provenances of *Populus trichocarps, Picea sitchensis* and *Pinus contorta* growing in Scotland, *Silvae Genetica* **25**, 49–59.

Cannell, M.G.R., Morgan, J. and Murray, M.B. (1988). Diameters and dry weights of tree shoots: effects of Young's modulus, taper, deflection and angle. *Tree Physiol.* **4**, 219–231.

Carlson, W.C. and Harrington, C.A. (1987). Cross-sectional area relationships in root systems of loblolly and short leaf pine. *Can. J. For. Res.* **17**, 556–558.

Charles-Edwards, D.A. (1982). *Physiological Determinants of Crop Growth.* Academic Press, New York.

Clarkson, D.T. (1985). Factors affecting mineral nutrient acquisition by plants. *Ann. Rev. Plant Physiol.* **36**, 77–115.

Clarkson, D.T. and Hanson, J.B. (1980). The mineral nutrition of higher plant. *Ann. Rev. Plant Physiol.* **31**, 239–298.

Cooper, A.J. and Thornley, J.H.M. (1976). Response of dry matter partitioning, growth, and carbon and nitrogen levels in the tomato plant to changes in root temperature: experiment and theory. *Ann. Bot.* **40**, 1139–1152.

Cooper, D. and Clarkson, D.T. (1989). Cycling of amino-nitrogen and other nutrients between shoots and roots in cereals—a possible mechanism integrating shoot and root in the regulation of nutrient uptake. *J. Exp. Bot.* **40**, 753–762.

Coutts, M.P. (1987). Developmental processes in tree root systems. *Can. J. For. Res.* **17**, 761–767.

Cowan, I.R. and Farquhar, G.D. (1977). Stomatal function in relation to leaf metabolism and environment. *Symp. Soc. Exp. Bot.* **31**, 471–505.

Davidson, R.L. (1969). Effect of root/leaf temperature differentials on root/shoot ratios in some pasture grasses and clover. *Ann. Bot.* **33**, 561–569.

Davies, W.J. and Zhang, J. (1991). Root signals and the regulation of growth and development of plants in drying soils. *Ann. Rev. Plant. Physiol.* **42**, 55–76.

Dean, T.J. and Long, J.N. (1986). Validity of constant-stress and elastic-instability principles of stem formation in *Pinus contorta* and *Trifolium pratense. Ann. Bot.* **58**, 833–840.

Dean, T.J. and Long, J.N. (1992). Influence of leaf area and canopy structure on size-density relations in even-aged lodgepole pine stands. *For. Ecol. Mgt* **49**, 109–117.

Denne, M.P. (1979). Wood structure and producing within the trunk and branches of *Picea sitchensis* in relation to canopy formation. *Can. J. For. Res.* **9**, 406–427.

Dewar, R.C. (1993). A root-shoot partitioning model based on carbon-nitrogen-water interactions and Münch phloem flow. *Funct. Ecol.* **7**, 356–368.

Dickson, R.E. (1991). Assimilate distribution and storage. In: *Physiology of Trees* (Ed. by A.S. Raghavendra), pp. 51–85. John Wiley & Sons Inc., Chichester.

Eamus, D. and Jarvis, P.G. (1989). The direct effects of increase in the global atmospheric CO_2 concentration on natural and commercial temperate trees. *Adv. Ecol. Res.* **19**, 1–55.

Ericsson, T. (1990). Dry matter partitioning at steady state nutrition. In: *Above-and Below-ground Interactions in Forest Trees and Acidified Soils.* Air Pollution Report, 32, pp. 236–243. Commission of European Communities and Swedish Agricultural University, Uppsala.

Espinosa, B.M.A., Perry, D.A. and Marshall, J.D. (1987). Leaf area-sapwood area relationship in adjacent young Douglas-fir stands with different early growth rates. *Can. J. For. Res.* **17**, 174–180.

Ewers, F.W. and Zimmerman, M.H. (1984). The hydraulic architecture of balsam fir (*Abies balsamea*). *Physiol. Plant.* **60**, 453–458.

Field, C. and Mooney, H.A. (1986). The photosynthesis–nitrogen relationship in wild plants. In: *On the Economy of Plant Form and Function* (Ed. by T.J. Givnish), pp. 25–55. Cambridge University Press, Cambridge.

Ford, E.D., Avery, A. and Ford, R. (1990). Simulation of branch growth in the *Pinaceae*: interactions of morphology, phenology, foliage productivity, and the requirement for structural support, on the export of carbon. *J. Theor. Biol.* **146**, 1–13.

Ford, R. and Ford, E.D. (1990). Structure and basic equations of a simulator for branch growth in the Pinaceae. *J. Theor. Biol.* **146**, 15–30.

Friend, A.D. (1991). Use of a model of photosynthesis and lead microenvironment to predict optimal stomatal conductance and leaf nitrogen partitioning. *Plant, Cell Environ.* **14**, 895–905.

Friend, A.D. (1993). PGEN: An integrated model of leaf photosynthesis, transpiration and conductance. *Ecol. Modelling* (in press).

Gamalei, Y. (1991). Phloem loading and its development related to plant evolution from trees to herbs. *Trees* **5**, 50–64.

Gholz, H.L. and Cropper, W.P. (1991). Carbohydrate dynamics in mature *Pinus elliottii* var. *elliotti* trees. *Can. J. For. Res.* **21**, 1742–1747.

Guaquinta, R.T. (1983). Phloem loading of sucrose. *Ann. Rev. Plant Physiol.* **34**, 347–387.

Gifford, R.M. and Evans, L.T. (1981). Photosynthesis, carbon partitioning, and yield. *Ann. Rev. Plant Physiol.* **32**, 485–509.

Givnish, T.J. (1986a). Optimal stomatal conductance, allocation of energy between leaves and roots, and the marginal cost of transpiration. In: *On the Economy of Plant Form and Function* (Ed. by T.J. Givnish), pp. 171–213. Cambridge University Press, Cambridge.

Givnish, T.J. (1986b). *On the Economy of Plant Form and Function.* Cambridge University Press, Cambridge.

Gleeson, S.K. (1993). Optimization of tissue nitrogen and root-shoot allocation. *Ann. Bot.* **71**, 23–31.

Grier, C.C. and Wareing, R.H. (1974). Conifer foliage mass related to sapwood area. *For. Sci.* **20**, 205–206.

Grime, J.P. and Hunt, R. (1975). Relative growth rate: its range and adaptive significance in a local flora. *J. Ecol.* **63**, 393–422.

Grime, J.P., Hodgson, J.G. and Hunt, R. (1988). *Comparative Plant Ecology: a Functional Approach to Common British Species.* Unwin Hyman, London.

Hanninen, H. (1990). Modelling the annual growth rhythm of trees: conceptual, experimental and applied aspects. *Silva Carelica* **15**, 35–45.

Harper, J.L. (1977). *Population Biology of Plants.* Academic Press, New York.

Hayman, D.S. (1974). Plant growth responses to vesicular-arbuscular mycorrhiza. VI. Effects of light and temperature. *New Phytol.* **73**, 71–80.

Hilbert, D.W. (1990). Optimization of plant root:shoot ratios and internal nitrogen concentration. *Ann. Bot.* **66**, 91–99.

Hilbert, D.W. and Reynolds, J.F. (1991). A model allocating growth among leaf proteins, shoot structure, and root biomass to produce balanced activity. *Ann. Bot.* **68**, 417–425.

Hilbert, D.W., Larigauderie, A. and Reynolds, J.F. (1991). The influence of carbon dioxide and daily photo-flux density on optimal leaf nitrogen concentration and root:shoot ratio. *Ann. Bot.* **68**, 365–376.

Hinckley, T.M. and Ceulemans (1989). Current focuses in woody plant water relations and drought resistance. *Ann. Sci. For.* **46**, 317–324.

Hirose, T. and Kitajima, K. (1986). Nitrogen uptake and plant growth. I. Effect of nitrogen removal on growth of *Polygonum cuspidatum*. *Ann. Bot.* **58**, 479–486.

Ho, L.C. (1988). Metabolism and compartmentation of imported sugars in sink organs in relation to sink strength. *Ann. Rev. Plant Physiol.* **39**, 355–378.

Hsiao, T.C. (1973). Plant response to water stress. *Ann. Rev. Plant Physiol.* **25**, 519–570.

Hunt, E.R. and Jaffe, M.J. (1980). Thigmomorphogenesis: the interaction of wind and temperature in the field on the growth of *Phaseolus vulgaris* L. *Ann. Bot.* **45**, 665–672.

Ingestad, T. (1979). Nitrogen stress in birch seedlings. II. N, K, P, Ca and Mg nutrition. *Physiol. Plant.* **45**, 149–157.

Ingestad, T. and Ågren, G.I. (1991). The influence of plant nutrition on biomass allocation. *Ecol. Appl.* **1**, 168–174.

Ingestad, T. and Lund, A.B. (1979). Nitrogen stress in birch seedlings. I. Growth technique and growth. *Physiol. Plant.* **45**, 137–148.

Jacobs, M.R. (1954). The effect of wind sway on the form and development of *Pinus radiata* D. Don. *Aust. J. Bot.* **2**, 35–51.

Jaffe, M.J. (1973). Thigmomorphogenesis: the response of plant growth and development to mechanical stimulation. *Planta (Berl.)* **114**, 143–157.

Johnson, I.R. (1985). A model for the partitioning of growth between the shoots and roots of vegetative plants. *Ann. Bot.* **55**, 421–431.

Johnson, I.R. and Thornley, J.H.M. (1987). A model of shoot:root partitioning with optimal growth. *Ann. Bot.* **60**, 133–142.

Jones, M.D., Durrall, D.M. and Tinker, P.B. (1991). Fluxes of carbon and phosphorus between symbionts in willow ectomycorrhizas and their changes with time. *New Phytol.* **119**, 99–106.

Kaipiainen, L. and Hari, P. (1985). Consistencies in the structure of Scots pine. In: *Crop Physiology of Forest Trees* (Ed. by P.M.A. Tigerstedt, P. Puttonen and V. Koski), pp. 32–37. Helsinki University Press, Finland.

Kaufmann, M.R. and Troendle, C.A., (1981). The relationship of leaf area and foliage biomass to sapwood conducting area in four subalpine forest tree species. *For. Sci.* **27**, 477–482.

Keane, M.G. and Watman, G.F. (1987). Leaf area-sapwood cross sectional area relationships in repressed stands of lodgepole pine. *Can. J. For. Res.* **17**, 205–209.

Keyes, M.R. and Grier, C.C. (1981). Above- and below-ground net production in 40-year-old Douglas fir stands on low and high productivity sites. *Can. J. For. Res.* **11**, 599–605.

King, D. and Loucks, O.L. (1978). The theory of tree bole and branch form. *Radiat. Environ. Biophys.* **15**, 141–165.

Kozlowski, T.T. (1992). Carbohydrate sources and sinks in woody plants. *Bot. Rev.* **58**, 107–222.

Kurz, W.A. (1989). Significance of shifts in carbon allocation patterns for long-term site productivity research. In: *Research Strategies for Long-term Site Productivity* (Ed. by. W.J. Dyck and C.A. Mees), pp. 149–164. IEA/BE A3 Report No. 8. Bulletin 152, Forest Research Institute, New Zealand.

Lambers, H. (1983). "The functional equilibrium", nibbling on the edges of a paradigm. *Neth. J. Agric. Sci.* **31**, 305–311.

Landsberg, J.J., Kaufmann, M.R., Binkley, D., Isebrands, J. and Jarvis, P.G. (1991). Evaluating progress towards closed forest models based on fluxes of carbon, water and nutrients. *Tree Physiol*, **9**, 1–15.

Lanner, R.M. (1976). Patterns of shoot development in *Pinus* and their relationship to growth potential. In: *Tree Physiology and Yield Improvement* (Ed. by M.G.R. Cannell and F.T. Last), pp. 223–243. Academic Press, London.

Larson, P.R. (1965). Stem form of young *Larix* as influenced by wind and pruning. *For. Sci.* **11**, 142–424.

Ledig, F.G. (1983). The influence of genotype and environment on dry matter distribution in plants. In: *Plant Research and Agroforestry* (Ed. by P.A. Huxley), pp. 427–454. ICRAF, Nairobi.

Linder, S. and Axelsson, B. (1982). Changes in carbon uptake and allocation patterns as a result of irrigation and fertilization in a young *Pinus sylvestris* stand. In: *Carbon Uptake and Allocation in Sub-alpine Ecosystems as a Key to Management.* (Ed. by R.H. Waring), pp. 38–44. Oregon State University, Forest Research Laboratory, Corvallis, OR.

Lloyd, D.G. (1980). Sexual strategies in plants. I. A hypothesis of serial adjustment of maternal investment during one reproductive season. *New. Phytol.* **86**, 69–79.

Long, J.N. and Smith, F.W. (1988). Leaf area-sapwood area relations of lodgepole pine as influenced by stand density and site index. *Can. J. For. Res.* **18**, 247–250.

Loomis, R.S., Rabbinge, R. and Ng, E. (1979). Explanatory models in crop physiology. *Ann. Rev. Plant Physiol.* **30**, 339–367.

Ludlow, A.R., Randle, T.J. and Grace, J.C. (1990). Developing a process-based growth model for Sitka spruce. In: *Process Modelling of Forest Growth Responses to Environmental Stress* (Ed. by R.K. Dixon, R.S. Meldahl, G.A. Ruark and W.G. Warren), pp. 249–262. Timber Press Inc., Oregon.

Ludlow, T. (1991). Seasonal allocation of carbon and nitrogen in Sitka spruce (*Picea sitchensis*). UK Forestry Commission Internal Report.

Luxmoore, R.J. (1991). A source-sink framework for coupling water, carbon and nutrient dynamics of vegetation. *Tree Physiol.* **9**, 267–280.

Maggs, D.H. (1961). Changes in the amount and distribution of increment induced by contrasting watering, nitrogen and environmental regimes. *Ann. Bot.* **25**, 353–361.

Mäkelä, A.A. (1986). Implications of the pipe model theory on dry matter partitioning and height growth in trees. *J. Theor. Biol.* **123**, 103–120.

Mäkelä, A.A. and Sievänen, R.P. (1987). Comparison of two shoot-root partitioning models with respect to substrate utilization and functional balance. *Ann. Bot.* **59**, 129–140.

Marshall J.D. and Waring, R.H. (1985). Predicting fine root production and turnover by monitoring root starch and soil temperature. *Can. J. For. Res.* **15**, 791–800.

Martin, F., Chemardin, M. and Gadel, P. (1981). Nitrate assimilation and nitrogen circulation in Austrian pine. *Physiol. Plant.* **53**, 105–110.

Mason, T.G. and Maskell, E.J. (1928). Studies on the transport of carbohydrates in the cotton plant. II. The factors determining the rate and direction of movement of sugars. *Ann. Bot.* **42**, 571–636.

Mattheck, C. (1990). Engineering components grow like trees. *Materialwiss Werkstoffteck* **21**, 143–168.

McDonald, A.J.S., Lohammar, T. and Ericsson, A. (1986). Growth response of step-decrease in nutrient availability in small birch (*Betula pendula* Roth.). *Plant, Cell Environ.* **9**, 427–432.

McLaughlin, S.B., McConathy, R.K., Barnes, R.L. and Edwards, N.T. (1980). Seasonal changes in energy allocation by white oak (*Quercus alba*). *Can. J. For. Res.* **10**, 379–388.

McMahon, T.A. and Kronauer, R.E. (1976). Tree structures: deducing the principle of mechanical design. *J. Theor. Biol.* **59**, 443–466.

Metzger, K. (1893). Der Win als massgebender Faktor für das Wachstum der Bäume. *Mündener Forstl.* **3**, 35–62.

Millard, P. and Proe, M.F. (1992). Storage and internal cycling of nitrogen in relation to seasonal growth of Sitka spruce. *Tree Physiol.* **10**, 33–43.

Miller, H.G. (1986). Carbon × nutrient interactions—the limitations to productivity. *Tree Physiol.* **2**, 373–385.

Milne, R. and Blackburn, P. (1989). The elasticity and vertical distribution of stress within stems of *Picea sitchensis*. *Tree Physiol.* **5**, 195–205.

Morgan, J. and Cannell, M.G.R. (1987). Structural analysis of tree trunks and branches: tapered cantilever beams subjected to large deflections under complex loading. *Tree Physiol.* **3**, 365–374.

Morgan, J. and Cannell, M.G.R. (1993). Shape of tree stems—a reexamination of the uniform stress hypothesis. *Tree Physiol.* (in press).

Münch, E. (1930). *Die Stoffbewegungen in er Pflanze.* Fisher, Jena.

Nadelhoffer, K.J., Aber, J.D. and Melillo, T.M. (1985). Fine roots, net primary production and soil nitrogen availability: a new hypothesis. *Ecology* **66**, 1377–1390.

Nambiar, E.S.K. and Fife, D.N. (1991). Nutrient retranslocation in temperate conifers. *Tree Physiol.* **9**, 185–207.

Oren, R., Wark, K.S. and Schulze, E.D. (1986). Relationships between foliage and conducting xylem in *Picae abies* (L.). Karst. *Trees* **1**, 61–69.

Pate, J.S. (1980). Transport and partitioning of nitrogenous solutes. *Ann. Rev. Plant Physiol.* **31**, 313–340.

Patrick, J.W. (1976). Hormone-directed transport of metabolites. In: *Transport and Transfer Processes in Plants* (Ed. by. I.F. Wardlaw and J.B. Passioura), pp. 433–446. Academic Press, New York.

Peel, A.J. (1974). *Transport of Nutrients in Plants.* Butterworths, London.

Priestly, C.A. (1962). Carbohydrates resources within the perennial plant. Commonwealth Bureau of Host and Plantation Crops. Tech. Comm. 27. Bureau of Trees and Plantation Crops, Wellingford, U.K.

Rastetter, E.B., Ryan, M.G., Shaver, G.R., Melillo, J.M., Nadelhoffer, K.J., Hobbie, J.E. and Aber, J.D. (1991). A general biogeochemical model describing the responses of the C and N cycles in terrestrial ecosystems to changes in CO_2 and climate, and N deposition. *Tree Physiol.* **9**, 101–126.

Reynolds, J.F. and Thornley, J.H.M. (1982). A shoot-root partitioning model. *Ann. Bot.* **49**, 585–597.

Rich, P.M., Helenurm, K., Kearns, D., Morse, S.R., Palmer, M.W. and Short, L. (1986). Height and diameter relationships for dicotyledonous trees and arborescent palms of Costa Rica tropical wet forest. *Bull. Torrey Bot. Club* **113**, 241–246.

Ritchie, G.A. and Dunlop, I.R. (1980). Root growth potential: its development and expression in forest tree seedlings. *NZ J. For. Sci.* **10**, 218–248.

Rogers, R. and Hinckley, T.M. (1979). Foliar weight and area related to current sapwood area in oak. *For. Sci.* **25**, 298–303.

Running, S.W. and Gower, S.T. (1991). FOREST-BGC, as a general model of forest ecosystem processes for regional applications. II. Dynamic carbon allocation and nitrogen budgets. *Tree Physiol.* **9** 147–160.

Sauter, J.J. and Kloth, S. (1986). Plasmodesmatal frequency and radial translocation rates in ray cells of popular (*Populus* × *canadensis* Moench, "robusta"). *Planta* **168**, 377–380.

Shinozaki, K., Yoda, K., Hozumi, K. and Kira, T. (1964). A quantitative analysis of plant form—the pipe model theory. I. Basic analysis. *Jap. J. Ecol.* **14**, 97–105.

Silverbush, M. and Barber, S.A. (1985). Sensitivity of simulated phosphorus uptake to parameters used by a mechanistic mathematical model. *Plant Soil* **74**, 93–100.

Telewski, W. and Jaffe, M.J. (1986). Thigmomorphogenesis: field and laboratory studies of *Abies frazeri* in response to wind or mechanical perturbation. *Physiol. Plant.* **66**, 211–218.

Thomas, T.H. (1985). Hormonal control of assimilate movement and compartmentation. In: *Plant Growth Substances* (Ed. by M. Bopp), pp. 305–359. Springer-Verlag, Berlin.

Thorne, J.H.M. (1985). Phloem unloading of C and N assimilates in developing seeds. *Ann. Rev. Plant Physiol.* **36**, 317–343.

Thornley, J.H.M. (1972a). A model to describe the partitioning of photosynthate during vegetative plant growth. *Ann. Bot.* **36**, 419–430.

Thornley, J.H.M. (1972b). A balanced quantitative model for root:shoot ratios in vegetative plants. *Ann. Bot.* **36**, 431–441.

Thornley, J.H.M. (1991). A transport-resistance model of forest growth and partitioning. *Ann. Bot.* **68**, 211–226.

Thornley, J.H.M. and Cannell, M.G.R. (1992). Nitrogen relations in a forest plantation—soil organic matter ecosystem model. *Ann. Bot.* **70**, 137–151.

Thornley, J.H.M. and Johnston, I.R. (1990). *Plant and Crop Modelling*. Oxford Science Publications, Oxford.

Trewavas, A.J. (1981). How do plant growth substances work? *Plant, Cell Environ.* **4**, 203–228.

Trewavas, A.J. (1991). How do plant growth substances work? II. *Plant, Cell Environ.* **14**, 1–12.

Tyree, M.T. (1988). A dynamic model for water flow in a single tree: evidence that models must account for hydraulic architecture. *Tree Physiol.* **4**, 195–217.

Tyree, M.T. and Sperry, J.S. (1988). Do woody plants operate near the point of catastrophic xylem dysfunction caused by dynamic water stress? Answers from a model. *Plant Physiol.* **88**, 574–580.

Valentine, H.T. (1985). Tree-growth models: derivations employing the pipe-model theory. *J. Theor. Biol.* **117**, 579–585.

Valinger, E. (1990). *Effects of Wind Sway on Stem Growth and Crown Development of Scots Pine Trees*. Swedish University of Agricultural Sciences, Umea.

van Bel, A.J.E. and Gamalei, Y.U. (1992). Ecophysiology of phloem loading in source leaves. *Plant, Cell Environ.* **15**, 265–270.

Wardlaw, I.F. (1990). The control of carbon partitioning in plants. *New Phytol.* **116**, 341–381.

Weinstein, D.A., Beloin, R.M. and Yanai, R.D. (1991). Modelling changes in red spruce carbon balance and allocation in response to interacting ozone and nutrient stresses. *Tree Physiol.* **9**, 127–146.

West, P.W., Jackett, D.R. and Sykes, S.J. (1989). Stress in, and the shape of, tree stems in forest monoculture. *J. Theor. Biol.* **140**, 327–343.

Whitehead, D., Edwards, W.R.N. and Jarvis, P.G. (1984a). Conducting sapwood area, foliage area and permeability in mature trees of *Picea sitchensis* and *Pinus contorta*. *Can. J. For. Res.* **14**, 940–947.

Whitehead, D., Jarvis, P.G. and Waring, R.H. (1984b). Stomatal conductance, transpiration and resistance to water uptake in a *Pinus sylvestris* spacing experiment. *Can. J. For. Res.* **14**, 692–700.

Wilson, J.B. (1988). A review of evidence on the control of shoot:root ratio, in relation to models. *Ann. Bot.* **61**, 433–449.

Wilson, B. and Archer, R. (1979). Tree design: some biological solutions to mechanical problems. *Bioscience* **29**, 293–298.

Zimmerman (1984). *Xylem Structure and the Ascent of Sap.* Springer-Verlag, Berlin, Heidelberg.

Zimmerman, M.H. and Brown, C.L. (1971). *Trees: Structure and Function.* Springer-Verlag, Berlin.

APPENDIX: THE RELATION BETWEEN FUNCTIONAL BALANCE AND WHOLE-PLANT MASS BALANCE

The rate of accumulation of plant carbon substrate (W_c, kg C) is the balance between shoot photosynthesis, incorporation into new plant structure (with associated growth respiration) and utilization for the maintenance of existing plant structure. At the whole-plant level, this may be expressed by the carbon balance equation (see e.g. Thornley and Johnson, 1990).

$$\frac{dW_c}{dt} = \sigma_c W_s - \frac{f_c}{Y}\frac{dW}{dt} - RW, \tag{A1}$$

where W_s (kg dry matter) is photosynthetically active shoot mass, σ_c (kg C per kg shoot dry matter per unit time) is the rate of carbon fixation per unit shoot mass, f_c (kg C per kg dry matter) is the fractional carbon content of new plant dry matter, Y is the efficiency of conversion of carbon substrate into new plant dry matter (W, kg total plant dry matter), and R (kg C per kg total plant dry matter) is maintenance respiration per unit plant mass.

Similarly, whole-plant nitrogen balance can be expressed as:

$$\frac{dW_n}{dt} = \sigma_n W_r - f_n\frac{dW}{dt}, \tag{A2}$$

where W_n (kg N) is nitrogen substrate, W_r (kg dry matter) is root mass actively involved in nitrogen uptake, σ_n (kg N per kg root dry matter per unit time) is the rate of nitrogen uptake per unit root mass, and f_n (kg N per kg dry matter) is the fractional nitrogen content of new plant dry matter. If W' denotes plant dry matter other than active shoot and root, such as stem wood, then total plant dry matter W is equal to $W_s + W_r + W'$, and we can define the shoot and root fractions by $f_s = W_s/W$ and $f_r = W_r/W$, respectively.

When self-shading and root competition are negligible, the specific uptake rates of carbon and nitrogen (σ_c and σ_n) are independent of shoot and root mass. The plant is then able to attain balanced exponential growth in which shoot mass, root mass, carbon substrate and nitrogen substrate all have the same constant relative growth rate μ; hence the shoot and root fractions (f_s and f_r) and the whole-plant concentrations of carbon and nitrogen substrate (C and N), defined by $C = W_c/W$ and $N = W_n/W$, are constant. Equations (A1) and (A2) then simplify (upon dividing by W) to give:

$$\sigma_c f_s = \mu\left(C + \frac{f_c}{Y}\right) + R \qquad (A3)$$

$$\sigma_n f_r = \mu(N + f_n). \qquad (A4)$$

While the relative growth rate (μ) may itself depend on the substrate concentrations (C and N) as well as on other factors such as plant water potential, it can be eliminated between equations (A3) and (A4) to give the following expression for the ratio of carbon fixation to nitrogen uptake, i.e. the ratio of shoot and root activities:

$$\frac{\sigma_c f_s}{\sigma_n f_r} = \frac{C + f_c/Y}{N + f_n} + \frac{R}{\sigma_n f_r} \qquad (A5)$$

This relation describes the equilibrium between the activities of the shoot and the root achieved in exponential growth. It is based solely on the mass balance of carbon and nitrogen at the whole-plant level, and is generally valid for any plant in exponential growth; it is independent of the relation between relative growth rate and the internal substrate status of the plant, and of the mechanism of internal substrate allocation between the shoot and the root.

Davidson's (1969) functional balance relation, equation (1) in the main text, is valid only if the right-hand side of equation (A5) is a constant, independent of the environment (i.e. of σ_c and σ_n). This is true if carbon and nitrogen substrates constitute either negligible or constant fractions of total plant carbon and nitrogen, and if maintenance respiration (R) is negligible in comparison to carbon fixation. In the case where C, N and R are negligible, equation (A5) simplifies to

$$\frac{\sigma_c f_s}{\sigma_n f_r} = \frac{f_c}{Y f_n} \qquad (A6)$$

which implies that shoot activity ($\sigma_c W_s$) is proportional to root activity ($\sigma_n W_r$); the constant of proportionality ($f_c/Y f_n$) depends on the C:N ratio of plant dry matter and on the efficiency with which carbon substrate is used for growth, but is independent of the environment (σ_c and σ_n).

Therefore, Davidson's (1969) statement of functional balance is an approximate expression of the mass balance of carbon and nitrogen at the whole-plant level, and provides a quantitative description of shoot:root responses only for plants in which (a) growth is exponential, (b) non-structural carbon and nitrogen constitute negligible or constant fractions of total plant carbon and nitrogen, and (c) maintenance respiration is negligible in comparison to carbon fixation. Assumptions (a) and (c) are invalid for trees, except small seedlings.

Physiologically, assumptions (b) is also unlikely to hold. In response to changes in σ_c σ_n an increase in C is often accompanied by a decrease in N and vice versa, leading to variations in the ratio of shoot:root activities (Thornley and Johnson, 1990). In that case, shoot:root responses to the environment cannot be predicted from equation (A5) without further information on the mechanisms governing internal substrate dynamics (i.e. the response of C and N to changes in σ_c and σ_n).

Module and Metamer Dynamics and Virtual Plants

P.M. ROOM, L. MAILLETTE and J.S. HANAN

I. SUMMARY

Plants are modular organisms and increasing attention is being paid to the dynamics of their constituent parts. There have been difficulties in handling spatial aspects of those dynamics but these are being overcome by computer models which generate realistic images of "virtual plants". Current models

ADVANCES IN ECOLOGICAL RESEARCH VOL. 25
ISBN 0–12–013925–1

have limited capacity to simulate responses to environmental conditions, a deficiency this review aims to help rectify by outlining the ecological significance of modules and metamers and by summarizing factors affecting and techniques for simulating their dynamics.

Theoretical analyses of branching dynamics are not well developed. There has been some work on segments in two-dimensional (2D) systems but metamers, modules, 3D systems, systems having more than one lateral branch per node, and the effects of damage have received little attention. The numbers, types and positions of new modules and metamers are determined by the states and positions of apical meristems. The states are controlled in turn by internal processes moderated by external conditions. Important internal factors include inherited patterns of apical dominance, phyllotaxis, resource partitioning, programmed abortion of apical meristems, inflorescence evocation, ageing and abscission. Important external factors include gravity, wind, temperature, photoperiod and the availability of light, water and nutrients.

Particular responses include etiolation when light intensity is low, opportunistic reiteration when resources become abundant in positions where they were previously scarce, and thigmomorphogenesis stimulated by mechanical disturbance such as wind. Competition modifies the availability of resources. The quality of light reflected from potential competitors triggers avoidance growth responses in some species. Pathogens, parasites and symbionts modify inherited patterns of resource partitioning. Loss of modules and metamers may be caused directly by extreme weather events or herbivores, or be due to abscission in response to shortage of resources caused by weather, competitors or parasites. Individuals which have lost structural units may exhibit tolerance or compensatory growth as well as chemical changes which affect the likelihood of further removal of biomass. A consistent theme is maintenance and restoration of appropriate root:shoot and source:sink ratios.

Different ways of modelling module and metamer dynamics are discussed and it is concluded that the L-system approach is the most versatile. Examples are given of how responses to particular internal and external factors have been simulated. The parameters needed for complete specification of internal and external determinants of module and metamer dynamics are summarized in a checklist. Potential uses of virtual plant models for research and education are discussed, as is the work needed for this potential to be realized. It is predicted that there will be increasing overlap between crop/physiological/biomass models and virtual plant models.

II. INTRODUCTION

Most studies of plant responses to their environments have been at the detailed level of physiology or the broader level of stand biomass (McGraw and Garbutt, 1990). There has been less work at the intermediate level of plant structural units (Table 1) but interest at this level is increasing. The ecological significance of the population and spatial dynamics of metamers and modules within ramets and genets was brought into modern focus in the 1970s (Harper, 1977; White, 1979). Sophisticated methods for dealing with population dynamics have been available for many years but techniques for handling spatial dynamics have been primitive until very recently. It was difficult to allow for the effects of changes in size and shape of proximal shoot units on the positions of more distal ones and there was no equivalent of population density graphs to provide summaries which were easy to interpret.

Developments in computer hardware and software are on the verge of making important new contributions to ecology by providing "virtual reality" tools which present complex spatial information in a way especially suited to human proficiency at processing visual detail (Moravec, 1988; Rheingold, 1992). In the case of plants, realistic, three-dimensional images can be created from growth rules obtained by observing real plants (Fig. 1) (Prusinkiewicz and Lindenmayer, 1990). The rules are incorporated into computer models which generate electronic abstractions or "virtual plants". The development of these virtual plants can be displayed as sequences of images which show the

Fig. 1. Spruce cones. An example of a realistic image based on a geometric model of cylindrical phyllotaxis. From *The Algorithmic Beauty of Plants* by Prusinkiewicz and Lindenmayer (1990), © 1990 D. Fowler and J. Hanan.

population and spatial dynamics of plant parts. Changes in the size and shape of parts, and consequent adjustments of whole structures, are illustrated by redrawing all parts in each time step. Such models can also keep track of the internal state of each module and metamer (e.g. age, dormancy, concentrations of hormones and nutrients) and display this information as different colours or shadings. Images displayed in rapid succession can show resources and metabolites flowing through the structure of a plant. Position-dependent interactions can also be explored, such as the interception of light by leaves or attempts by two plants to occupy the same space. By dealing with populations and positions of parts as well as the more traditional biomass, these models are creating new opportunities for interfacing with herbivore models which deal with numbers and positions of animals.

At present, virtual plant models assume mostly constant environmental conditions and they have little ability to simulate responses to changing conditions and external stimuli. Some models can simulate populations of plants in which individuals differ stochastically from one another but they have very limited ability to simulate how neighbouring individuals interact.

The aim of this review is to consolidate information relevant to making virtual plant models more realistic and useful. The review is concerned with the number and positions of plant parts as they become visible externally, persist and die in ecological time. Limited attention is given to roots. Though roots are modular, they are not metameric, and their dynamics are less predictable than the dynamics of shoots due to soil being less uniform than air. The review is confined to the Kingdom Planta, though many of the issues addressed apply to branching algae, fungi and sessile colonial animals. Relevant publications were found by searching on-line databases and reference lists in papers and by requesting preprints from authors and editors of journals up to July 1992. Some 2500 citations were stored and cross-referenced in a bibliographic database.

III. TERMINOLOGY

This review generally follows the usage of terms found in Esau (1977), Harper (1977), Hallé et al. (1978), White (1984) and Bell (1991). Some terms have a history of imprecise or inconsistent use. A node, for example, has often been defined as the point of attachment of a leaf to a stem even though leaf attachments occupy short lengths of stem and not "points". Imprecision on this detail carries forward into definitions of internode and metamer. If a node is to remain a *point* defined in relation to attachment of a *leaf* rather than an axillary bud, if each branch is not to have a leafless metamer at its base, and if an internode is to be coincident with the stem component of a metamer, then a node must be positioned at the distal limit of attachment of a leaf, immediately proximal to the axillary bud subtended by the leaf. This is consistent with Kurihara *et al.* (1978), illustrated in White (1984), but it is inconsistent with the three definitions of metamer given by Bell (1991) which place axillary buds in the same metamers as the leaves which subtend them.

To avoid ambiguity we define structural components in Table 1 and Figs. 2 to 5 and present a glossary of other terms in the Appendix. We use "individual" in the sense of an observably coherent structural entity to avoid the practical difficulties of determining what constitutes a physiologically integrated or genetically uniform individual. Our use of "segment" is consistent with MacDonald (1983) but differs from that of Tomlinson (1978) and McMahon and Kronauer (1976) who use it to denote what we refer to as an axis (Table 1). We use "module" in the broad sense of the product of a single apical meristem (White, 1984), rather than the more restricted sense of the determinate, sympodial unit originally defined by French workers as "l'article" (Hallé *et al.*, 1978). In clonal species, a new ramet may be formed by each new metamer (e.g. *Salvinia molesta* Mitchell), each new module (e.g. *Iris* spp.), or only certain types of new modules (Fig. 3).

IV. FUNDAMENTALS

A. Ecology

Being sessile, plants can only explore their environment by growth and they are chronically exposed to weather, competitors and herbivores. Their success, and that of many sessile animals, is related to their early evolution of growth by reiteration of a few basic types of structural unit (Harper, 1985). Growth towards or away from stimuli, regrowth after damage, and the production of multiple reproductive organs, are facilitated by the potential of each vegetative meristem to give rise to organs of different types and the partial autonomy of each module (Harper *et al.*, 1986). Modular growth is at the heart of clonal increase and confers potential immortality on clonal genets (Watkinson and White, 1985).

The overall shape of a plant is largely determined by the timing, topology and geometry of production and loss of metamers (Porter, 1983). Small changes in parameters such as metamer length, branching angle and proportion of buds remaining suppressed are amplified by reiteration to result in large effects on crown shape (Gottlieb, 1986). This makes plants inherently plastic and able to alter their phenotypes in response to environmental cues (Schmid, 1992). Different patterns of growth result in different branching patterns and foraging strategies (Hutchings and de Kroon, 1994). Thus, there are direct links between module and metamer dynamics, the parameters of branching growth, architecture, the ability to capture resources, productivity and fitness.

Table 1

Definitions of components of individual plants in order of size: components following metamer may contain many metamers; components following module may contain many modules

Component	Definition
Node	The most distal point of the junction between a stem and a leaf (or coincident whorl of leaves), just proximal to any subtended axillary bud(s) (Kurihara *et al.*, 1978)
Axillary/apical meristem	A cell or group of cells, specialized for mitosis, initiating or at the apex of a shoot. An axillary meristem becomes an apical meristem as soon as it starts to produce a shoot
Bud	An unextended, partly developed, shoot having at its summit the apical meristem which produced it; an unexpanded metamer or group of metamers (Bell, 1991)
Internode	A portion of stem between nodes i.e. from immediately distal to the junction of a leaf with the stem to the same position with respect to the next most distal leaf. The basal internode of a branch starts at the node of the leaf from whose axil the branch grew
Metamer	An internode, the axillary bud(s) at its proximal end and the leaf or leaves at its distal end but not any shoots resulting from growth of axillary buds (Kurihara *et al.*, 1978)
Segment	The one or more metamers between nodes at which successive branches are attached. Equivalent to an edge in graph theory terminology (MacDonald, 1983)
Unit of growth	A morphologically discrete growth increment, the result of one episode of rhythmic growth by a module, i.e. extension of the preformed contents of a previously dormant apical bud followed by growth of neoformed leaves (if any) and formation of a new, dormant, apical bud (= unit of extension of Hallé *et al.*, 1978)
Unit of morphogenesis	The product of a single episode of mitotic activity of an apical meristem having rhythmic growth (Hallé, 1986), extending from proximal to the oldest of a set of neoformed leaves to the most distal of the next distal set of preformed leaves
Shoot	A young stem which has grown from a single axillary/apical meristem and the leaves and buds which it carries; the young portion of a module
Short shoot	A shoot having shorter/fewer internodes which exploits ambient conditions and often bears flowers
Long shoot	A shoot having longer/more internodes which explores for new resources
Module	The product of one meristem; in shoots: a set of metamers originating from one axillary/apical bud (White, 1984); the smallest unit of morphology capable of producing daughter units and/or seeds (Maillette, 1982b)

Table 1 (continued)

Component	Definition
Axis	A sequence of units of growth in the same general direction from one (monopodial) or more (sympodial) meristems
Branch	An axis other than the main stem plus any subordinate axes it bears
Ramet	The unit of clonal growth, capable of an independent existence if severed from the parent plant (Harper, 1977)
Individual	A physically coherent, structural individual (Vuorisalo and Tuomi, 1986)
Architectural unit	The complete set of axis types and their relative arrangements found in a species—cannot be seen until an individual is old enough to have expressed its architectural model in full (Barthélémy, 1991)

There are trade-offs between different uses of limiting resources (Tuomi *et al.*, 1988; Chapin *et al.*, 1990) and a plant's responses to environmental conditions are constrained by its growth form, pattern of resource partitioning, photosynthetic rate, intrinsic growth rate, degree of physiological integration and the abundance and positions of meristems (Maschinski and Whitham, 1989). The structural dynamics and architectures we observe today are unlikely to be optimal for any particular circumstances because they represent compromises resulting from the past actions of many, sometimes opposing, selective forces (Bloom *et al.*, 1985). In addition, the diversity of plant architectures that coexist is unequivocal evidence that there are many different answers to the question of survival. There is no single optimum for particular circumstances because phyllotaxis, leaf shape and branching pattern interact with and compensate for each other (Niklas, 1987).

B. Demography

New metamers are produced by apical meristems. New modules are initiated when apical meristems form daughter meristems in the axils of leaves. These axillary meristems become the apical meristems of new modules. The dynamics of meristems, metamers, modules, ramets and individual plants are necessarily linked because of their physical and parent–offspring relationships. The population density of metamers in a stand is the product of the population density of individuals and the mean numbers of ramets per individual, modules per ramet and metamers per module. At each level of the structural hierarchy:

$$N_t = N_0 + \text{Births} - \text{Deaths} + \text{``Immigration''} - \text{``Emigration''} \quad (1)$$

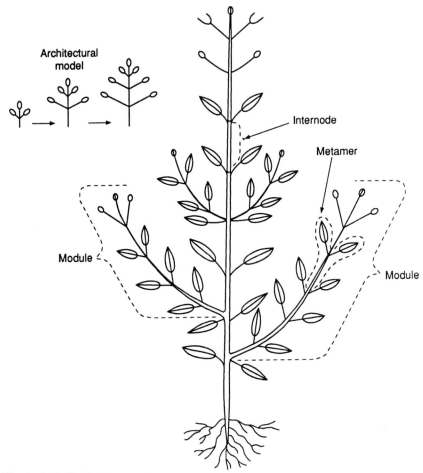

Fig. 2. An individual annual herb showing structural components and the architectural model. Each module is also a unit of growth and a unit of morphogenesis. The architectural model is that of Rauh (Hallé *et al.*, 1978), having an orthotropic main axis producing distichous, orthotropic, lateral shoots.

where N_t is population density at time t and N_0 is the initial population density. Depending on the structural unit, births may result from growth or germination, deaths may result from senescence, abscission or destruction (Harper, 1977), "immigration" may result from grafting and "emigration" from clonal fragmentation.

Within a module, the number of metamers can increase only at the distal end and therefore in arithmetic fashion. New metamers are added to a module one at a time during continuous growth, or in groups during the expansion phase of rhythmic growth (Fig. 4). Metamer numbers may de-

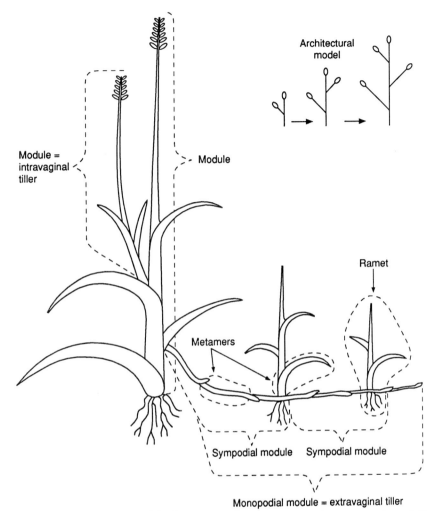

Fig. 3. An individual clonal herb or grass showing structural components and the architectural model which is that of Attims (Hallé *et al.*, 1978). The orthotropic main axis produces alternating lateral shoots which may be orthotropic (intravaginal tillers) or initially plagiotropic (extravaginal tillers/stolons) followed by continued plagiotropy if stolon growth is monopodial or orthotropy if sympodial growth produces new ramets. If stolon growth is monopodial, each complete stolon is a module and each ramet formed from an axillary bud is a module. If stolon growth is sympodial, each plagiotropic section of stolon plus the ramet formed by orthotropic growth is a module.

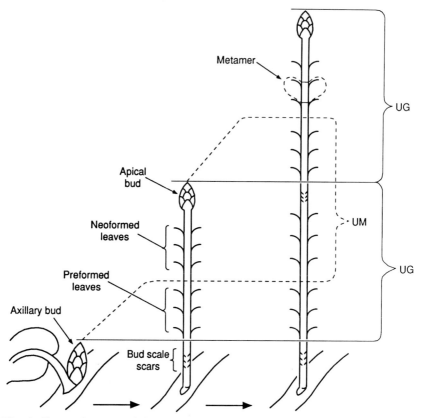

Fig. 4. Shoot of a perennial plant which has rhythmic growth: only the petioles of leaves are shown. The whole shoot is a single module which, after one and two episodes of growth from dormant initial axillary bud or apical bud to dormant apical bud, consists of one and two units of growth (UG) and has a basal metamer aged 9 and 18 plastochrons respectively. The unit of morphogenesis (UM) is the structure resulting from a single episode of cell division in the apical meristem.

crease arithmetically, either singly at the distal end of a module or in a group if a more proximal metamer dies and takes more distal, dependent metamers with it. A module may give rise to several new modules simultaneously, one at each axillary bud, and if the process is repeated the number of modules and the sum of their metamers increase exponentially. Rapid early growth, whether exponential or otherwise, becomes asymptotic at physiological or resource limits as typified by the logistic equation:

$$\frac{dN}{dt} = rN\frac{(K-N)}{K}$$

(2)

where N is the number of modules or metamers in an individual or a stand

and K is the carrying capacity in relation to stands or the maximum sustainable number of modules or metamers in relation to individuals (Bazzaz and Harper, 1977). The same equation applies when N is the number of ramets, individuals or genets in a stand.

In stands of aclonal plants, small differences in vigour or time of establishment are rapidly transformed by competition into skewed size distributions of a few dominant and many subordinate individuals (Pacala and Weiner, 1991). After a period of competitive growth and irrespective of population density, stands of the same age have the same biomass according to the Law of Constant Final Yield (Kira *et al.*, 1953). Variation in the size of individuals is commonly due to variation in the number rather than the size of metamers and in such cases the Law applies to metamer population density as well as to biomass. When the intensity of competition is high, density-dependent mortality of individuals takes place as the smallest die and the others continue to grow. This may or may not occur according to the "Self-thinning Rule" (Yoda *et al.*, 1963; Lonsdale, 1990). When individual variation in size is due to variation in the number rather than the size of metamers, the relationship applies to metamer population density as well as to biomass.

Amongst clonal plants the Law of Constant Final Yield probably applies and self-thinning of independent physiological units appears to occur (Eriksson, 1989) but not smaller units such as ramets or modules (Kays and Harper, 1974). Internal nutrient cycling and control of spatial arrangement by apical dominance adjust the natality and mortality of ramets and prevent intraclonal self-thinning of ramets (Cain, 1990).

C. Patterns of Branching

When a seed germinates, an apical meristem starts to construct a shoot system by adding metamers to the single, initial module to form a main axis or trunk. The apical meristem follows one of four alternative paths (Hallé, 1986):

(i) Continued growth without multiplication.
(ii) Production of additional meristems of equal potential.
(iii) Production of additional meristems of unequal potential.
(iv) Production of additional meristems of mixed potential.

Subsequent growth, abortion and suppression of apical and axillary meristems gives rise to a diverse array of branching patterns which has been analysed in various ways. Prusinkiewicz and Lindenmayer (1990) noted that there are four basic patterns of branching:

(i) Terminal: main apex terminates, all lateral apices terminate.
(ii) Sympodial: main axis terminates, some lateral apices continue.
(iii) "Monopodial": main apex continues, all lateral apices terminate. (Note,

elsewhere we use monopodial in the conventional sense of an axis produced by a single apical meristem irrespective of branching habit.)
(iv) Polypodial: main apex continues, some lateral apices continue.

   The resulting axes have been classified according to their potential for
further differentiation (e.g. trunk, branch, long shoot, short shoot) and
Barthélémy (1991) reported that no more than seven of these categories have
been found in a single species. This insight has been used in building some
virtual plant models (de Reffye et al., 1988).
   Branching in clonal plants is basically two-dimensional (2D) in the horizontal plane, although it is also 3D if ramets add modules in the vertical plane.
Branching in aclonal plants is 3D. Branching in 2D represents greater investment in foraging for patchy resources and in pre-empting substrate, in contrast to branching in 3D which allows more metamers and modules to occupy
an area of substrate, greater rates of metamer population increase per unit of
biomass accumulated, and more efficient interception of light (Niklas, 1987).
Two contrasting 2D branching strategies have been recognized: a guerilla
strategy having little branching and more extensive linear foraging, and a
phalanx strategy having much branching and more complete occupation of
surface area to exclude competitors (Lovett Doust, 1981). The guerilla strategy results in lower rates of metamer and module population increase and
smaller populations per unit of biomass because of greater investment in long
internodes.
   Theoretical studies relating abstract, drainage-basin and other branching
patterns to the general properties of space are mainly restricted to 2D systems
having one lateral axis per node (MacDonald, 1983), although Thornley and
Johnson (1990) describe some studies of 3D systems. Several schemes classify
segments or axes into orders to calculate bifurcation ratios but most of these
schemes are centripetal and not suited to structures like plants which grow
distally. A centrifugal scheme for axes (Gravelius, 1914) has been applied to
simulations of plant growth in which orders increased from zero (Prusinkiewicz and Lindenmayer, 1990) or one (de Reffye et al., 1988) at the main axis.
Branch order analyses of plants are few because ordering ignores ontogeny,
allowing axes to be of the same order even if they originated at different ages
of the plant and parent segment (Steingraeber and Waller, 1986). Ordering
also fails to distinguish between monopodial and sympodial axes and takes no
account of how many metamers are present in segments. We agree with the
conclusion of Waller and Steingraeber (1985) that comparisons of bifurcation
ratios between branches within individuals, between species and between
habitats are meaningless.
   Under constant conditions, the lengths and diameters of segments and
branching angles are consistent throughout individual 2D plants, while in 3D
plants distal segments are usually shorter, thinner and have larger branching

angles than proximal segments (Bell, 1991). Branching angles and diameters of segments optimal for fluid flow within a branching system can be calculated from fluid dynamics theory. Some plants have branching patterns similar to theoretical optima but most do not, probably due to the action of other selective forces such as gravity, competitors and the need to minimize self-shading (MacDonald, 1983).

The mathematics of 2D branching involving more than one lateral axis per node appear to have been little studied. In *Salvinia molesta*, few branches were formed when nitrogen availability was low but there were as many as three lateral axes per node when nitrogen was abundant. Formation of the first metamer of each lateral axis was delayed by apical dominance until the parent axis, or immediately adjacent older lateral axis, had produced two metamers distal to the node. When all axillary buds became axes and all axes added metamers at the same rate, the population of metamers y in an individual was given by:

$$y_x = 2y_{x-1} - y_{x-n-2} \qquad (3)$$

where x is the number of metamers in the main axis (i.e. plastochron index physiological age) and n is the number of lateral axes per node (Room, 1983).

Mathematical aspects of 3D patterns have received even less attention than 2D systems. Despite extensive work relating phyllotaxis to light interception, vascular architecture and the structure of composite flowers, the implications of phyllotaxis for metamer and module dynamics remain largely unexplored (Kirchoff, 1984). At the empirical level of gross morphology, Corner (1949) proposed the following two rules for branches:

(i) Axial conformity: the stouter an axis, the larger and more complicated are its appendages.

(ii) Diminution of ramification: the more branching, the smaller the terminal branches and their appendages.

In similar vein, Shinozaki *et al.* (1964) proposed a "pipe-stem model" on the basis that the quantity of leaves above a given horizontal plane is proportional to the sum of the cross-sectional areas of all branches at that plane.

In a more detailed analyses of growth by tropical trees, 24 branching programs or "architectural models" were identified by Hallé *et al.* (1978). Bell (1991) found another three programs amongst lianas and suggested that more might be found amongst herbs. Six of the programs are based on modules whose apical meristems eventually become sexual and stop growing (Halle, 1986). Eighteen of the programs were classified by Porter (1989a) according to whether the main and lateral axes were monopodial or sympodial, orthotropic or plagiotropic, hapaxanthic or pleonanthic. He found that of the 64 potential combinations, 27 have been observed (some of his 18 programs conforming to more than one combination), 27 are impossible and 10 are possible but have not been observed. Both he and Fisher (1992) noted

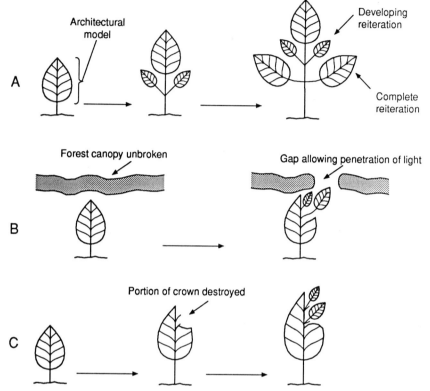

Fig. 5. Three categories of reiteration of an architectural model: A, metamorphic; B, opportunistic; C, traumatic.

that the programs do not correlate with taxonomic or ecological classifications and the latter pointed out that during ageing of complex trees, the programs are not predictive of crown shape because patterns of reiteration (Fig. 5) and intercalation are at present unpredictable. There do not seem to have been any studies of the significance of the different programs for the dynamics of modules and metamers.

The broad implications of physical constraints on the evolution of 3D structural patterns were explored by Niklas (1987) using simulations. He showed that the sequence of vegetative architectures in the fossil record was consistent with progressive modification of phyllotaxis and branching pattern to gain increased interception of light without exceeding mechanical limits to shape and size. Though the underlying geometry, physics and mathematics are straightforward, much remains to be done to bring about a synthesis of how spatial dimension, number of axes per node, branch angle, phyllotaxis, fluid dynamics, light interception, mechanical strength and the characteristics of architectural models interact to constrain the dynamics of metamers and modules.

D. Monitoring Techniques

The dynamics of structural units can be assessed at three levels of increasing detail and difficulty: population, topology and geometry. Population assessment by counting is generally straightforward, although keeping track of which parts have been counted on large plants can be difficult (Harper, 1977). Cohorts of plant parts may be marked with colour-coded threads or plastic rings placed around petioles or stolons (e.g. Maillette, 1985). Topology, together with population size, can be recorded in plant maps and the contexts in which structural units are produced and lost can be determined by comparison of maps drawn at intervals (Room, 1988). Mapping is a 2D exercise and can be problematical for large 3D plants which may need separate maps for different vertical layers or for each lateral branch. Maps take time to prepare the first time but can be constructed to accommodate future sampling dates so saving time, allowing errors to be detected and corrected, and reducing the amount of plant handling.

Measurements of the geometric positions of metamers can capture information on population size and on topology if the locations of all nodes and apices are recorded. There is currently no quick and highly accurate way to assess the geometry of 3D plants, the main problem being that measurements cannot all be made by line-of-sight from a single reference point. A useful method for plants up to about 1 m tall involves placing the end of an articulated arm at each point of interest. A sensor for angular displacement at each articulation sends measurements to a computer which calculates the 3D coordinates of the end of the arm. Twenty measurements can be made per minute with a precision of 3 mm (Lang, 1990). For larger plants, infra-red beam surveying equipment (e.g. Monmos, Sokkia Co. Ltd, Japan, 0·1 mm resolution) might be used to determine the position of the end of a large pointer bearing two reflective targets. However, positioning the pointer and holding it steady would probably be difficult and time-consuming. To speed up data collection there is a need for some form of portable, non-destructive, 3D scanning device or a 3D version of the 2D system used in sports medicine for digitizing video images.

V. INTERNAL FACTORS

A. States of Apical Meristems

Whether new metamers and modules will be produced in particular positions, and of what types, depends on the state of apical meristems. A meristem may be vegetative or reproductive, active (dividing and/or elongating) or inactive (suppressed by apical dominance or seasonally dormant), or dead (Maillette, 1982a). Hormones are probably involved in mediating most state changes but there is considerable uncertainty about mechanisms (Salisbury and Ross, 1992). Meristems become reproductive in response to changes in such factors as the age of a plant, the size of the supporting shoot, or the abundance of resources, and in response to external signals such as day length (Reid et al., 1991). Growth by reproductive modules is analogous to that of vegetative modules. Change to the reproductive state is irreversible, and ultimately fatal to an apical meristem, so that most of a shoot structure usually originates from vegetative meristems. Reproduction reduces the number of apical meristems able to generate photosynthetic tissue which in turn may reduce subsequent growth and survival (Carlsson and Callaghan, 1990).

Transition of an apical meristem from not dividing to dividing indicates that it is completing construction of a metamer. The process results in one or more leaves and axillary meristems. The orientation of axillary meristems determines phyllotaxis and branching angle. The orientation is usually species-specific, under strong genetic control and, in the case of phyllotaxis, correlated with vascular architecture (Esau, 1977).

Once axillary meristems have been formed, they are frequently held suppressed by apical dominance from the parent apical meristem until it has grown some distance away (Cline, 1991). Plants with strong apical dominance have conical crowns with well developed lower branches while weak apical dominance results in rounded or flat crowns, the upper branches developing more than the lower ones (Lavarenne et al., 1971; Ladipo et al., 1991). Weak apical dominance early in life followed by strong dominance by lateral apices results in broad and round crowns (Brown et al., 1967; Remphrey and Davidson, 1991). Patterns of apical dominance vary between species and in response to light intensity and quality, temperature, and nutrient supply (Cline, 1991). Suppressed meristems become active when dominant apices are lost or weakened and those plants which lack suppressed meristems, such as palms, are entirely dependent on the survival of continually active meristems (Tomlinson et al., 1989). Leader shoots in the upper portions of crowns often grow faster than shoots lower down and are said to exert apical control. This implies that apical meristems may have more complex influences on the meristems of daughter branches than simple suppression via apical dominance (Bell, 1991). Apical dominance and apical control are examples of correlative inhibition (Goebel, 1900).

Seasonal dormancy is common in perennial plants and results in rhythmic growth (see Fig. 4). It is usually triggered by photoperiod in anticipation of annual seasons of low temperature or lack of water. Some species maintain rhythmic growth even under constant conditions (Thiébaut et al., 1990). Meristem activity is continuous in only 20% of tropical trees (Hallé et al.,

1978). Dormant buds of some species include all of the next growing season's metamers preformed, while other species produce additional, neoformed metamers after the preformed ones have expanded (Hayes *et al.*, 1989). Abortion or programmed death of meristems is integral to some patterns of growth (Mueller, 1988). After the abortion of an apical meristem has halted the growth of a determinate module, an axillary meristem may grow into a subapical module which replaces the parent module as axis leader, to be replaced by another in turn and so on to form a sympodial axis. In contrast, a monopodial axis is produced by a single apical meristem. Single individuals can have both monopodial and sympodial axes (Callaghan *et al.*, 1986).

B. Age

Age may be measured in real time, physiological time based on heat accumulated between upper and lower threshold temperatures, or physiological time expressed as the plastochron index, i.e. the number of new metamers created in sequence along an axis.

Age is not a straightforward concept for whole modular organisms because of the continued birth and death of constituent parts. Old genets, individuals or ramets, for example, may be composed entirely of young modules, in which case physiological status will not be proportional to age. Moreover, size tends not to be proportional to age because individuals of the same age but grown under different conditions have different numbers and sizes of parts (Watkinson and White, 1985). Indeed, size is usually a better predictor of an individual's fitness than age (Herrera, 1991). If an individual always retains some of its original metamers as is usual in aclonal plants, ageing can have significant effects on distal growth by reducing inflorescence production, growth rate, and mean shoot length, causing a switch from continuous to rhythmic growth, and stimulating metamorphic reiteration (see Fig. 5) (Barthélémy *et al.*, 1989; Morrisson and Myerscough, 1989; Peng and Krieg, 1991). In addition, the accumulation of deleterious mutations in apical meristems can be significant in long-lived plants (Klekowski and Godfrey, 1989).

The concept of age is less problematical for branches, axes and modules than it is for whole modular organisms because the former are as old as their oldest metamers. Even so, the number of structural units within them will usually reflect past conditions for growth rather than age. Apical dominance can decline as branches become older (de Reffye *et al.*, 1991) and, in *Rhododendron maximum* L., the probabilities of branches surviving, growing, flowering or dying vary with branch age and size (McGraw, 1989). The growth rates of modules may decrease or increase as they age (Caesar and Macdonald, 1983; Fetcher and Shaver, 1983).

The concept of age is conventional for metamers. Their initial performance, in photosynthesis for example (Harper, 1989a), generally declines with age and resources may be withdrawn from them for re-use elsewhere, leaves may be shed, and remaining tissues in trunks and main branches may become specialized for mechanical support and translocation.

C. Integration

Modules, branches and root–trunk–branch sectors behave as semiautonomous physiological units in many species (Watson, 1986). The arrangement of these units is determined by vascular architecture. Dicots often have vertical connections from roots to leaves, and between leaves directly above each other along a shoot (i.e. of the same orthostichy), but limited horizontal connections between orthostichies (Shinozaki et al., 1964; Kirchoff, 1984; Price and Hutchings, 1992). This may result in sectoral asymmetry when some sectors have better access to resources and can add more structural units than others. In contrast, monocots have extensive vascular interconnections which promote pooling of solutes and symmetrical growth (Esau, 1977). In gymnosperm and dicot trees, vascular constrictions near branch junctions allocate water unequally among branches (Tyree and Ewers, 1991). Amongst clonal plants, the phalanx growth form is usually associated with greater integration than the guerilla growth form (Schmid and Bazzaz, 1987). Nonetheless, integration in phalanx Schizachyrium scoparium (Michx.) Nash is limited by the rapid disintegration of connections between ramets (Welker and Briske, 1992).

The internal management of nutrients and water helps determine the rates at which structural units are added and lost. Efficient mobilization of resources from senescing parts and movements to and from storage are conducive to faster growth and reduced losses during periods of adversity (Larsson et al., 1991; Helmisaari, 1992). Evergreen species, for example, seem to have poor nutrient retrieval prior to leaf abscission, limited efficiency of internal nutrient use and storage, and limited ability to regrow lost structures (Jonasson, 1989).

VI. EXTERNAL FACTORS

A. Conditions

1. Temperature

The rates at which apical meristems form new metamers and initiate new modules change with temperature in a predictable manner (Room, 1988; Sackville Hamilton and Harper, 1989) and cumulative temperature can trigger developmental switches necessary for appropriate phenology, as in low temperature vernalization (Hänninen, 1990). High temperatures can weaken apical dominance and increase internode length (Moe, 1990; Erwin *et al.*, 1991). Extreme temperatures cause mortality, with flower buds, flowers and young fruits being especially sensitive (Konsens *et al.*, 1991; Reddy *et al.*, 1991).

2. Gravity and Wind

Terrestrial plants must support their own weight and any extra loads from wind, rain, snow or epiphytes (Vincent, 1990). Trees maintain a safety margin against extra loads by typically reaching only one-quarter of the height at which they buckle under their own weight, although there is considerable inter- and intraspecific variation (King, 1991).

Wind increases transpiration, cools meristems and reduces rates of growth (Grace, 1977). Wind, rain and other mechanical stimulation activate thigmo-morphogenetic responses which typically result in new modules and metamers being unusually short and broad (Jaffe, 1980). Shaking, even for very short periods, can halve stem elongation and reduce the birth rate and size of leaves (Rees and Grace, 1981). Dwarfing responses result in greater stem strength, reduced wind resistance and consequent lower risk of mechanical collapse (Tateno, 1991). Strong wind causes trees to become gnarled and twisted and often carries particles of soil or snow which cause lesions leading to reduced growth, delayed flowering and increased mortality (Armbrust, 1968; Fryrear, 1975).

At higher elevation, conditions are both windier and colder. The plastochron increases and leaf longevity drops in herbaceous species but tends to increase in conifers (Ewers and Schmid, 1981; Diemer *et al.*, 1992). Near the tree line, normally single-stemmed trees exposed to the elements become multi-stemmed and rely on banks of arrested meristems to regrow after especially unfavourable weather kills above-ground modules (Fanta, 1981; Maillette, 1987).

3. Chemicals

At high concentrations, materials such as salt and heavy metals kill metamers and individuals, while at sublethal concentrations they cause reduced growth rates, miniaturized plants and longer-lived leaves (Blits and Gallagher, 1991). Biologically active chemicals may also originate from organisms such as parasitic plants, sap-sucking insects, neighbouring plants, pathogens or symbionts (Whittaker and Feeny, 1971; Waller, 1987) but their effects on metamer dynamics are generally not well documented. Metamers become distorted when secretions from nematodes, mites, insects and fungi stimulate abnormal cell enlargement or uncontrolled cell divisions to form galls (Bell, 1991). Witches brooms are formed when fungi and mistletoes induce excessive proliferation of modules. *Arceuthobium* spp. and other mistletoes also cause reduced internode expansion, conversion of short shoots to long shoots and failure to initiate flowers (Kuijt, 1969). Reproduction is affected when the fungus *Epichloe typhina* and some viruses induce early abortion of inflorescences in grasses, greatly increased tillering, and increased tolerance of heavy grazing (Burdon, 1987).

B. Resources and Competition

1. Light

Light is the primary resource fuelling photosynthesis and growth, thereby generating demand for nutrients and water. Consequently, light tends to have greater effects on the dynamics of metamers and modules than other resources (Slade and Hutchings, 1987a; Boutin and Morisset, 1988; Denslow *et al.*, 1990). Growth rates are commonly proportional to the intensity of photosynthetically active radiation (PAR) between compensation and saturation intensities (Canham, 1988; Mitchell and Woodward, 1988). Variations in the intensity of PAR trigger temporary adjustments of photosynthetic efficiency and leaf inclination as well as more permanent changes such as production of sun or shade leaves and opportunistic reiteration of branches (see Fig. 5) (Boutin and Morisset, 1988; Elias, 1990). Leaves growing in the sun tend to be shorter-lived than those in the shade (Maillette, 1982a; Xu *et al.*, 1990).

In low light intensities plants commonly become etiolated, producing petioles and internodes which are longer and leaves having larger surface areas (Atkinson, 1984). Etiolation allows more rapid discovery and exploitation of spaces receiving more light but may result in shoots unable to support their own weight (Holbrook and Putz, 1989). Not all plants become etiolated and some reduce rhizome length, increase petiole strength, or appear to have no response (Lovett Doust, 1987; de Kroon and Knops, 1990). Flowering and seed production are usually delayed and reduced under low light intensities (Boutin and Morisset, 1989). The negative effects of shading are lessened in clonal species when unshaded ramets subsidize connected ones which are shaded (Slade and Hutchings, 1987a). Shade-adapted trees and shrubs tend to have sparsely branched crowns with horizontal leaves that enhance the potential for capture of sunflecks, or laterally spreading, layered crowns which forage for small gaps (King, 1990). In contrast, emergent tree species favour growth in height so as to quickly reach the canopy (Kohyama and Hotta, 1990).

In addition to being a resource, light conveys signals to many plants. Photoperiod, often in conjunction with thermoperiod, is a reliable indicator of seasonality and has well documented effects on phenology, growth, and reproduction (Downs and Bevington, 1981). The spectral composition of light transmitted through or reflected by leaves is different from that of sunlight. Under a plant canopy, the ratio of red:far-red wavelengths is reduced (Thompson and Harper, 1988; Methy *et al.*, 1990). This can stimulate production of longer petioles and internodes, and reduced branching, leading to more rapid growth through shaded spaces (Hunt *et al.*, 1990; Ballaré *et al.*, 1990). In other species, growth may be retarded with the result that branches do not penetrate further into shaded spaces, freeing the resources they would have consumed for opportunistic reiteration elsewhere. Light reflected from leaves can cause similar responses which result in branches avoiding becoming shaded at all (Ballaré *et al.*, 1987). Shade-adapted plants appear to respond less to far-red light than sun-adapted species (Morgan and Smith, 1979).

2. Nutrients

In shoots, high nutrient levels increase not only the rates of production of modules and metamers but also their death rates, resulting in faster turnover (Schmid *et al.*, 1990). Low nutrient availability reduces rates of production of modules, metamers, flowers and inflorescences and can shorten leaf longevity (Room, 1988; Hocking and Pinkerton, 1991). Responses to nutrient availability vary among species with different rates of uptake, internal recycling and storage of nutrients. In *Acer pseudoplatanus* L., canopy development in the spring depends initially on stored nitrogen but afterwards on current nitrogen supply. Leaf mortality occurs prematurely on poor soils but extra foliage is added on rich soils (Millard and Proe, 1991). The proliferation of roots is generally proportional to the local abundance of nutrients (Fitter and Stickland, 1991).

Responses to nutrient levels can also vary because of differences in the plasticity of modular growth, especially in clonal herbs. Some clonal herbs show little change in growth form in response to nutrient availability (Alpert, 1991). However, stolons of *Ranunculus repens* L. and *Glechoma hederacea* L. in nutrient-poor microsites have long internodes and sparse branching for extensive foraging and stolons in nutrient-rich microsites have short internodes and heavy branching for intensive foraging (Ginzo and Lovell, 1973; Slade and Hutchings, 1987a,b). This response can be modified within a stolon by subsidy from longer-established ramets to distal ramets, i.e. acropetal translocation. Parent ramets in rich microsites can subsidize daughter ramets in resource-poor microsites, thereby enhancing stolon growth (Slade and Hutchings, 1987a,b,c). Translocation which is exclusively acropetal results in resource movement only from parent to daughter stolon, not from sister to sister stolon, and results in sectorial growth enhancement.

3. Water

During periods of water shortage the rates of production, and sometimes the sizes, of new modules and metamers decline rapidly (Moran *et al.*, 1989; Shumway *et al.*, 1991). Rates of leaf and metamer mortality usually increase more slowly (Tanner, 1983) but can increase rapidly if the plant sheds branches to reduce the demand for water (Kozlowski, 1973). In grasses, sudden water stress increases leaf death rate but gradual stress may decrease it (Reekie and Redmann, 1991). In regions with distinct wet and dry seasons, many plants have regular phenologies coupled to predictable rainfall, whereas in deserts, growth and development tend to be opportunistic in response to rain (Boeken, 1989). Roots are usually less affected by water shortage than shoots, resulting in a decrease in shoot:root ratio during drought (Reid *et al.*, 1991). Too much water can be as detrimental as too little. Flooding reduces oxygen supply to roots, severely decreasing root and shoot growth and causing death if prolonged (Laan and Blom, 1990; Pezeshki, 1991).

4. Competition

Crowding reduces the resources available to each individual, causing many of the responses to shortages described above. Mean individual biomass is less than in uncrowded plants, rates of module and metamer production and reproduction are reduced, and phenology is altered.

Competition involves a scramble to occupy space in which to capture resources. When two modules in the same or different individuals grow towards a vacant space:

(i) both may stop growing or change their orientation as they detect each other's resource depletion zone;

(ii) one may be inhibited and the other may continue growing;

(iii) neither may be affected by the other (Franco, 1986).

Response (i) produces a clear separation between neighbouring modules within and between plants (Jones and Harper, 1987a,b); response (ii) gives a hierarchy of dominance-suppression; response (iii) will result in collisions and may give the impression of a type (i) outcome if abrasion causes metamers to be lost (Franco, 1986).

When a space is already occupied:

(iv) an approaching module may stop growing as it detects the resource depletion zone of the incumbent;

(v) the approaching module may continue growing but remain subordinate or outgrow the incumbent through being more efficient at resource acquisition (Campbell *et al.*, 1992);

(vi) within an individual, architectural rules and module longevity may ensure that old modules die before new ones enter the same spaces (Harper, 1977).

Responses (i) and (iv) result in plants which are taller than uncrowded individuals and have reduced biomass, smaller stem diameters, fewer leaves and fewer, shorter branches higher on the main stem (Weiner and Thomas, 1992). Asymmetry resulting from any of responses (i) to (v) can be used as a measure of interference within and between neighbours (Franco, 1986).

C. Removal of Biomass

We use "removal of biomass" rather than "damage" to avoid the latter's connotations of reduced vigour and fitness which are not universal outcomes (Paige, 1992). Biomass is sometimes "removed" when tissues are killed *in situ*.

The agents of biomass removal include herbivorous animals, parasitic plants, pathogens, fire, wind and hail. The biotic agents are usually more specific in the types and positions of tissues they attack. The literature relating to biomass removal is large and a number of authors have discussed response patterns, mostly in connection with herbivory. Few publications deal with modular responses even though these are central to linking the population dynamics of plants and herbivores (Dirzo, 1984; Haukioja, 1991). The broad picture is that plants usually respond to restore physiological balance and to increase passive defences. Response patterns of practical significance can be discovered using simple experimental procedures (Jameson, 1963) but comprehensive studies have yet to be attempted for any species.

1. Responses to Biomass Removal

Plants do not heal in the sense of restoring tissues, instead they compartmentalize (wall-off) or abscise damaged tissues (Shigo, 1984), grow new metamers to replace those destroyed, and modify chemical pathways to change the impact of current and future encounters with agents of biomass removal (Rhoades, 1985). High levels of investment in toxins and spines appear to have evolved in response to severe injury by herbivores, especially when intraspecific competition and abiotic stress were intense (Mattson *et al.*, 1988; Grubb, 1992). We expect analogous investment in mechanical strengthening and regrowth to have evolved where removal of biomass by abiotic agents was frequent and severe.

The nature of responses is broadly related to growth form. For example, many grasses maintain the potential for rapid regrowth by having meristems below the impact zones of fire and grazers, and by storing reserves in roots. In contrast, many dicot herbs have meristems in exposed positions and many allocate some resources to defence at all times, while woody plants grazed by ground-dwelling herbivores only invest in defences until they have grown to a safe height (Bryant *et al.*, 1983).

Plants "tolerate" (i.e. do not respond to) the removal of structures which are about to be shed (Stephenson, 1981) and may grow faster if such structures are removed as soon as they become metabolic sinks. When the biomass removed would have contributed to future accumulation of assimilates, plants usually grow faster to "compensate" for the loss. "Under-", "equal-" and "over-" compensation correspond to a plant's size and fecundity becoming less than, equal to, or greater than intact controls (Maschinski and Whitham, 1989). "Instantaneous" compensation occurs when metabolites are redirected as soon as sinks are removed, or when remaining leaves increase the rate of photosynthesis after partial defoliation. This has been distinguished from "time-dependent" compensation in which removal of sinks delays the onset of metabolite limitation of growth (Hearn and Room, 1979). Instantaneous compensatory expansion of previously suppressed buds has been seen as soon as 30 minutes after decapitation of shoots (Hillman, 1984). When a shoot apex is removed, adjustments of correlative inhibition follow which become evident as "traumatic reiteration", due to axillary buds being released from apical dominance, and "metamorphosis" (= "dedifferentiation") due to stronger, more orthotropic growth of lateral branches released from apical control (see Fig. 5) (Bell, 1991).

High regrowth potential of shoots may be indicated by a high root:shoot ratio indicating storage of resources in roots and high root regrowth potential may be indicated by a low root:shoot ratio (van der Meijden *et al.*, 1988). In defoliated sunflowers and grasses, the ratio was restored by greater growth of shoots at the expense of roots (Mariko and Hogetsu, 1987; Oesterheld, 1992). The ratio may be adjusted rapidly by abscission of branches following drought (Savile, 1976) and by abscission of roots following heavy pruning of shoots (Addicott and Lyon, 1973).

Under-compensation may result irrespective of the quantity and frequency of biomass removal (Belsky, 1986) but over-compensation is common at low to moderate intensities of removal, especially amongst grasses (McNaughton, 1983; Seastedt, 1985). In apple trees, an arithmetic increase in the proportion of buds removed by pruning caused a geometric increase in growth of vegetative buds and a more globular canopy (Porter, 1989b). The time needed for equal- and over-compensation varies from days to infinity. Disbudded *Salvinia molesta* took 28 days to achieve the same number of ramets as intact controls (Julien and Bourne, 1986) and *Solanum carolinense* L. took 13 weeks to produce as many seeds as intact controls when all fruit were removed

(Solomon, 1983). The number of flowering buds in fruit trees was still depressed a year after pruning (Porter, 1989b) and replacement of pruned roots took from 25 days for seedlings to several months for mature trees (Geisler and Ferree, 1984).

Removal of biomass may cause new metamers to be smaller as in root-pruned bonsai trees. In contrast, shoot-pruning of fruit trees in winter generally results in larger fruit (Loreti and Pisani, 1990). The schedule of growth may alter, as in defoliated cocoa shoots which changed from rhythmic to continuous extension (Vogel, 1975) and in *Impatiens balsamina* L. where destruction of axillary flower buds caused indeterminate shoots to become determinate and flower (Nanda and Purohit, 1966). Removal of flowers may alter sexuality, as in *Pinus cembroides* Zucc. which is usually a monoecious tree but becomes a male-only shrub following destruction by caterpillars of the apical shoots that usually bear female flowers (Whitham and Mopper, 1985). Pruning at all intensities reduces fruit yield in small trees but, in larger trees, moderate pruning increases yield by shifting the balance between vegetative and reproductive meristems (Mika, 1986). Defoliation generally reduces seed production, especially in fruit with direct vascular connections to damaged leaves, but can increase it depending on timing and intensity (Jameson, 1963).

Traumatic reiteration may result in some tissues becoming more protected when shrubs produce highly branched canopies (Vesey-Fitzgerald, 1973) and grasses produce densely packed tillers (McNaughton, 1984). Other induced defences include abscission, compartmentalization, nutrient withdrawal, lignification, and synthesis of poisons, repellents and adhesives (Berryman, 1988) but only abscission and compartmentalization have direct effects on metamer and module dynamics. Patterns of abscission depend on the causative agent, for example, all organs infected by pathogens are usually shed whereas insects usually cause shedding when they attack flowers and fruit but not leaves (Addicott and Lyon, 1973). In compartmentalization, wounds are isolated with barriers of cork, periderm, gum, lignin or suberin. The structural but not physiological integrity of metamers is preserved, which is important to large, long-lived species (Shigo, 1984).

A critical factor affecting chemical responses may be whether dominance relations amongst modules are disrupted. In *Betula pubescens* Ledeb., removing biomass from leaves reduced the quality of remaining shoots as food for herbivores but removing buds increased it (Haukioja *et al.*, 1990). In *Salvinia molesta*, destruction of buds by a beetle increased concentrations of nitrogen in remaining tissues which in turn stimulated population growth of the beetle (Room and Thomas, 1985). Chemical responses may spread between metamers which are not nearest neighbours and may be restricted to single branches or other semiautonomous physiological units depending on vascular architecture (Watson and Casper, 1984).

2. Nature of Biomass Destroyed

The impact of biomass removal depends on the topological position and function of tissues destroyed: removal of material by galling of leaves and stems has least impact; defoliation, root and shoot destruction intermediate impact; and phloem/cambium/sapwood destruction in the main stem and root crown most impact (Mattson *et al.*, 1988). This ranking recognizes the paramount importance of vascular connections between roots and shoots but topological position is significant in less extreme situations when the death of one metamer inevitably results in the loss of those distal to it. Removal of material which includes an apical meristem, or the expanding leaves of a terminal metamer, triggers adjustments of correlative inhibition. This is particularly well documented for fruit trees where removal of apical meristems results in growth of more new metamers and modules than removal of the same biomass and number of axillary meristems (Mika, 1986).

The role, condition and potential of tissues removed affect responses. Removal of young leaves, for example, is likely to reduce lifetime productivity of a plant more than removal of older leaves (Harper, 1989b). In *Jurinea mollis* Ascherson, destruction of basal rosette meristems resulted in multiple stalks and up to three times more seeds than controls, whereas removal of flower heads resulted in axillary flower heads and fewer seeds, and removal of fruit stimulated no response and could result in zero seeds (Inouye, 1982). In *Salvinia molesta*, removal of buds stimulated traumatic reiteration and equal-compensation (Julien and Bourne, 1986) whereas removal of leaves and vascular tissues resulted in no reiteration and under-compensation (Julien and Bourne, 1988). In *Abutilon theophrasti* Medic., 75% defoliation caused over-compensation in seed production in bright light but under-compensation when light was limiting (Lee and Bazzaz, 1980). These cases suggest that over- and equal-compensation are possible when biomass is removed from sinks, but not sources, of limiting resources.

3. Constraints on Responses

The number and positions of meristems can place limits on responses. In the extreme case, herbivory may have selected against unitary architecture because genets can recover from biomass removal more completely by having modular growth (Haukioja, 1991). Three levels of capability for traumatic reiteration have been recognized based on the availability of meristems for regrowth (de Castro e Santos, 1980):

(i) None, e.g. palms with a single meristem.
(ii) Limited, e.g. due to axillary buds having a limited lifespan.
(iii) Partial, all others.

Physiological autonomy of plant parts restricts the availability of meristems for regrowth according to their positions. The shoots and roots of most plants have sufficient autonomy that regrowth is restricted to regions close to the site of biomass removal (Geisler and Ferree, 1984; Thomas and Watson, 1988).

The timing of biomass removal imposes constraints because seasonal and reproductive cycles preclude some responses at some times. Suppressed meristems available for compensatory growth are more numerous late in the growing season but time is short and nutrients are usually less abundant than at the beginning of the season (Maschinski and Whitham, 1989). There are also constraints associated with life history. Annuals, for example, have less time than perennials for accumulating reserves and for compensatory growth (van der Meijden *et al.*, 1988), whilst evergreens accumulate fewer reserves which might be used for compensation than deciduous trees.

VII. SIMULATION

A. Types of Models

Models of plant growth have been classified in several different ways on the basis of how they simulate processes. In crop models, Whisler *et al.* (1986) contrasted empirical with mechanistic treatments of environmental effects. Amongst models of branching, Waller and Steingraeber (1985) distinguished between spatial or non-spatial, stochastic or deterministic, and stationary or non-stationary types. Bell (1986) recognized blind, sighted and self-regulatory control of branching. Françon (1991) grouped graphical models according to the algorithms used: empirical assembly, fractal, combinatoric, production system, or stochastic. Unfortunately, many models use mixtures of different approaches for different processes. For the purpose of this review, we found it most useful to group models on the level of spatial detail they addressed and then to concentrate on how specific phenomena were simulated.

1. Levels of Spatial Detail

Models of module and metamer dynamics have been built at the population, topological, and geometric levels, each succeeding level incorporating the features of the previous one. Population dynamics alone have been simulated using difference equations (Room, 1983), matrices (Carlsson and Callaghan, 1991), Markov chains (Maillette, 1990), and differential equations (Maillette, 1992). Such models ignore topology and they allow little scope for geometric analysis beyond determination of the fates of parts in broad spatial regions such as upper or lower portions of the canopy (Maillette, 1982b). They have the advantage of simple mathematical formulation (Jeffers, 1978) and can be useful for the verification of statistics produced by virtual plant models. A number of crop models work at the level of module and metamer populations, using biomass parameters to calculate crop yield by multiplication (Whisler *et al.*, 1986).

Topological models record the parent–offspring relationships of structural components, arranging the components into correct branching structures. This allows the movement of materials through the developing structure to be followed, enabling the simulation of hormonal control mechanisms (Frijters, 1978), resource partitioning (Borchert and Honda, 1984) and accumulation of induced defences.

In geometric models, the spatial position and orientation of each structural component is tracked. This allows the simulation of position-specific interactions such as collisions between branches, interception of light by leaves and bending of branches due to gravity. Geometric models also provide the information necessary for realistic images of virtual plants to be produced (see Fig. 1) (de Reffye *et al.*, 1988; Prusinkiewicz and Lindenmayer, 1990). Positions are recalculated in each time step because the positions of distal structures are affected by changes in position and orientation of proximal structures. The spatial coordinates of structural units are found by measurements which are subject to natural variability. This variability can be simulated by random variation of parameter values. Deterministic models abstract from this variability by using mean values, allowing the contribution of particular parameters to be examined without being obscured by random variation (Fisher, 1992).

2. Control of Development

Growth is a continuous process but simulation models operate in discrete time steps which make it convenient to simulate the discrete addition of structural units. Consequently, the structural unit considered by a model, the time taken for a unit to appear, and the time step used by the model are usually related. Models which operate at the level of segments or modules (Honda, 1971), or entire branch complexes (Remphrey and Powell, 1987), usually take time steps from one season to the next. Models which operate at the level of metamers (de Reffye *et al.*, 1989) usually have time steps of days, hours or plastochrons. A close approximation to continuous growth can be achieved by solving for very small time steps differential equations which describe metamer elongation (Prusinkiewicz *et al.*, 1993).

The simulation of modular and metamer dynamics requires replication of the behaviour of apical meristems and their products (White, 1984). This behaviour can be conceptualized as being controlled by inherent "rules of growth" characteristic of a species (Lindenmayer, 1982; Bell, 1986). The dynamics of a virtual plant are produced by the successive application of rules appropriate to the state of each structural component. The states are determined by internal factors, such as hormone flow, component age, size and position, and external factors, such as resource availability, day length, and toxins injected by insects. While an ideal virtual plant model will take both internal and external sets of factors into account, modelling efforts to date have focused on either one or the other.

Models that focus on internal factors have rules of growth derived solely from analysis of the structural and temporal patterns of module and metamer dynamics. Environmental conditions in force during the period of observation become implicit in the rules. These models operate in physiological time using time steps of plastochrons (Barlow, 1989). All geometric models that we are aware of take this approach. Crop models provide the best examples of simulations focusing on external factors (Whisler *et al.*, 1986; Hanks and Ritchie, 1991). They use environmental conditions and resources as driving variables for the physiological processes of development. Rules typically describe the changing balance of resources such as nitrogen and water, determine the production of photosynthate, then partition it among activities such as the production of new metamers. In order to determine the rules, plant development must be measured in a wide range of conditions. The models operate in physical time, with daily or hourly steps, since conditions and resource availability are recorded in those units. Some models of this type have incorporated metamer dynamics at the topological but not geometrical level. Although the concepts and techniques are applicable to virtual plant models, details of modelling physiological processes are beyond the scope of this review.

Control of the changing states of apical meristems and structural units can be achieved either by application of observed probabilities of state changes ("implicit" models) or by explicit simulation of hypothesized control mechanisms ("explicit" models). The use of probabilities provides a convenient abstraction from the operational details of internal and external factors and is especially useful if a species has highly variable morphology (de Reffye *et al.*, 1989; Cain, 1990). It produces models that are descriptive of branching structure rather than explanatory, although the use of probabilities conditional on particular internal or external events can extend the range of the model as far as the modeller is prepared to measure probabilities. Models which use this approach include the analytical population models described earlier, as well as stochastic models such as those simulating modular development of clones (Bell and Tomlinson, 1980). In the stochastic approach, a computer generated "random" number is compared with the state-change probability to determine if an event such as production of a new module occurs.

The most detailed stochastic 3D models trace meristem activity at a metamer to metamer level, using probabilities for death, flowering, dormancy, and growth of new metamers (de Reffye *et al.*, 1989). A module of a given order is characterized by its growth rate relative to the main stem and its "dimension", the maximum number of time steps in which the meristem state will be checked. The system has been successful in creating 3D geometric models of a wide range of plant architectures. An alternative stochastic approach captures meristem activity in aggregate. Numbers, locations and yearly growth increments of new and developing modules are related to annual variation in leader growth by empirical formulae determined through regression analysis (Renshaw, 1985; Remphrey and Powell, 1987).

Deterministic models, in which a given state defines completely the rule to be applied, allow the effects of parameters to be explored without the noise generated by stochasticity. Honda (1971), for example, showed that small changes in branch angles and lengths could produce a wide variety of tree-like forms, leading to a number of more detailed models of tree architecture (Fisher, 1986).

Explicit simulation of control mechanisms gives the most scope for providing explanatory models of dynamics. Information flow between components in the modelled structure can be used to simulate signals controlling phenomena such as apical dominance and floral initiation (Frijters 1978; Prusinkiewicz and Lindenmayer, 1990). If an apex exerting apical control dies, or a branch acting as a sink for assimilates is removed, the mechanisms of the model should generate the appropriate response automatically.

3. An Evaluation

While the models reviewed here are well suited to their respective purposes, two systems stand out as having the most potential to become virtual plant models having a wide range of generic uses for ecologists: AMAP (Atelier pour la Modélisation de l'Architecture des Plantes) originated by de Reffye (1981) and L-systems initiated by Lindenmayer (1968). Both simulate 3D growth of any kind of plant in physiological time. AMAP uses stochastic mechanisms while L-systems, although originally deterministic for modelling internal control mechanisms, can incorporate stochastic mechanisms and duplicate AMAP simulations (Françon, 1991). Unlike AMAP, L-systems allow parameter values and growth rules to be changed easily, feedbacks to be simulated as flows of information through plant structure, and operation in real time by including appropriate equations within rules (Prusinkiewicz and Lindenmayer, 1990). Thus L-systems are inherently more versatile and hold greater promise than AMAP.

4. L-systems

Plant components are represented in L-systems by letters, different types and states of structural units being assigned different letters. A sequence of letters, with branches delimited by brackets, represents an entire plant. The order of letters in the sequence determines the topological connections in the structure, while special symbols represent relative geometry. Development is simulated by a process of rewriting. In each time step, every letter in the sequence is replaced according to a rule specific to that letter. The replacement may be a new letter representing a different state of the same component, a group of letters representing new components produced in this time step, or a blank representing the death of the component. The resulting string of letters and blanks represents the state of the plant in the next time step. The string of letters may be interpreted geometrically to create a 3D image. The rewriting process is repeated for each time step until the desired stage of development is reached. For example, consider the following set of rules:

$$A \rightarrow I[+K]B$$
$$B \rightarrow I[-K]A$$
$$K \rightarrow L$$

A and B represent different states of the apical meristem, I represents an internode, and K and L represent two stages of development of a leaf. The symbols + and − control geometry, placing the components which follow the symbol to the left and right, respectively. The brackets [and] enclose branches. The symbol → indicates that the letter on the left is transformed into the string of symbols on the right when the rule is applied. The first rule says that an apex in state A transforms itself into state B while producing an internode and a branch on the left containing a leaf. If this set of rules is applied to a "seed" or starting string A, representing the initial apex, the following series of strings will be produced in four time steps:

$$A$$
$$I[+K]B$$
$$I[+L]I[-K]A$$
$$I[+L]I[-L]I[+K]B$$
$$I[+L]I[-L]I[+L]I[-K]A$$

which represents the growth of a monopodial shoot with alternating leaves.

In deterministic L-systems one rule exists for each different letter. In stochastic L-systems, more than one rule may exist for a letter, and each will have an associated probability of application. In context-sensitive L-systems, the application of a rule is not only dependent on the letter under consideration, but also on the values of neighbouring letters in the sequence. This makes it possible to pass information both up and down the developing structure, providing an effective means for simulating internal factors such as hormone and assimilate flow.

B. Internal Factors

1. States of Apical Meristems

The basic approaches to control of state changes are explicit simulation as described above and stochastic application of probabilities. Explicit models can use information flow through the plant structure to simulate control of internode elongation, apical dominance and flower initiation in inflorescences (Frijters, 1978; Prusinkiewicz and Lindenmayer, 1990). Stochastic models may employ conditional probabilities to link their behaviour to internal factors. Mechanisms such as apical dominance and control may be simulated by making probabilities and developmental delays dependent on age and position of a bud (Room, 1983), or on the life or death of the leader apex (Callaghan et al., 1990). Alternatively, the basic probabilities can be assumed to reflect these factors if the number of buds developing are assigned stochastically (Cain et al., 1991) or determined by the stochastically assigned type of the parent module (Remphrey et al., 1983).

When a new metamer, segment or module is created, its geometric parameters must be assigned, not necessarily using the same approach taken to control meristem state. Models that express meristem activity stochastically may use deterministic geometric parameters (Bell et al., 1979), while other models are completely stochastic (Cain et al., 1991), and deterministic simulations of meristem state may incorporate variability in their geometric parameters. The key geometric parameters for simulating the 2D geometry of clonal plants are segment length and branching angle, both of which may vary with the type and position of the apical meristem producing the module (Bell et al., 1979; Room, 1983), or with distance from the meristem exerting apical control (Callaghan et al., 1990).

The geometry of 3D plants is more complex, requiring parameters to fix a metamer's orientation relative to its parent axis. In the series of models based on Honda (1971), trigonometric formulae were derived to capture different geometric patterns of tropical trees, such as asymmetric branching in *Tabebuia rosea* D.C. (Borchert and Honda, 1984), and the shift from radial to dorsiventral symmetry in *Neea amplifolia* Donn. (Fisher and Weeks, 1985). The latter model is capable of either stochastic or deterministic simulation of branching angles and lengths, while earlier models are strictly deterministic, the lengths of new segments being defined by applying a contraction ratio to parent segment lengths. Models based on de Reffye's approach (de Reffye et al., 1989) capture module length through the stochastically determined number of metamers produced in the life of an apical meristem. Individual metamer lengths and branching angles may be assigned either deterministically or stochastically.

In 2D models of clonal growth, branching may alternate from side to side, an effect which may be captured by module type (Bell *et al.*, 1979) or be derived stochastically (Remphrey *et al.*, 1983). In 3D models, the phyllotactic patterns of branches and leaves are typically captured using a divergence angle parameter (Honda, 1971). The more intricate phyllotactic patterns found in flower heads and fruits have been simulated using descriptive, geometry-based models (Fowler *et al.*, 1989) as well as explanatory algorithms based on contact, collision and reaction-diffusion models (Thornley and Johnson, 1990; Fowler *et al.*, 1992).

2. Age

If the ages of structural units are required for rules controlling senescence, they can be incremented in each time step of a model (Room, 1988). While such parameters could be used in the simulation of metamorphosis and de-differentiation, we are not aware of any model that does this explicitly.

3. Resource Sharing

Honda *et al.* (1981) simulated declining frequency of higher order branching by assigning different "flow rates" of a bifurcation-determining factor to daughter branches. A similar technique was applied in a model of *Tabebuia rosea* (Borchert and Honda, 1984), with growth flux additionally limited by a sigmoidal function simulating the effect of declining root:shoot ratio as a tree grows. Apical control was simulated by a feedback mechanism favouring flux to the leader, which also provided control of death and abscission of lower laterals.

An alternative approach is to calculate assimilate supply based on the available photosynthetic area (Ford *et al.*, 1990) and then partition it to control further development. Mechanistic crop models incorporate similar mechanisms to partition the available photosynthate to various sinks, either globally (Jones *et al.*, 1980) or by priority within branch strata (Mutsaers, 1984). More detailed models of root:shoot partitioning can be found in the crop modelling literature (Whisler *et al.*, 1986). For the purposes of virtual plant models, roots could be modelled either as a black box or by incorporation of an existing model. We are not aware of any models which simulate limiting nutrients being recycled from senescent metamers.

C. External Factors

The key to simulating the effects of an external factor is to provide a mechanism for detecting its strength and rules specifying the appropriate response. In deterministic models, the response may be an apical state change or the release of a signal that will percolate through the plant structure. In stochastic

models, responses require that a set of probabilities appropriate for the external condition be assigned (Harper and Bell, 1979; Room, 1988).

1. Conditions

Temperature is the usual driving variable for timing developmental events in
crop models, usually supplied as daily minimum and maximum values
(Whisler et al., 1986; Hanks and Ritchie, 1991). The simplest models base the
appearance of metamers on the accumulation of thermal time above empirically determined thresholds (Porter, 1984). These thresholds may be different
for each new metamer (McMaster et al., 1991). More complex environmental
interactions can be included by using a threshold associated with the most
limiting resource (Whisler et al., 1986), or by reducing rates from a theoretical
maximum by resource stress factors (McKinion et al., 1989).

Gravity-related curvature of branches has been simulated by modifying
metamer-to-metamer angles using a model of bending based on the theory of
material resistance (de Reffye, 1983). The parameters included Young's
Modulus for the plant being simulated as well as internode length and thickness. Taking the opposite approach, branch thickness can be determined
using power-beam taper equations (McMahon and Kronauer, 1976) that fit
branch bending to an observed deflection profile (Ford and Ford, 1990). A
more abstract simulation of bending due to gravity, growth habit, physical
disturbance and phototropism uses direction and susceptibility-to-bending
parameters to determine the angle at each node (Aono and Kunii, 1984; de
Reffye et al., 1989; Prusinkiewicz and Lindenmayer, 1990).

2. Resources and Competition

Resource handling by crop models has been described previously. With the
addition of the empirical relationships between plant density and development, these models simulate intraspecific competition as a matter of course.

Clonal foraging for nutrients can be modelled by adjusting growth direction
towards areas of low branch density or high resource availability (Cohen,
1967; Greene, 1989), by modifying growth rates to match patchy resources
(Sutherland and Stillman, 1988), or by restricting development of branches if
others already exist in the direction of extension (Remphrey et al., 1983). The
latter approach has been used in simulations of clonal competition (Bell,
1985) and responses to shading (Fisher and Weeks, 1985; Ford and Ford,
1990). Existing branches and other obstacles may be avoided by testing for
potential collisions prior to calculating the direction and extent of growth
(Dabadie, 1991; Prusinkiewicz and McFadzean, 1992).

Direct effects of shading can be modelled by assigning light thresholds
below which meristems will die (Dabadie, 1991). The reduction of light due to

shading can be found by tracing back along a sample of light rays to detect earlier intersections with plant parts and other objects. Alternatively, the photosynthate production of a single leaf may be related to the number of leaves above it (Bell, 1986). If the requirements for space and light are considered as the driving forces of shoot development, a combination of routines for intersection-testing and determination of light availability can be used as the basis for modelling growth (Greene, 1989).

3. Removal of Biomass

Simulating responses to biomass removal requires that information be passed through the plant structure. For example, growth flux can be redistributed to simulate traumatic reiteration after pruning (Borchert and Honda, 1984) (Fig. 6), and adjustment of bud states and probabilities along an axis can be

Fig. 6. Simulated traumatic reiteration after pruning. Sample output from an L-system implementation of the model from Borchert and Honda (1984). Apical control and reduction of lateral branching were simulated by interactive mechanisms controlling distribution of "metabolic flux" for 12 time steps. Segment thickness reflects the maximum flux through a branch. On the left is a non-pruned branch, on the right is a branch with main stem pruned at the fifth node after eight time steps.

used to model response to biomass removal by insects (Room, 1988). Information on plant structure has been incorporated into crop models to allow the appropriate distribution of biomass removal by pests (Jones *et al.*, 1980; Whisler *et al.*, 1986). In these cases, the mechanistic nature of the models is relied on to provide the appropriate responses, although none that we are aware of computes allocations to defence.

C. Checklist of Simulation Parameters

Table 2 is intended as an exhaustive list of parameters for simulating metamer and modular dynamics and only a subset may be relevant for a particular case. Depending on the species being modelled, there may be a need for different parameter values for each possible combination of state parameters. Some items represent the lowest common denominator of more complex parameters presented in the literature. For example, the progression of angles from metamer to metamer along a module may define the combined plagiotropic/orthotropic nature of the module, the phyllotactic pattern of the leaves and responses to gravity, light, and wind.

Table 2 is in four sections. The meristem parameters capture control aspects of the developmental process. Metamer/module parameters capture physical properties of structural units. The positions of new meristems determines the topology of an individual plant, while lengths and angles capture details of its geometry. Organ parameters give further details of metamer construction, playing a large role in driving physiology-based simulations through elements such as a leaf's role in photosynthesis or a fruit's role in yield. Environmental parameters capture the conditions for growth and the resources available through time. These will be modified to some extent by a plant's development: nutrients in the soil will be depleted and the plant will create a microclimate in its own canopy. Thus, it may be necessary to recalculate environmental parameters for each component individually.

VIII. FUTURE DEVELOPMENTS

We are entering a new era in ecology as developments in computer science provide tools increasingly suited to handling the complexity of ecological interactions. The most likely impacts in the near future will be on questions involving feedbacks and substantial spatial aspects.

In plant ecology, the ability to simulate module and metamer dynamics will provide new insights into relationships between morphology, environment and productivity. Understanding of intra- and interspecific competition for light and its role in community dynamics will be transformed. Iterations will identify combinations of spacing, agronomic inputs and pruning which result in crop canopies optimal for light interception, penetration of pesticides and

Table 2
A checklist of parameters affecting metamer and module dynamics

Parameters	Note number*
Internal to the plant	
Meristem	
State	
(vegetative, reproductive, dormant, senescent, dead, age,	
topological position, geometrical position)	1
Mechanism and timing of state changes	
(explicit inherent control, probability)	2
(environmental responses)	3
Type of metamer/module produced	1
Rate of metamer/module production	4, 5
(plastochron vs. temperature and meristem state)	6, 7
Resource sink type, strength and priority	7
Morphogen source/signal type and strength	
Metamer and Module	
State	
(vegetative, reproductive, senescent, dormant, dead, age,	
topological position, geometrical position)	1, 8
Timing and mechanism of state changes	1
(explicit inherent control, probability)	
(environmental responses)	9
Shape, size, and growth function	10
(internode length, nodal diameter, module growth habit)	
Number of leaves and axillary meristems	1
Arrangement of leaves and axillary meristems	10
Orientation relative to distal neighbour	
(branching angle, divergence angle, deflection angle, tropism and	
disturbance effects)	10, 11
Resource source/sink/storage properties and priorities	7
Resource and morphogen transmission properties	12
Levels of secondary chemicals	
Physical properties	
(density, Young's modulus)	13
Organ (leaf, flower, fruit)	
State	
(senescent, dead, age, topological position, geometrical position,	7, 8, 14
stage of development e.g. bud to flower to fruit)	
Timing and mechanism of state changes	1
(lineage, interaction, probability)	
(environmental responses)	
Orientation and positioning relative to metamer	15
Shape, size, and growth function	15
Resource source/sink/storage properties and priorities	7
Morphogen source properties	
Levels of secondary chemicals	
Other properties	7
(density, seeds per fruit, other yield properties)	

Table 2 (continued)

Parameters	Note number*
External to the plant	
Gravity (constant), wind	13
Temperature	6, 7
Water availability	7
Light availability	7
Nutrient availability	7
Space availability	
(objects in the environment,	16
row spacing/population density of individuals,	7
soil structure)	17
Timing, degree, and location of disturbance	
Timing, type, location and amount of biomass removed	18

* Notes
1. Widely used in metameric models.
2. Explicit simulation—Honda *et al.* (1981), Prusinkiewicz and Lindenmayer (1990), Honda *et al.* (1981); probability—Bell *et al.* (1979), de Reffye *et al.* (1989).
3. Remphrey *et al.* (1983), Room (1988), Ford and Ford (1990), Dabadie (1991), Prusinkiewicz and McFadzean (1992).
4. Used in all models.
5. May be expressed as a probability—de Reffye *et al.* (1989).
6. Widely used for phenology, e.g. Ford and Ford (1990), McMaster *et al.* (1991).
7. Widely used in physiological crop models, e.g. Whisler *et al.* (1986).
8. e.g. leaf senescence by age in many crop models.
9. Greene (1989).
10. Widely used in geometric models.
11. Aono and Kunii (1984), Prusinkiewicz and Lindenmayer (1990), Dabadie (1991).
12. Honda *et al.* (1981); Prusinkiewicz and Lindenmayer (1990).
13. McMahon and Kronauer (1976), de Reffye (1976), Ford and Ford (1990)
14. Prusinkiewicz and Lindenmayer (1990).
15. Prusinkiewicz and Hanan (1989), de Reffye *et al.* (1989).
16. Dabadie (1991), Prusinkiewicz and McFadzean (1992).
17. Whisler *et al.* (1986), Diggle (1988).
18. Room (1988).

harvestability. New ideotypes or target "designer plants" will be suggested to plant breeders and genetic engineers by simulations of plants with modified internode lengths, branching angles and leaf shapes. Simulations of biomass removal will indicate the intensity of grazing needed for optimal tillering and productivity of pastures. For weeds, architectures sensitive to biomass removal of particular kinds will be identified for use in ranking target species for biological control and to specify the ideal agents to be searched for.

A "virtual laboratory" which allows non-programmers to experiment with virtual plants has already been developed (Prusinkiewicz and Lindenmayer, 1990). In addition to becoming useful in research, virtual plants will become valuable components of software for education, training and decision-support. They will be used to teach the importance of timing agronomic inputs and pest control in relation to the developmental stages of crops. When interfaced with pest population models and expert systems, they will assist growers to decide whether probable loss of yield warrants application of control measures.

To allow these capabilities to be achieved, the main tasks outstanding are:

(i) To make virtual plants responsive to their physical environments in real, rather than physiological, time.
(ii) To make virtual plants responsive to their biotic environments, especially neighbouring plants, to allow accurate simulation of plant populations.
(iii) To devise more efficient methods of measuring the structure of plants, e.g. automated scanning.

It would also be useful for model verification if boundary conditions could be defined through a theoretical analysis of growth in 2D or 3D, number of axes per node, syllepsis/prolepsis, branch angle, phyllotaxis and characteristics of architectural models interact to constrain the dynamics of metamers and modules. Some of the responses relevant to items (i) and (ii), especially those concerned with rates of growth, have already been defined and incorporated into crop models and it should be straightforward to adapt them to virtual plant models. Indeed, it is likely that overlap between crop and virtual plant models will increase until there is no clear boundary between the two.

ACKNOWLEDGEMENTS

We are grateful to Mr John Whiteman for helping to obtain reprints, to a number of colleagues who shared unpublished information and manuscripts, and to an anonymous reviewer who made valuable comments on a draft of this paper. The Cooperative Research Centre for Tropical Pest Management paid for Dr Maillette to work in Brisbane and she received grants from the Natural Sciences and Engineering Research Council of Canada and the Universities of Calgary and Québec à Trois-Rivières.

REFERENCES

Addicott, F.T. and Lyon, J.L. (1973). Physiological ecology of abscission. In: *Shedding of Plant Parts* (Ed. by T.T. Kozlowski), pp. 85–124. Academic Press, New York.
Alpert, P. (1991). Nitrogen sharing among ramets increases clonal growth in *Fragaria chiloensis. Ecology* **72**, 69–80.
Aono, M. and Kunii, T.L. (1984). Botanical tree image generation. *IEEE Comput. Graphics & Applics.* **4**, 10–34.
Armbrust, D.V. (1968). Windblown soil abrasive injury to cotton plants. *Agron. J.* **60**, 622–625.

Atkinson, C.S. (1984). Quantum flux density as a factor controlling the rate of growth, carbohydrate partitioning and wood structure of *Betula pubescens*. *Ann. Bot.* **54**, 397–412.

Ballaré, C.L., Sanchez, R.A., Scopel, A.L., Casal, J.J. and Ghersa, C.M. (1987). Early detection of neighbour plants by phytochrome perception of spectral changes in reflected sunlight. *Plant, Cell Environ.* **10**, 551–557.

Ballaré, C.L., Scopel, A.L. and Sanchez, R.A. (1990). Far-red radiation reflected from adjacent leaves: An early signal of competition in plant canopies. *Science* **247**, 329–332.

Barlow, P.W. (1989). Meristems, metamers and modules and the development of shoot and root systems. *Bot. J. Linn. Soc.* **100**, 255–279.

Barthélémy, D. (1991). Levels of organization and repetition phenomena in seed plants. *Acta Biotheor.* **39**, 309–323.

Barthélémy, D., Edelin, C. and Hallé, F. (1989). Some architectural aspects of tree ageing. *Ann Sci. For.* **46**, 194–198.

Bazzaz, F.A. and Harper, J.L. (1977). Demographic analysis of the growth of *Linum usitatissimum*. *New Phytol.* **78**, 193–207.

Bell, A.D. (1985). On the astogeny of six-cornered clones: an aspect of modular construction. In: *Studies on Plant Demography: a Festschrift for John L. Harper* (Ed. by J. White), pp. 187–207. Academic Press, London.

Bell, A.D. (1986). The simulation of branching patterns in modular organisms. *Philos. Trans. R. Soc. Ser. B* **313**, 143–160.

Bell, A.D. (1991) *Plant Form. An Illustrated Guide to Flowering Plant Morphology*. Oxford University Press, Oxford.

Bell, A.D. and Tomlinson, P.B. (1980). Adaptive architecture in rhizomatous plants. *Bot. J. Linn. Soc.* **80**, 125–160.

Bell, A.D., Roberts, D. and Smith, A. (1979). Branching patterns: the simulation of plant architecture. *J. Theor. Biol.* **81**, 351–375.

Belsky, A.J. (1986). Does herbivory benefit plants? A review of the evidence. *Am. Nat.* **127**, 870–892.

Berryman, A.A. (1988). Towards a unified theory of plant defense. In: *Mechanisms of Woody Plant Defences Against Insects* (Ed. by W.J. Mattson, J. Levieux and C. Bernard-Dagan), pp. 39–55. Springer Verlag, New York.

Blits, K.C. and Gallagher, J.L. (1991). Morphological and physiological responses to increased salinity in marsh and dune ecotypes of *Sporobolus virginicus* (L.) Kunth. *Oecologia* **87**, 330–335.

Bloom, A.J., Chapin, F.S. and Mooney, H.A. (1985). Resource limitation in plants— an economic analogy. *Ann. Rev. Ecol. Syst.* **16**, 363–392.

Boeken, B. (1989). Life histories of desert geophytes: the demographic consequences of reproductive biomass partitioning patterns. *Oecologia* **80**, 278–283.

Borchert, R. and Honda, H. (1984). Control of development in the bifurcating branch system of *Tabebuia rosea*: a computer simulation. *Bot. Gaz.* **145**, 184–195.

Boutin, C. and Morisset, P. (1988). Study of the phenotypic plasticity of *Chrysanthemum leucanthemum*: 1. Growth, biomass allocation and reproduction. *Can. J. Bot.* **66**, 2285–2298.

Boutin, C. and Morisset, P. (1989). Study of the phenotypic plasticity of *Chrysanthemum leucanthemum*. II. Demography. *Can. J. Bot.* **67**, 977–983.

Brown, C.L., McAlpine, R.G. and Kormanik, P.P. (1967). Apical dominance and form in woody plants: a reappraisal. *Am. J. Bot.* **54**, 153–162.

Bryant, J.P., Chapin, F.S. and Klein, D.R. (1983). Carbon/nutrient balance of boreal plants in relation to vertebrate herbivory. *Oikos* **40**, 357–368.

Burdon, J.J. (1987). *Diseases and Plant Population Biology*. Cambridge University Press, Cambridge.

Caesar, J.C. and MacDonald, A.D. (1983). Shoot development in *Betula papyrifera*. II. Comparison of vegetative and reproductive short shoot growth. *Can. J. Bot.* **61**, 3066–3071.

Cain, M.L. (1990). Models of clonal growth in *Solidago altissima*. *J. Ecol.* **78**, 27–46.

Cain, M.L., Pacala, S.W. and Silander, J.A. Jr (1991). Stochastic simulation of clonal growth in the tall goldenrod, *Solidago altissima*. *Oecologia* **88**, 477–485.

Callaghan, T.V., Headley, A.D., Svensson, B.M., Lixian, L., Lee, J.A. and Lindley, D.K. (1986). Modular growth and function in the vascular cryptogam *Lycopodium annotinum*. *Proc R. Soc. Ser. B* **228**, 195–206.

Callaghan, T.V., Svensson, B.M., Bowman, H., Lindley, D.K. and Carlsson, B.A. (1990). Models of clonal plant growth based on population of dynamics and architecture. *Oikos* **57**, 257–269.

Campbell, B.D., Grime, J.P. and Mackey, J.M.L. (1992). Shoot thrust and its role in plant competition. *J. Ecol.* **80**, 633–641.

Canham, C.D. (1988). Growth and canopy architecture of shade-tolerant trees: response to gaps. *Ecology* **69**, 786–795.

Carlsson, B.A. and Callaghan, T.V. (1990). Effects of flowering on the shoot dynamics of *Carex bigelowii* along an altitudinal gradient in Swedish Lapland. *J. Ecol.* **78**, 152–165.

Carlsson, B.A. and Callaghan, T.V. (1991). Simulation of fluctuating populations of *Carex bigelowii* tillers classified by type, age and size. *Oikos* **60**, 231–240.

de Castro e Santos, A. (1980). Essai de classification des arbres tropicaux selon leur capacité de réitération. *Biotropica* **12**, 187–194.

Chapin, F.S., Schulze, E.D. and Mooney, H.A. (1990). The ecology and economics of storage in plants. *Ann. Rev. Ecol. Syst.* **21**, 423–447.

Cline, M.G. (1991). Apical dominance. *Bot. Rev.* **57**, 318–358.

Cohen, D. (1967). Computer simulation of biological pattern generation process. *Nature, Lond.* **216**, 246–248.

Corner, E.J.H. (1949). The Durian theory of the origin of the modern tree. *Ann. Bot.* **13**, 367–414.

Dabadie, P. (1991). Contribution à la modélisation et simulation de la croissance des vegetaux. PhD Thesis. Montpellier.

Denslow, J.S., Schultz, J.C., Vitousek, P.M. and Strain, B.R. (1990). Growth responses of tropical shrubs to treefall gap environments. *Ecology* **71**, 165–179.

Diemer, M., Korner, C. and Prock, S. (1992). Leaf life spans in wild perennial herbaceous plants: a survey and attempts at a functional interpretation. *Oecologia* **89**, 10–16.

Diggle, A.J. (1988), ROOTMAP—a model in three-dimensional coordinates of the growth and structure of fibrous root systems. *Plant and Soil* **105**, 169–178.

Dirzo, R. (1984). Herbivory: a phytocentric overview. In: *Perspectives on Plant Population Ecology* (Ed. by R. Dirzo and J. Sarukhan), pp. 141–165. Sinauer Associates, Sunderland, MA.

Downs, R.S. and Bevington, J.M. (1981). Effects of temperature and photoperiod on growth and dormancy of *Betula papyrifera*. *Am. J. Bot.* **68**, 795–800.

Elias, P. (1990). Effects of the canopy position on shoots and leaves in various trees of an oak hornbeam forest. *Biologia, Bratisl.* **45**, 31–42.

Eriksson, O. (1989). Seedling dynamics and life histories in clonal plants. *Oikos* **55**, 231–238.

Erwin, J.E., Heins, R.D. and Moe, R. (1991). Temperature and photoperiod effects on *Fuchsia hybrida* morphology. *J. Am. Soc. Hort. Sci.* **116**, 955–960.

Esau, K. (1977). *Anatomy of Seed Plants.* John Wiley & Sons, New York.

Ewers, F.W. and Schmid, R. (1981). Longevity of needle fascicles of *Pinus longaeva* (bristlecone pine) and other North American pines. *Oecologia* **51**, 107–115.

Fanta, J. (1981). *Fagus sylvatica* and the *Aceri fagetum* of the alpine timberline in central European Mountains. *Vegetatio* **44**, 13–24.

Fetcher, N. and Shaver, G.R. (1983). Life histories of tillers of *Eriophorum vaginatum* in relation to tundra disturbance. *J. Ecol.* **71**, 131–148.

Fisher, J.B. (1986). Branching patterns and angles in trees. In: *On the Economy of Plant Form and Function* (Ed. by T.J. Givnish), pp. 493–524. Cambridge Univ. Press, Cambridge.

Fisher, J.B. (1992). How predictive are computer simulations of tree architecture? *Int. J. Plant. Sci.* **153 (suppl.)**, 137–146.

Fisher, J.B. and Weeks, C.L. (1985). Tree architecture of *Neea* Nyctaginaceae geometry and simulation of branches and the presence of two different models. *Bull. Mus. Natl. Hist. Nat. Sect. B, Adansonia Bot. Phytochim.* **7**, 385–402.

Fitter, A.H. and Stickland, T.R. (1991). Architectural analysis of plant root systems: 2. Influence of nutrient supply on architecture in contrasting plant species. *New Phytol.* **118**, 375–382.

Ford, E.D., Avery, A. and Ford, R. (1990). Simulation of branch growth in the Pinaceae: interactions of morphology, phenology, foliage productivity, and the requirement for structural support, on the export of carbon. *J. Theor. Biol.* **146**, 15–36.

Ford, R. and Ford, E.D. (1990). Structure and basic equations of a simulator for branch growth in the Pinaceae. *J. Theor. Biol.* **146**, 1–14.

Fowler, D.R., Hanan, J. and Prusinkiewicz, P. (1989). Modelling spiral phyllotaxis. *Comput. Graph.* **13**, 291–296.

Fowler, D.R., Prusinkiewicz, P. and Battjes, J. (1992). A collision-based model of spiral phyllotaxis. *Comput. Graph.* **26**, 361–368.

Franco, M. (1986). The influence of neighbours on the growth of modular organisms with an example from trees. *Phil. Trans. R. Soc. Ser. B* **313**, 209–226.

Françon, J. (1991). Sur la modélisation informatique de l'architecture et du développment des végétaux. In: *L'Arbre. Biologie et Développement* (Ed. by C. Edelin), pp. 231–249. Naturalia Monspeliensia, Montpellier.

Frijters, D. (1978). Mechanisms of developmental integration of *Aster novae-angliae* L. and *Hieracium murorum* L. *Ann. Bot.* **42**, 561–575.

Fryrear, D.W. (1975). Estimating seedling survival from wind erosion parameters. *Trans. Am. Soc. Agric. Engrs* **18**, 888–891.

Geisler, D. and Ferree, D.C. (1984). Response of plants to root pruning. *Hort. Rev.* **6**. 155–188.

Ginzo, H.D. and Lovell, P.H. (1973). Aspects of the comparative physiology of *Ranunculus bulbosus* L. and *R. repens* L. *Ann. Bot.* **37**, 753–764.

Goebel, K. (1990). *Organography of Plants, Especially of the Archegoniatae and Spermaphyta.* Part 1. Clarendon Press, Oxford.

Gottlieb, L.D. (1986). The genetic basis of plant form. *Phil. Trans. R. Soc. Ser. B* **313**, 197–208.

Grace, J. (1977). *Plant Response to Wind.* Academic Press, London.

Gravelius, H. (1914). *Flusskunde.* Goschen, Berlin.

Greene, N. (1989). Voxel space automata: modeling with stochastic growth processes in voxel space. *Comput. Graph.* **23**, 175–184.

Grubb, P.J. (1992). A positive distrust of simplicity—lessons from plant defences and from competition among plants and among animals. *J. Ecol.* **80**, 585–610.

Hallé, F. (1986). Modular growth in seed plants. *Phil. Trans. R. Soc. Ser. B* **313**, 77–88.

Hallé, F., Oldeman, R.A.A. and Tomlinson, P.B. (1978). *Tropical Trees and Forests: an Architectural Analysis.* Springer-Verlag, Berlin.

Hanks, J.H. and Ritchie, J.T. (1991). *Modelling Plant and Soil Systems.* American Society of Agronomy Inc., Madison, Wisconsin.

Hänninen, H. (1990). Modelling bud dormancy release in trees from cool and temperate regions. *Acta For. Fenn.* **213**, 1–47.

Harper, J.L. (1977). *Population Biology of Plants.* Academic Press, London.

Harper, J.L. (1985). Modules, branches, and the capture of resources. In: *Population Biology and Evolution of Clonal Organisms* (Ed. by J.B.C. Jackson, L.W. Buss and R.E. Cook), pp. 1–33. Yale University Press, New Haven.

Harper, J.L. (1989a). Canopies as populations. In: *Plant Canopies: their Growth, Form and Function* (Ed. by G. Russell, B. Marshall and P.G. Jarvis), pp. 105–128. Cambridge University Press, Cambridge.

Harper, J.L. (1989b). The value of a leaf. *Oecologia* **80**, 53–58.

Harper, J.L. and Bell, A.D. (1979). The population dynamics of growth form in organisms with modular construction. In: *Population Dynamics* (Ed. by R.M. Anderson, B.D. Turner and L.R. Taylor), pp. 29–52. Blackwell Scientific Publications, Oxford.

Harper, J.L., Rosen, B.R. and White, J. (1986). Preface. The growth and form of modular organisms. *Phil. Trans. R. Soc. Ser. B* **313**, 3–5.

Haukioja, E. (1991). The influence of grazing on the evolution, morphology and physiology of plants as modular organisms. *Phil. Trans. R. Soc. Ser. B* **333**, 241–248.

Haukioja, E., Ruohomaki, K., Senn, J., Suomela, J. and Walls, M. (1990). Consequences of herbivory in the mountain birch (*Betula pubescens ssp. tortuosa*): Importance of the functional organization of the tree. *Oecologia* **82**, 238–247.

Hayes, P.A., Steeves, T.A. and Neal, B.R. (1989). An architectural analysis of *Shepherdia canadensis* and *Shepherdia argentea*: patterns of shoot development. *Can. J. Bot.* **67**, 1870–1877.

Hearn, A.B. and Room, P.M. (1979). Analysis of crop development for cotton pest management. *Protection Ecol.* **1**, 265–277.

Helmisaari, H.S. (1992). Nutrient retranslocation within the foliage of *Pinus sylvestris*. *Tree Physiol.* **10**, 45–58.

Herrera, C.M. (1991). Dissecting factors responsible for individual variation in plant fecundity. *Ecology* **72**, 1436–1448.

Hillman, J.R. (1984). Apical dominance. In: *Advanced Plant Physiology* (Ed. by M.B. Wilkins), pp. 127–148. Pitman, London.

Hocking, P.J. and Pinkerton, A. (1991). Response of growth and yield components of linseed to the onset or relief of nitrogen stress at several stages of crop development. *Fld Crops Res.* **27**, 83–102.

Holbrook, N.M. and Putz, F.E. (1989). Influence of neighbors on tree form: effects of lateral shade and prevention of sway on the allometry of *Liquidambar styraciflua* (sweet gum). *Am. J. Bot.* **76**, 1740–1749.

Honda, H. (1971). Description of the form of trees by the parameters of the tree-like body: effects of the branching angle and the branch length on the shape of the tree-like body. *J. Theor. Biol.* **31**, 331–338.

Honda, H., Tomlinson, P.B. and Fisher, J.B. (1981). Computer simulation of branch interaction and regulation by unequal flow rates in botanical trees. *Am. J. Bot.* **68**, 569–585.

Hunt, P.G., Matheny, T.A. and Kasperbauer, M.J. (1990). Cowpea yield response to light reflected from different colored mulches. *Crop Sci.* **30**, 1292–1294.

Hutchings, M.J. and de Kroon, H. (1994). Foraging in plants: the role of morphological plasticity in resource acquisition. *Adv. Ecol. Res.* **25**, 160–237.

Inouye, D.W. (1982). The consequences of herbivory: a mixed blessing for *Jurinea mollis* (Asteraceae). *Oikos* **39**, 269–272.

Jaffe, M.J. (1980). Morphogenetic responses of plants to mechanical stimuli or stress. *BioScience* **30**, 239–243.

Jameson, D.A. (1963). Responses of individual plants to harvesting. *Bot. Rev.* **29**, 532–594.

Jeffers, J.N.R. (1978). *An Introduction to Systems Analysis: with Ecological Applications*. Edward Arnold, London.

Jonasson, S. (1989). Implications of leaf longevity, leaf nutrient reabsorption and translocation for the resource economy of five evergreen plant species. *Oikos* **56**, 121–131.

Jones, J.W., Brown, L.G. and Hesketh, J.D. (1980). COTCROP: a computer model for cotton growth and yield. In: *Predicting Photosynthesis for Ecosystem Models* (Ed. by J.D. Hesketh and J.W. Jones), pp. 209–241. CRC Press, Boca Raton, FL.

Jones, M. and Harper, J.L. (1987a). The influence of neighbours on the growth of trees. I. The demography of buds in *Betula pendula*. *Proc. R. Soc. Ser. B* **232**, 1–18.

Jones, M. and Harper, J.L. (1987b). The influence of neighbours on the growth of trees. II. The fate of buds on long and short shoots in *Betula pendula*. *Proc. R. Soc. Ser. B* **232**, 19–33.

Julien, M.H. and Bourne, A.S. (1986). Compensatory branching and changes in nitrogen content in the aquatic weed *Salvinia molesta* in response to disbudding. *Oecologia* **70**, 250–257.

Julien, M.H. and Bourne, A.S. (1988). Effects of leaf-feeding by larvae of the moth *Samea multiplicalis* Guen. (Lep., Pyralidae) on the floating weed *Salvinia molesta*. *J. Appl. Entomol.* **106**, 518–526.

Kays, S. and Harper, J.L. (1974). The regulation of plant and tiller density in a grass sward. *J. Ecol.* **62**, 97–105.

King, D.A. (1990). Allometry of saplings and understorey trees of a Panamanian forest. *Funct. Ecol.* **4**, 27–32.

King, D.A. (1991). Tree size. *Natl Geog. Res. Explor.* **7**, 342–351.

Kira, T., Ogawa, H. and Shinozaki, K. (1953). Intraspecific competition among higher plants. I. Competition–density–yield interrelationships in regularly dispersed populations. *J. Inst. Polytech. Osaka Cy Univ.* **D4**, 1–16.

Kirchoff, B.K. (1984). On the relationship between phyllotaxy and vasculature: a synthesis. *Bot. J. Linn. Soc.* **89**, 37–51.

Klekowski, E.J. and Godfrey, P.J. (1989). Ageing and mutation in plants. *Nature, Lond.* **340**, 389–391.

Kohyama, T. and Hotta, M. (1990). Significance of allometry in tropical saplings. *Funct. Ecol.* **4**, 515–522.

Konsens, I., Ofir, M. and Kigel, J. (1991). The effect of temperature on the production and abscission of flowers and pods in snap bean (*Phaseolus vulgaris* L.). *Ann. Bot.* **67**, 391–400.

Kozlowski, T.T. (1973). *Shedding of Plant Parts*. Academic Press, New York.

de Kroon, H. and Knops, J. (1990). Habitat exploration through morphological plasticity in two chalk grassland perennials. *Oikos* **59**, 39–49.

Kuijt, J. (1969). *The Biology of Parasitic Flowering Plants*. University of California Press, Berkeley.

Kurihara, H., Kuroda, H. and Kinoshita, O. (1978). Morphological bases of shoot growth to estimate tuber yields with special reference to phytomer concept in potato plant. *Jap. J. Crop Sci.* **47**, 690–698.

Laan, P. and Blom, C.W.P.M. (1990). Growth and survival responses of *Rumex* spp. to flooded and submerged conditions: the importance of shoot elongation, underwater photosynthesis and reserve carbohydrates. *J. Exp. Bot.* **41**, 775–784.

Ladipo, D.O., Leakey, R.R.B. and Grace, J. (1991). Clonal variation in a four year old plantation of *Triplochiton scleroxylon* K. Schum, and its relation to the predictive test for branching habit. *Silvae Genet.* **40**, 130–135.

Lang, A.R.G. (1990). An instrument for measuring canopy structure. *Remote Sens. Revs.* **5**, 61–71.

Larsson, C.M., Larsson, M., Purves, J.V. and Clarkson, D.T. (1991). Translocation and cycling through roots of recently absorbed nitrogen and sulfur in wheat (*Triticum aestivum*) during vegetative and generative growth. *Physiol. Plant.* **82**, 345–352.

Lavarenne, S., Champagnat, P. and Barnola, P. (1971). Croissance rythmique de quelques végétaux ligneux de régions tempérées cultivés en chambres climatiques à température élevée et constante et sous diverses photopériodes. *Bull. Soc. Bot. Fr.* **118**, 131–162.

Lee, T.D. and Bazzaz, F.A. (1980). Effects of defoliation and competition on growth and reproduction of the annual plant *Abutilon theophrasti*. *J. Ecol.* **68**, 813–821.

Lindenmayer, A. (1968). Mathematical models for cellular interactions in development, Parts I and II. *J. Theor. Biol.* **18**, 280–315.

Lindenmayer, A. (1982). Developmental algorithms: lineage versus interactive control mechanisms. In: *Developmental Order: Its Origin and Regulation* (Ed. by S. Subtelny and P.B. Green), pp. 219–245. Alan R. Liss, New York.

Lonsdale, W.M. (1990). The self-thinning rule: dead or alive? *Ecology* **71**, 1373–1388.

Loreti, F. and Pisani, P.L. (1990). Structural manipulation for improved performance in woody plants. *Hort. Sci.* **25**, 64–74.

Lovett Doust, L. (1981). Population dynamics and local specialization in a clonal perennial (*Ranunculus repens*). I. The dynamics of ramets in contrasting habitats. *J. Ecol.* **69**, 743–755.

Lovett Doust, L. (1987). Population dynamics and local specialization in a clonal perennial (*Ranunculus repens*). III. Responses to light and nutrient supply. *J. Ecol.* **75**, 555–568.

MacDonald, N. (1983). *Trees and Networks in Biological Models*. Wiley, Chichester.

Maillette, L. (1982a). Structural dynamics of silver birch. I. The fates of buds. *J. Appl. Ecol.* **19**, 203–218.

Maillette, L. (1982b). Structural dynamics of silver birch. II. A matrix model of the bud population. *J. Appl. Ecol.* **19**, 219–238.

Maillette, L. (1985). Modular demography and growth patterns of two annual weeds (*Chenopodium album* L. and *Spergula arvensis* L.) in relation to flowering. In: *Studies in Plant Demography: a Festschrift for John L. Harper*. (Ed. by J. White), pp. 239–255. Academic Press, London.

Maillette, L. (1987). Effects of bud demography and elongation patterns on *Betula cordifolia* near the tree line. *Ecology* **68**, 1251–1261.

Maillette, L. (1990). The value of meristem states, as estimated by a discrete time Markov chain. *Oikos* **59**, 235–240.

Maillette, L. (1992). Seasonal model of modular growth in plants. *J. Ecol.* **80**, 123–130.

Mariko, S. and Hogetsu, K. (1987). Analytical studies of the response of sunflowers (*Helianthus annuus*) to various defoliation treatments. *Ecol. Res.* **2**, 1–17.

Maschinski, J. and Whitham, T.G. (1989). The continuum of plant responses to herbivory: the influence of plant association, nutrient availability, and timing. *Am. Nat.* **134**, 1–19.

Mattson, W.J., Lawrence, R.K., Haak, R.A., Herms, D.A. and Charles, P.J. (1988). Defensive strategies of woody plants against different insect-feeding guilds in relation to plant ecological strategies and intimacy of association with insects. In: *Mechanisms of Woody Plant Defenses Against Insects: Search for Pattern* (Ed. by W.J. Mattson, J. Levieux and C. Bernard-Dagan), pp. 3–38. Springer-Verlag, New York.

McGraw, J.B. (1989). Effects of age and size on life histories and population growth of *Rhododendron maximum* shoots. *Am. J. Bot.* **76**, 113–123.

McGraw, J.B. and Garbutt, K. (1990). The analysis of plant growth in ecological and evolutionary studies. *Trends Ecol. Evolut.* **5**, 251–254.

McKinion, D.N., Baker, D.N., Whisler, F.D. and Lambert, J.R. (1989). Application of the GOSSYM/COMAX system to cotton crop management. *Agric. Syst.* **31**, 55–65.

McMahon, T.A. and Kronauer, R.E. (1976). Tree structures: deducing the principle of mechanical design. *J. Theor. Biol.* **59**, 443–466.

McMaster, G.S., Klepper, B., Rickman, R.W., Wilhelm, W.W. and Willis, W.O. (1991). Simulation of shoot vegetative development and growth of unstressed winter wheat. *Ecol. Model.* **53**, 189–204.

McNaughton, S.J. (1983). Compensatory plant growth as a response to herbivory. *Oikos* **40**, 329–336.

McNaughton, S.J. (1984). Grazing lawns: animals in herds, plant form, and coevolution. *Am. Nat.* **124**, 863–886.

Methy, M., Alpert, P. and Roy, J. (1990). Effects of light quality and quantity on growth of the clonal plant *Eichhornia crassipes*. *Oecologia* **84**, 265–271.

Mika, A. (1986). Physiological responses of fruit trees to pruning. *Hort. Rev.* **8**, 337–378.

Millard, P. and Proe, M.F. (1991). Leaf demography and the seasonal internal cycling of nitrogen in sycamore (*Acer pseudoplatanus* L.) seedlings in relation to nitrogen supply. *New Phytol.* **117**, 587–596.

Mitchell, P.L. and Woodward, F.I. (1988). Response of three woodland herbs to reduced photosynthetically active radiation and low red to far-red ratio in shade. *J. Ecol.* **76**, 807–825.

Moe, R. (1990). Effect of day and night temperature alternations and of plant growth regulators on stem elongation and flowering of the long day plant *Campanula isophylla* Moretti. *Scientia Hort.* **43**, 291–306.

Moran, M.S., Pinter, P.J. Jr, Clothier, B.E. and Allen, S.G. (1989). Effect of water stress on the canopy architecture and spectral indices of irrigated alfalfa. *Remote Sens. Environ.* **29**, 251–262.

Moravec, H. (1988). *Mind Children*. Harvard University Press, Cambridge, MA.

Morgan, D.C. and Smith, H. (1979). A systematic relationship between phytochrome-controlled development and species habitat, for plants grown in simulated natural radiation. *Planta* **145**, 253–258.

Morrisson, D.A. and Myerscough, P.G. (1989). Temporal regulation of maternal investment in populations of the perennial legume *Acacia suaveolens*. *Ecology* **70**, 1629–1938.

Mueller, R.J. (1988). Shoot tip abortion and sympodial branch reorientation in *Brownea ariza* Leguminosae. *Am. J. Bot.* **75**, 391–400.

Mutsaers, H.J.W. (1984). KUTUN: a morphogenetic model for cotton. *Agric. Syst.* **14**, 229–257.

Nanda, K.K. and Purohit, A.N. (1966). Experimental induction of apical flowering in an indeterminate plant *Impatiens balsamina* L. *Naturwissenschaften* **54**, 230.

Niklas, K.J. (1987). Evolution of plant shape: design constraints. *Trends Ecol. Evolut.* **1**, 67–72.

Oesterheld, M. (1992). Effect of defoliation intensity on aboveground and belowground relative growth rates. *Oecologia* **92**, 313–316.

Pacala, S.W. and Weiner, J. (1991). Effects of competitive asymmetry on a local density model of plant interference. *J. Theor. Biol.* **149**, 165–180.

Paige, K.N. (1992). Overcompensation in response to mammalian herbivory: from mutualistic to antagonistic interactions. *Ecology* **73**, 2076–2085.

Peng, S. and Krieg, D.R. (1991). Single leaf and canopy photosynthesis response to plant age in cotton. *Agron. J.* **83**, 704–708.

Pezeshki, S.R. (1991). Root responses of flood tolerant and flood sensitive tree species to soil redox conditions. *Trees (Berl.)* **5**, 180–186.

Porter, J.R. (1983). A modular approach to analysis of plant growth. II. Methods and results. *New Phytol.* **94**, 191–200.

Porter, J.R. (1984). A model of canopy development in winter wheat. *J. Agric. Sci.* **102**, 383–392.

Porter, J.R. (1989a). Modules, models and meristems in plant architecture. In: *Plant Canopies: their Growth, Form and Function* (Ed. by G. Russell, B. Marshall and P.G. Jarvis), pp. 143–160. Cambridge University Press, New Rochelle, New York.

Porter, J.R. (1989b). Pruning, canopy architecture and plant productivity. In: *Manipulation of Fruiting*. (Ed. by C.J. Wright), pp. 293–304. Butterworth and Co., Sutton Bonnington.

Price, E.A.C. and Hutchings, M.J. (1992). The causes and developmental effects of integration and independence between different parts of *Glechoma herderacea* clones. *Oikos* **63**, 376–386.

Prusinkiewicz, P. and Hanan, J. (1989). Lindenmayer systems, fractals, and plants. *Lecture Notes in Biomathematics 79*. Springer Verlag, New York.

Prusinkiewicz, P. and Lindenmayer, A. (1990). *The Algorithmic Beauty of Plants*. Springer-Verlag, New York.

Prusinkiewicz, P. and McFadzean, D. (1992). Modelling plants in environmental context. In: *Proceedings of the Fourth Annual Western Computer Graphics Symposium*, pp. 47–51. University of Alberta Press, Edmonton.

Prusinkiewicz, P., Hammel, M. and Mjolsness, E. (1993). Animation of plant development. In: *Proceedings of SIGRAPH '93* Anaheim 1993, 351–360.

Reddy, V.R., Baker, D.N. and Hodges, H.F. (1991). Temperature effects on cotton canopy growth, photosynthesis, and respiration. *Agron. J.* **83**, 699–704.

Reekie, E.G. and Redmann, R.E. (1991). Effects of water stress on the leaf demography of *Agropyron desertorum*, *Agropyron dasystachyum*, *Bromus inermis*, and *Stipa viridula*. *Can. J. Bot.* **69**, 1647–1654.

Rees, D.J. and Grace, J. (1981). The effect of wind and shaking on the water relations of *Pinus contorta*. *Physiol. Plant* **51**, 222–228.

de Reffye, P. (1976). Modélisation et simulation de la verse du caféier, à l'aide de la théorie de la résistance des matériaux. *Café Cacao Thé* **20**, 251–272.

de Reffye, P. (1981). Modèle mathématique aléatoire et simulation de la croissance et de l'architecture de *Caffeier robusta*. 2e partie. Etude de la mortalité des méristèmes plagiotropes. *Café Cacao Thé* **25**, 219–230.

de Reffye, P. (1983). Modèle mathématique aléatoire et simulation de la croissance et de l'architecture de *Caffeier robusta*. 4e partie. Programmation sur microordinateur du trace en trois dimensions de l'architecture d'un arbre. Application au caféier. *Café Cacao Thé* **27**, 3–20.

de Reffye, P., Edélin, C., Françon, J., Jaeger, M. and Puech, C. (1988). Plant models faithful to botanical structure development. *Comput. Graph.* **22**, 151–158.

de Reffye, P., Lecoustre, R., Edélin, C. and Dinouard, P. (1989). Modelling plant growth and architecture. In: *Cell to Cell Signalling: From Experimental to Theoretical Models*, pp. 237–246. Academic Press, London.

de Reffye, P., Dinouard, P. and Barthélemy, D. (1991). Modélisation et simulation de l'architecture de l'orme du Japon *Zelkova serrata* (Thunb.) Makino (Ulmaceae): la notion d'axe de référence. In: *L'Arbre, Biologie et Développement* (Ed. by C. Edelin), pp. 252–266. Naturalia Monspeliensia, Montpellier.

Reid, D.M., Beall, F.D. and Pharis, R.P. (1991). Environmental cues in plant growth and development. In: *Growth and Development*, vol. X, pp. 65–181. Academic Press, London.

Remphrey, W.R. and Davidson, C.G. (1991). Spatiotemporal distribution of epicormic shoots and their architecture in branches of *Fraxinus pennsylvanica*. *Can. J. For. Res.* **22**, 336–340.

Remphrey, W.R. and Powell, G.R. (1987). Crown architecture of *Larix laricina* saplings: an analysis of higher order branching. *Can. J. Bot.* **65**, 268–279.

Remphrey, W.R., Neal, B.R. and Steeves, R.A. (1983). The morphology and growth of *Arctostaphylos uva-ursi* (bearberry): an architectural model simulating colonizing growth. *Can. J. Bot.* **61**, 2451–2458.

Renshaw, E. (1985). Computer simulation of Sitka spruce: spatial branching models for canopy growth and root structure. *IMA J. Math. Appl. Med. Biol.* **2**, 183–200.

Rheingold, H. (1992). *Virtual Reality*. Martin, Secker & Warburg, London.

Rhoades, D.F. (1985). Offensive-defensive interactions between herbivores and plants: their relevance in herbivore population dynamics and ecological theory. *Am. Nat.* **125**, 205–238.

Room, P.M. (1983). "Falling apart" as a lifestyle: the rhizome architecture and population growth of *Salvinia molesta*. *J. Ecol.* **71**, 349–365.

Room, P.M. (1988). Effects of temperature, nutrients and a beetle on branch architecture of the floating weed *Salvinia molesta* and simulations of biological control. *J. Ecol.* **76**, 826–848.

Room, P.M. and Thomas, P.A. (1985). Nitrogen and establishment of a beetle for biological control of the floating weed salvinia in Papua New Guinea. *J. Appl. Ecol.* **22**, 139–156.

Sackville Hamilton, N.R. and Harper, J.L. (1989). The dynamics of *Trifolium repens* in a permanent pasture: I. The population dynamics of leaves and nodes per shoot axis. *Proc. R. Soc. Ser. B* **237**, 133–173.

Salisbury, F.B. and Ross, C.W. (1992). *Plant Physiology*, 4th edn. Wadsworth Publishing Company, Belmont, California.

Savile, D.B.O. (1976). Twig abscission in maples (section *Rubra: Acer rubrum* and *A. saccharinum*) as a defense against water stress. *Can. Fld Nat.* **90**, 184–185.

Schmid, B. (1992). Phenotypic variation in plants. *Evolut. Trends Plants* **6**, 45–60.

Schmid, B. and Bazzaz, F.A. (1987). Clonal integration and population structure in perennials: effects of severing rhizome connections. *Ecology* **68**, 2016–2022.

Schmid, B. and Bazzaz, F.A. (1990). Plasticity in plant size and architecture in rhizome derived vs. seed derived *Solidago* and *Aster*. *Ecology* **71**, 523–535.

Schmid, B., Miao, S.L. and Bazzaz, F.A. (1990). Effects of simulated root herbivory and fertilizer application on growth and biomass allocation in the clonal perennial *Solidago canadensis*. *Oecologia* **84**, 9–15.

Seastedt, T.R. (1985). Maximisation of primary and secondary productivity by grazers. *Am. Nat.* **126**, 559–564.

Shigo, A.L. (1984). Compartmentalization: a conceptual framework for understanding how trees grow and defend themselves. *A. Rev. Phytopath.* **22**, 189–214.

Shinozaki, K., Yoda, K., Hozumi, K. and Kira, T. (1964). A quantitative analysis of plant form—the pipe model theory. I. Basic analyses. *Jap. J. Ecol.* **14**, 97–105.

Shumway, D.L., Steiner, K.C. and Abrams, M.D. (1991). Effects of drought stress on hydraulic architecture of seedlings from five populations of ash. *Can. J. Bot.* **69**, 2158–2164.

Slade, A.J. and Hutchings, M.J. (1987a). Clonal integration and plasticity in foraging behaviour in *Glechoma hederacea*. *J. Ecol.* **75**, 1023–1036.

Slade, A.J. and Hutchings, M.J. (1987b). An analysis of the cost and benefits of physiological integration between ramets in the clonal perennial herb *Glechoma hederacea*. *Oecologia* **73**, 425–431.

Slade, A.J. and Hutchings, M.J. (1987c). An analysis of the influence of clone size and stolon connections between ramets on the growth of *Glechoma hederacea*. *New Phytol.* **106**, 759–771.

Solomon, B.P. (1983). Compensatory production in *Solanum carolinense* following attack by a host-specific herbivore. *J. Ecol.* **71**, 681–690.

Steingraeber, D.A. and Waller, D.M. (1986). Non-stationarity of tree branching patterns and bifurcation rations. *Proc R. Soc. B* **228**, 187–194.

Stephenson, A.G. (1981). Flower and fruit abortion: proximate causes and ultimate functions. *A. Rev. Ecol. Syst.* **12**, 253–279.

Sutherland, W.J. and Stillman, R.A. (1988). The foraging tactics of plants. *Oikos* **52**, 239–244.

Tanner, E.V.J. (1983). Leaf demography and growth of the tree fern *Cyatea pubescens* Mett. ex Kuhn in Jamaica (West Indies). *Bot. J. Linn. Soc.* **87**, 213–228.

Tateno, M. (1991). Increase in lodging safety factor of thigmomorphogenically dwarfed shoots of mulberry tree. *Physiol. Plant.* **81**, 239–243.

Thiébaut, B., Comps, B. and Teissier, Du Cros E. (1990). Development of the axis in trees: annual growth, syllepsis, and prolepsis in the European beech (*Fagus sylvatica*). *Can. J. Bot.* **68**, 202–210.

Thomas, L.P. and Watson, M.A. (1988). Leaf removal and the apparent effects of architectural constraints on development in *Capsicum annuum*. *Am. J. Bot.* **75**, 840–843.

Thompson, L. and Harper, J.L. (1988). The effect of grasses on the quality of transmitted radiation and its influence on the growth of white clover *Trifolium repens*. *Oecologia* **75**, 343–347.

Thornley, J.H.M. and Johnson, I.R. (1990). *Plant and Crop Modelling*. Clarendon Press, Oxford.

Tomlinson, P.B. (1978). Branching and axis differentiation in tropical trees. In: *Tropical Trees as Living Systems*. (Ed. by P.B. Tomlinson and M.H. Zimmerman), pp. 187–207. Cambridge University Press, Cambridge.

Tomlinson, P.B., Takaso, T. and Rattenbury, J.A. (1989). Developmental shoot morphology in *Phyllocladus* (Podocarpaceae). *Bot. J. Linn. Soc.* **99**, 223–248.

Tuomi, J., Niemela, P., Chapin, F.S., Bryant, J.P. and Siren, S. (1988). Defensive responses of trees in relation to their carbon/nutrient balance. In: *Mechanisms of Woody Plant Defenses Against Insects: Search for Pattern* (Ed. by W.J. Mattson, J. Levieux and C. Bernard-Dagan), pp. 57–72. Springer-Verlag, New York.

Tyree, M.T. and Ewers, F.W. (1991). Tansley review no. 34: The hydraulic architecture of trees and other woody plants. *New Phytol.* **119**, 345–360.

van der Meijden, E., Wija, M. and Verkaar, H.J. (1988). Defence and regrowth, alternative plant strategies in the struggle against herbivores. *Oikos* **51**, 355–363.

Vesey-Fitzgerald, D.F. (1973). Animal impact on vegetation and plant succession in Lake Manyara National Park, Tanzania. *Oikos* **24**, 314–325.

Vincent, J.F.V. (1990). Fracture properties of plants. *Adv. Bot. Res.* **17**, 235–287.

Vogel, M. (1975). Recherche du détérminisme du rythm de croissance du cacaoyer. In: *Café Cacao Thé* **19**, 265–290.

Vuorisalo, T. and Tuomi, J. (1986). Unitary and modular organisms: criteria for ecological division. *Oikos* **47**, 382–385.

Waller, D.M. and Steingraeber, D.A. (1985). Branching and modular growth: theoretical models and empirical patterns. In: *Population Biology and Evolution of Clonal Organisms* (Ed. by J.B.C. Jackson, L.W. Buss and R.E. Cook), pp. 225–257. Yale University Press, New Haven, CT.

Waller, G.R. (Ed.). (1987). *Allelochemicals: Role in Agriculture and Forestry*. American Chemical Society, Washington, DC.

Watkinson, A.R. and White, J. (1985). Some life-history consequences of modular construction in plants. *Phil. Trans. R. Soc. Ser. B* **313**, 31–50.

Watson, M.A. (1986). Integrated physiological units in plants. *Trends Ecol. Evolut.* **1**, 119–123.

Watson, M.A. and Casper, B.B. (1984). Morphometric constraints on patterns of carbon distribution in plants. *A. Rev. Ecol. Syst.* **15**, 233–258.

Weiner, J. and Thomas, S.C. (1992). Competition and allometry in three species of annual plants. *Ecology* **73**, 648–656.

Welker, J.M. and Briske, D.D. (1992). Clonal biology of the temperate, caespitose, graminoid *Schizachyrium scoparium*: a synthesis with reference to climate change. *Oikos* **63**, 357–365.

Whisler, F.D., Acock, B., Baker, D.N., Fye, R.E., Hodges, H.F., Lambert, J.R., Lemmon, H.E., McKinion, J.M. and Reddy, V.R. (1986). Crop simulation models in agronomic systems. *Adv. Agron.* **40**, 141–208.

White, J. (1979). The plant as a metapopulation. *A. Rev. Ecol. Syst.* **10**, 109–145.

White, J. (1984). Plant metamerism. In: *Perspectives on Plant Population Ecology* (Ed. by R. Dirzo and J. Sarukhán), pp. 15–47. Sinauer Associates, Sunderland, MA.

Whitham, T.G. and Mopper, S. (1985). Chronic herbivory: impacts on architecture and sex expression of pinyon pine. *Science* **228**, 1089–1091.

Whittaker, R.H. and Feeny, P.P. (1971). Allelochemics: chemical interactions between species. *Science* **171**, 757–770.

Xu, G., Ninomiya, I. and Ogino, K. (1990). The change of leaf longevity and morphology of several tree species grown under different light conditions. *Bull. Ehime Univ. For.* **28**, 35–44.

Yoda, K., Kira, T., Ogawa, H. and Hozumi, H. (1963). Self-thinning in overcrowded pure stands under cultivated and natural conditions. *J. Inst. Polytech. Osaka Cy Univ.* **14**, 107–129.

APPENDIX: GLOSSARY

abortion   programmed, predictable death of a meristem

abscission   cutting off from self, followed by shedding

acropetal   towards the apex

acrotony   distal branches more developed than proximal ones

apical control   suppression of lateral branch growth by the trunk apex or by branches growing closer to the apex

apical dominance   suppression of axillary meristems/buds by the presence of a shoot apex

apposition growth   apical meristem continues to grow but is replaced as the main apex by growth of a more vigorous axillary meristem (no meristem abortion involved)

articulated growth   rhythmic growth resulting in shoot units separated by morphological discontinuities (e.g. leaf scars)

axil   the space immediately distal to the attachment of a leaf to a stem

basipetal   towards the base

basitony   proximal branches more developed than distal ones

caulomer   the unit of sympodial growth formed by one apical meristem (= French "L'article")

cladoptosis   shedding of shoots, stems and branches

compensatory growth   growth to replace what would have existed in the absence of tissue loss

continuous growth   growth without visible rhythm, producing shoots without distinct articulations

continuous ramification   every node is the origin of an axis of the next order

correlative inhibition   inhibition of growth by an organ on another part of a plant

decurrent habit   strong apical dominance over axillary buds and weak apical control of lateral branch growth once axillary buds have developed so that lateral branches eventually outgrow the main axis, results in poorly defined trunk and spherical crown

dedifferentiation   change to a level of differentiation characteristic of a younger ontogenetic stage

determinate   axial growth determined early in development to be unable to continue indefinitely, usually due to inflorescence evocation or programmed abortion of the apical meristem

deterministic model   generates exactly the same outputs each time it is run with a given set of inputs

distal   further from the junction between root and shoot

epinasty   the effect of one branch on another which determines their final orientation

excurrent habit   weak apical dominance and strong apical control of lateral branches, leads to a pronounced trunk and conical crown

geometry   the arrangement of objects in space in relation to a fixed, external point

hapaxanthic   determinate by developing terminal reproductive structures

hypopodium   the basal internode of a new sylleptic branch having a single node; presence indicates sylleptic growth

indeterminate   monopodial axial growth able to continue indefinitely, any flowers being produced laterally

intercalation   crown infilling by metamorphosis of lateral branches

mesotony   branches more developed in the median rather than proximal or distal parts of a trunk or branch

metamorphosis   change in meristem potential from plagiotropic to orthotropic

monopodial   growing as a single and continuous axis arising from a single meristem

orthotropic   growing vertically

orthostichy   vertical series of leaves or leaf scars connected to the same vascular bundle

parastichy   a series of leaves or leaf scars in order of origin

phyllotaxis   leaf arrangement on a stem

phytomer   a structural unit in a plant = metamer

plagiotropic   growing away from the vertical, horizontally or obliquely

plastochron   the time interval between production of two adjacent metamers in a shoot

plastochron index   a measure of developmental age in plastochrons, i.e. the number of metamers between a given location on a shoot and the shoot apex

pleonanthic   producing lateral reproductive structures as the main axis continues to grow

polypodial   main apex continues, some lateral apices continue

prolepsis   branching which is delayed by an inactive period of the axillary buds

proximal   closer to the junction between root and shoot

rhythmic growth   alternating periods of rapid and slow growth or dormancy, usually in seasonal environments

stochastic model   generates different outputs each time it is run with a given set of inputs because internal workings are affected by random variations

syllepsis   branches grow with no dormancy of axillary buds, results in more regular internode lengths and branch spacing than prolepsis

sympodial   *by substitution*: apex terminates due to apical meristem aborting or becoming reproductive, forward growth is continued by activity of a lateral meristem; *by apposition*: modules are indeterminate but growth slows and they become subordinate to axillary shoots. The subordinate apical meristem often produces an erect short shoot while the new leader remains plagiotropic

thigmomorphogenesis   growth response to mechanical contact pressure

topology   the relative arrangements of objects with respect to each other; which parts are connected to which others

Foraging in Plants: the Role of Morphological Plasticity in Resource Acquisition

M.J. HUTCHINGS and H. DE KROON

ADVANCES IN ECOLOGICAL RESEARCH VOL. 25
ISBN 0–12–013925–1

Triffids were, admittedly, a bit weird—but that was, after all, just because they were novelties. People had felt the same about novelties of other days—about kangaroos, giant lizards, black swans. And, when you came to think of it, were triffids all that much queerer than mudfish, ostriches, tadpoles, and a hundred other things? The bat was an animal that had learned to fly: well, here was a plant that had learned to walk—what of that?

(from John Wyndham, *The Day of the Triffids*, 1951)

I. SUMMARY

Resources which are essential for plant growth are usually heterogeneously distributed both in space and in time within the habitat. As a result of plasticity in the modular construction of branches ("spacers") in response to resource availability, plants are capable of placing leaves and root tips ("resource-acquiring structures") non-randomly within their environment. Selective placement of organs of resource uptake actively modifies the potential for resource acquisition, and is interpreted as a consequence of foraging behaviour.

There is ample evidence that plants are capable of placing resource-acquiring structures selectively within the habitat. Orthotropic (vertical) shoots commonly reduce spacer length and increase lateral branch formation under more favourable light conditions. These morphological modifications promote more effective placement of leaves in the high light zone at the top of a vegetation canopy. In many species increased lateral root formation and growth increases local root surface area in patches with high nutrient content. The morphological responses to resource supply from the environment are usually purely responses to local patch quality, and they are normally unaffected by the conditions experienced by other parts of the plant. In contrast to orthotropic stems, plagiotropic (horizontal) stems of clonal plants exhibit a variety of responses to resource availability, and the magnitude of response is usually much smaller than that seen in orthotropic shoots. Rather than providing the plant with a capacity to place ramets selectively in favourable patches within a heterogeneous environment, plagiotropic stems appear to be important for the continuous exploration of new habitat space.

Foraging behaviour which enhances the probability of locating resource-rich patches does not appear to be profitable in all environments. In habitats of inherently low resource availability, in which resources become available in the form of ephemeral pulses, physiological plasticity has been viewed as a profitable adjunct to morphological plasticity for the acquisition of resources. It has been proposed that morphological plasticity will be more restricted in plants which are characteristic of such conditions. Evidence is accumulating that shoots of shade-tolerant species are indeed morphologically less plastic than shoots of shade-intolerant species. However, it has not been conclusively shown that the root morphology of species from infertile soils is less plastic than that of species from fertile soils.

Physiological studies have demonstrated that the photosynthetic apparatus has the ability to utilize ephemeral sunflecks efficiently, and that plasticity in nutrient uptake rates promotes efficient capture of nutrient pulses. Few comparative studies have been carried out to date, but the available data show only a weak tendency for greater physiological plasticity in species from habitats with low light or low nutrient availability. In resource-poor habitats, the ability of resource-acquiring structures to remain viable during periods of resource depletion seems to be at least as important for the acquisition of resource pulses as a high level of physiological plasticity. In general, the relative importance of morphological *versus* physiological plasticity for resource acquisition in heterogeneous environments has yet to be assessed. We conclude that, to date, support is as yet incomplete for the contention that species from habitats which differ in resource supply possess different resource acquiring syndromes.

The ultimate aim of research into foraging should be to understand the benefits and disadvantages of different resource-acquiring syndromes, and their ecological and evolutionary consequences. Achievement of this goal requires evaluation of foraging behaviour in terms of its costs and benefits. There is also a need for an increase in information on the developmental and physiological traits—and the genetics of the traits—which underlie plasticity.

II. INTRODUCTION

All plant species require the same limited number of essential resources from their environment, namely sunlight, carbon dioxide, water and a number of elements in nutrient form, obtained, at least in higher plants, predominantly from the soil. The availability of these resources from the environment is patchy both in space and time. Therefore, the difficulty in obtaining sufficient resources varies spatially and temporally. Although there are differences between species in the levels of resources which are needed for growth, and although the ratios in which resources are required may differ between species, all plants have the same basic problems of resource acquisition to solve. This review is a discussion of the solutions employed by plants for enhancing the acquisition of essential resources in spatially and temporally heterogeneous habitats.

The possession of morphological plasticity and the ability to produce branched structures by the proliferation of modules are attributes which free plants from some of the limitations imposed by sessility. Resource-acquiring structures (these have also been referred to as "feeding sites" (Bell, 1984) and "mouths" (Watson, 1984)) such as leaves, root tips and, in clonal plants, ramets, are projected into the environment upon branches ("spacers"; Bell, 1984). When the part of the spacer proximal to a resource-acquiring structure stops extending, the position of that resource-acquiring structure becomes fixed. Evidence will be presented showing that many plants do not place their resource-acquiring structures in random positions within their environment, but instead place them selectively in patches of greater resource supply. Such morphological responses to resource availability may actively modify the potential of the plant for resource acquisition and they are interpreted as manifestations of foraging behaviour.

In the following section of this review we define and delimit the issues which will be discussed. In Sections IV–VI we document the foraging activities of the spacers and resource-acquiring organs of higher plants, and the foraging activities of fungi. These sections also review what is known about the proximal physiological and developmental mechanisms which control foraging activities. In Section VII we discuss integration—the extent to which foraging behaviour is coordinated throughout the whole plant. Physiological plasticity in the rates of resource acquisition by leaves and roots has been viewed as a profitable adjunct to foraging, and this topic is discussed in Section VIII. In the penultimate section (IX), we summarize the hypotheses which have been presented in the literature on foraging syndromes as adaptive traits in given environmental settings. These hypotheses are then evaluated, given the empirical evidence presented earlier. This section culminates in the formulation of a number of questions which remain to be solved. We conclude the review (Section X) by briefly outlining an agenda for a research programme on plant foraging, which includes proposals for empirical and mathematical studies on the costs and benefits of foraging behaviour.

III. DEFINING THE SCOPE OF THE REVIEW: ON FORAGING, PLASTICITY AND GROWTH

Description of plant responses to environmental quality in terms of foraging is a relatively recent development. To our knowledge, Bray (1954) was the first to use the term in this way when he described the search patterns of roots for nutrients in the soil. Later on, Grime and his co-workers (e.g. Grime, 1979; Grime et al., 1986) established common usage of the term in the vocabulary of plant ecology. Here we define foraging as "the processes whereby an organism searches, or ramifies within its habitat, which enhance its acquisition of essential resources" (modified after Slade and Hutchings, 1987a). We believe that this definition fits the subject as it is perceived by animal ecologists and behaviourists. Foraging in plants is accomplished by morphological plasticity. As such it is one aspect of plant "behaviour", a term which has been widely used by plant ecologists in recent years as an alternative to phenotypic plasticity (Silvertown and Gordon, 1989).

Bradshaw (1965) has defined plasticity thus: "Plasticity is shown by a genotype when its expression is able to be altered by environmental influences. The change that occurs can be termed the response. Since all changes in the characters of an organism which are not genetic are environmental, plasticity is applicable to all intragenotypic variability." In response to different environmental conditions, morphological plasticity can generate different patterns of spacer production and hence different patterns in the placement of resource-acquiring structures. This review concentrates on analysis and interpretation of these patterns and their consequences for future resource acquisition.

Our definition emphasizes that placement of resource-acquiring structures is an essential component of foraging. It must be realized that under certain conditions a lack of selective placement may be an important element of a foraging syndrome; habitat heterogeneity may be such that selective placement will not contribute to an increase in resource acquisition. It is probable that in many cases plants use a combination of morphological and physiological techniques to acquire resources efficiently, with the balance of importance between these techniques depending on the type of habitat occupied. In analogy with animal foraging theory we postulate that the economics (i.e. the costs and benefits) of different resource-acquiring behaviours are reflected in the performance of the plant, and that in a given environmental setting the behaviour conferring the highest long-term resource gain will have a selective advantage.

In addition to plasticity in spacer morphology, plasticity in a number of other plant characteristics, such as root:shoot ratio, leaf and root turnover rates, and storage capability, will also influence resource acquisition or resource retention. Although the consequences of such variation can be investigated using economic principles (Bloom *et al.*, 1985; Chapin *et al.*, 1990) we do not regard variation in these characteristics as aspects of foraging behaviour, because the positioning of plant parts is not involved. They are thus largely omitted from this discussion.

Foraging is thus concerned with the placement of resource-acquiring structures within the surroundings of the plant, and its implications for future resource acquisition. Growth, by contrast, refers to the consequences of resource acquisition for the production of new modules or ramets and the

concomitant increase in plant size. The distinction between growth and foraging must be clear. Consider two environments, one providing few resources
and thus affording only slow growth, the other providing ample resources and
supporting rapid growth. After a given length of time, two plants with identical genotypes will have achieved different biomasses in the two environments.
If this is all that distinguishes them and if, in the course of time, the plant in
poor conditions would come to resemble the plant in good conditions in both
biomass *and* morphology, there would be no evidence that they have responded to the different conditions by foraging in different ways. They only
differ in the extent to which they have grown. If the plants have responded to
their environments by foraging, however, they will differ from each other in
the morphology of their spacers and in the patterns of placement of their
resource-acquiring structures. For example, plants of *Abutilon theophrasti*
grown under high and low light conditions had very different shoot heights
when compared at the same plant weight (Rice and Bazzaz, 1989). This is an
indication that shoots foraged in these environments in different ways.

In analogy with the literature on foraging in animals, Kelly (1990) has
suggested that responses in plants should only be accepted as foraging activities if they have taken place *prior to* resource uptake (see also Oborny, 1991).
We regard such a limitation in the use of the concept as unnecessarily restric-
tive and difficult to apply because, unlike the situation in animals, resource
uptake in plants is usually a continuous process. Moreover, resource uptake is
often the only means by which a plant can sense the level of resource supply
from its environment (the use of light *quality* detection is a notable exception,
as will be discussed below). Rather than emphasizing the timing of foraging
responses relative to the timing of resource uptake, we focus on the importance for future resource acquisition by the plant, of the ability to produce
changes in its modular construction in different environments.

IV. THE FORAGING ACTIVITIES OF SHOOTS

Shoots do not all grow towards the strongest source of light. Those which do
are usually referred to as orthotropic, although strictly they should be described as positively orthophototropic. However, many species produce stem
homologues, such as stolons and subterranean rhizomes, which are oriented
at some angle, rather than vertically (plagiotropically) or strictly horizontally
(diagravitropically; see Salisbury and Marinos, 1985). In this section we
review morphological plasticity in stems and their homologues. We discuss
the effects of variation in a number of environmental factors on shoot morphology. This discussion is followed by a consideration of the proximal mechanisms regulating morphological plasticity. We conclude with an examination
of the extent to which shoots will be capable of selective placement in the

more favourable parts of a patchy environment as a consequence of the morphological responses which have been discussed.

A. Response to Photon Flux Density

The shoots of higher plants are constructed from a number of repeated building blocks or metamers (White, 1979). A metamer consists of a node, an internode, an axillary meristem and a leaf. The components of metamers may not all be equally conspicuous. For example, the leaf may be rudimentary, and may drop soon after the metamer is formed, or the internode may be compressed. Metamers are usually arranged in a monopodial axis, or module. Each module is terminated by an apex, which may be living, dead, vegetative or sexual. Axillary (or lateral) meristems may grow out and form new modules, but in many plants they remain dormant for long periods.

The most consistent plastic responses to availability of light are internode elongation (etiolation) and reduced branching (a lower proportion of axillary meristems grow out) at lower flux densities (e.g. Bazzaz and Harper, 1977; Ford and Diggle, 1981; Givnish, 1982, 1986; Smith, 1983; Jones, 1985; Maillette, 1985, 1986; Ellison, 1987; Menges, 1987; Ellison and Niklas, 1988). The etiolation response commonly takes place without a corresponding increase in shoot weight, and thus involves a change in shoot allometry, with etiolated internodes having lower weight per unit length (e.g. Ogden, 1970; Hutchings, 1986; Foggo, 1989; Rice and Bazzaz, 1989). The tendency to increase in height by etiolation varies between species with orthotropic shoots (Grime, 1966; Hara et al., 1991), being more pronounced in shade-intolerant than in shade-tolerant species. Although increased height may promote overtopping of neighbours and improve access to light in some communities, it will not enable herbaceous species, or species of low stature, to avoid shade in woodlands (Fitter and Ashmore, 1974; Frankland and Letendre, 1978).

Height extension in response to low light availability has also been observed in some graminoids which lack a true orthotropic stem. For example, in *Carex flacca* (Cyperaceae) and *Brachypodium pinnatum* (Poaceae), shading stimulates leaf elongation and leaf sheath elongation respectively (de Kroon and Knops, 1990). When vegetation is dense, these responses could lift leaf blades into the high light zone of the canopy, enabling both more efficient photosynthesis and greater shading of neighbours. Increase in leaf sheath length can be protracted in time in *B. pinnatum*, with the degree of elongation depending on the local light regime, allowing a fine-scaled, efficient response to local canopy conditions. *B. pinnatum* and *C. flacca* also form plagiotropic rhizomes, the mean length of which does not change in *B. pinnatum* and decreases in *C. flacca* under low photon flux density, although the variations in rhizome lengths are large for both species under all growing

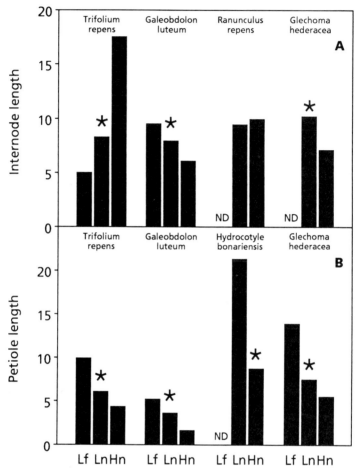

Fig. 1. Responses of (A) stolon internode length and (B) petiole length of some herbaceous stoloniferous species to photon flux density (PFD) and light quality. Treatments are: Lf, low photon flux density, filtered (low R/FR ratio); Ln, low photon flux density, neutral (high R/FR ratio); Hn, high photon flux density, neutral. *Trifolium repens* data from Thompson and Harper (1988): PAR under low photon flux density was 32% of PAR under high photon flux density; R/FR ratios in neutral and filtered light were approximately 1·0 and 0·2, respectively. Filtered light was produced by shade from a *Holcus lanatus* canopy. *Galeobdolon luteum* data from Mitchell and Woodward (1988): PAR under low photon flux density was 12% of PAR under high photon flux density; R/FR ratios in neutral and filtered light were approximately 0·85 and 0·40, respectively. *Ranunculus repens* internode data from Lovett Doust (1987): PAR under low photon flux density was 66% of PAR under high photon flux density. The results shown here are for plants of *R. repens* from a woodland population growing under high nutrient availability. *Hydrocotyle bonariensis* petiole data from Evans (1992): PAR under low photon flux density was 25% of PAR under high photon flux density. Results for severed plants growing under high nutrient and high water availability.

conditions. As in *C. flacca*, the main vertical shoots of *Glaux maritima*, a member of the Primulaceae, showed clear etiolation responses to low light, producing longer internodes, although the horizontal subterranean offshoots which produce hibernacles at their ends were shorter (Jerling, 1988).

Several species with plagiotropic stems etiolate under low light availability (Hutchings and Turkington, 1993), while others do not show a significant response (Fig. 1). In *Glechoma hederacea*, a shade-tolerant stoloniferous species, internode and petiole lengths increase and branching is reduced when photon flux density is low (Slade and Hutchings, 1987b; E.A.C. Price and M.J. Hutchings, 1993, unpublished). However, the total length of primary stolons hardly responds to light level (Hutchings and Slade, 1988). This is also true in *Ranunculus repens*, although in this species the lengths of individual internodes are unresponsive to light level (Lovett Doust, 1987). There is a fall in weight per unit length of stolon in both *G. hederacea* and *R. repens* when photon flux density is reduced. A species which is closely related to *G. hederacea*, *Lamiastrum galeobdolon* (syn. *Galeobdolon luteum*), also has longer stolon internodes and petioles under lower light levels (Mitchell and Woodward, 1988; Dong, 1993), as does *Hydrocotyle bonariensis* (Evans, 1992). In *Trifolium repens*, internode lengths are unresponsive or even shorter at low light levels, while stolon branching declines dramatically and vertical petioles etiolate as light supply declines (Solangaarachchi and Harper, 1987; Thompson and Harper, 1988). Thompson (1993) recently showed that the responses of *T. repens* depend on the actual photon flux densities that are applied. At relatively low light, an increase in flux density may result in an increase in stolon internode and petiole length but at higher light supply an increase in flux density may result in shorter structures.

Tropical vines use the support of host trees to reach high light zones in the top of the forest canopy. Vines employ a number of solutions to reach a suitable support. Slender leafless shoots of *Monstera gigantea* seedlings have been shown to grow preferentially towards "the darkest sector of the horizon", thus increasing the probability of encountering a nearby tree trunk up

---

Fig. 1. (continued)
Glechoma hederacea internode data from Slade and Hutchings (1987b): PAR under low photon flux density was 25% of PAR under high photon flux density. Data for primary stolons only. *Glechoma hederacea* petiole data from E.A.C. Price and M.J. Hutchings (1993, unpublished): in the Ln treatment, PAR was 20% of PAR under high photon flux density and R/FR ratio was 0·9; in the Lf treatment PAR was 2·5% of PAR under high photon flux density and R/FR ratio was 0·45. Shading was imposed by competition with *Lolium perenne*. Data for primary stolons only. All mean lengths in cm except for *Trifolium repens* internodes, which are given in mm. Asterisks indicate significant differences between treatments at least at $p < 0.05$. ND is not determined.

which the vine can grow (Strong and Ray, 1975). In deep shade, the tropical liana *Ipomoea phillomega*, and many other vine species (Ray, 1992) produce flagellar shoots with long internodes, high elongation rates and rudimentary, short-lived leaves (Peñalosa, 1983). Such shoots typically circumnutate, sweeping the terminal part of the shoot through the air as the result of an endogenous growth-related rhythm (Putz and Holbrook, 1991). Perception of nearby support can modify the rotational movement into an ellipse with the long axis oriented towards the support (Tronchet, 1977, cited in Putz and Holbrook, 1991). Circumnutation thus improves the chances of finding a suitable support. Once shoots have reached a support and commenced ortho-tropic growth, the slender stems may alter their form to produce shorter internodes. They also show slow elongation and eventually, upon reaching a zone of high photon flux density, develop large, long-lived leaves (Peñalosa, 1983; Ray, 1992). Similar morphological responses to open and shaded habi-tats are shown by the scrambling stem internodes of *Rubia peregrina* (Navas and Garnier, 1990).

B. Response to Light Quality

In terms of illumination, the habitats of most plants are patchy, consisting of areas where leaves can be illuminated by direct or diffuse sunlight, and areas where the light they receive first passes through the leaves of overhanging competing species. Light transmitted through leaves has a lower ratio of red/far-red (R/FR) light than unfiltered light (Holmes and Smith, 1975; Holmes, 1976, 1981; Hutchings, 1976). The R/FR ratio may fall from approximately 1·15 in full sunlight to as little as 0·05 under a dense canopy of vegetation. When plants are exposed to a R/FR ratio lower than that found in unfiltered light, stems develop a greater degree of apical dominance. This is expressed as an increase in stem elongation rate and an inhibition of the growth of lateral buds. Responsiveness differs between species (Grime, 1966, 1979; Child *et al.*, 1981; Morgan, 1981), with shade-tolerant species responding less than shade-intolerant ones (Morgan and Smith, 1979; Smith, 1982; Corré, 1983; Holmes, 1983). Light which has passed through plant leaves also has a lower proportion of blue light than is found in unfiltered light. Supplementing the blue light incident on a plant has been shown to induce the production of shorter internodes (Thomas, 1981; Quail, 1983; Schafer and Haupt, 1983). Thus, plants may be expected to produce longer internodes when beneath a canopy transmitting a low level of blue light.

Both in species with orthotropic and plagiotropic shoots, the largest effects of light quality on stem elongation are seen below the leaves of plants trans-mitting the lowest R/FR ratio (Smith and Holmes, 1977; Thompson and Harper, 1988). Although much of the information collected on these effects is for dicotyledonous species, similar effects are seen in grasses (Deregibus *et al.*, 1983; Casal *et al.*, 1985, 1987; Skálová and Krahulec, 1992). Supplementation of FR light stimulates shoot length elongation in grasses, with the response being most marked in the leaf sheaths. At the same time tillering is suppressed. That these changes can be induced by alterations to the light environment caused by plant density is shown by ingenious work by Deregibus *et al.* (1985) and Casal *et al.* (1985), in which red-light-emitting diodes were placed at the bases of plants of the grass species *Sporobolus indicus* and *Paspalum dilatatum*, increasing the R/FR ratio. Plants treated in this way produced more tillers than control plants grown at the same density (Fig. 2), thus aggravating local competition for light. It has been suggested that the greater apical dominance induced in control plants when density is high (and therefore R/FR ratio low) could promote escape from competitive situations if the energy saved on tillering was instead allocated to more or longer rhizomes.

Several recent studies (Ballaré *et al.*, 1987, 1988, 1990, 1991a; Casal *et al.*, 1990) have demonstrated that an increase in internode length can be caused by changes in the quality of the light reflected from the stems and foliage of neighbouring plants. As with light which has passed through a canopy of leaves, this reflected light has a low R/FR ratio. Changes in stem morphology can be induced by this reflected light well before the presence of the neighbouring plants causes measurable shading. This observation has led to the hypothesis that the change in the quality of light reflected from neighbouring plants may serve as an early warning of impending competition for light, allowing the plant to develop an appropriate morphological response before competition for light becomes intense. Similarly, *Portulaca oleracea*, a prostrate species with many scrambling stems which develop from a single rooted origin, grows preferentially away from shade and areas with high FR levels when placed in a heterogeneous light environment (Novoplansky *et al.*, 1990a). That light quality is more important as a signal than light quantity, is shown by the observation that seedlings of *P. oleracea* became recumbent preferentially in the direction of light with a higher R/FR ratio, rather than higher photon flux density (Novoplansky, 1991). A strong tendency to grow away from sectors of the horizon with a low R/FR ratio may also underlie observations by Strong and Ray (1975) showing that a tropical vine on the forest floor grows preferentially towards darkness rather than towards lighter sectors of its horizon that presumably have a low R/FR ratio.

We conclude this section with an exception to the general response of etiolation to low light levels and low R/FR ratios. The extreme densities which can be developed in pure stands of bryophytes produce very low light fluxes close to the surface of a moss turf (During, 1990). Severe stem etiolation might be expected in such colonies, accompanied by substantial shoot mortality, but many mosses do *not* etiolate when given light with a low R/FR ratio, and their mortality rates are low (Bates, 1988). There seems to be a

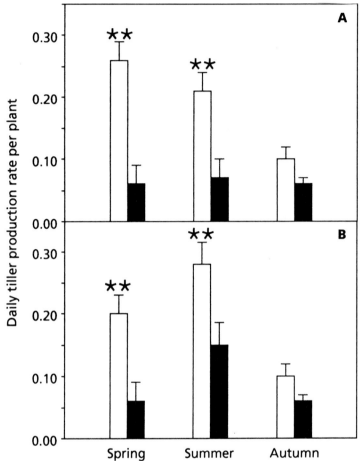

Fig. 2. Mean (± SE) daily tiller appearance rates for plants of (A) *Paspalum dilatatum* and (B) *Sporobolus indicus*. Plants were either grown with additional red light provided to the base of plants by red-light-emitting diodes (open bars) or controls (closed bars). Significant treatment differences ($p < 0.01$) are indicated by asterisks. After Deregibus *et al.* (1985).

clear benefit in the high shoot densities of moss colonies. Because shoot height is very even, these colonies have a smooth surface, producing a boundary layer within which all of the shoots are "immersed". Under such circumstances, evaporation of water from the colony is limited by the boundary layer area of the colony rather than by the total area of leaves. For mosses that lack a cuticle, the advantage of reduced water loss outweighs the disadvantage of reduced irradiance, giving the highest productivity when shoots grow in dense stands.

C. Response to Soil Nutrient Status and Water Supply

Many of the studies relevant to this topic have been made on species with stolons and rhizomes. Experiments by Wareing (1964) and McIntyre (1976) on the rhizomatous *Agropyron (Elymus) repens* showed that many axillary buds do not grow when nitrogen is in low supply, but that provision of nitrogen stimulates the growth of laterals and the production of more branches (McIntyre, 1965; Williams, 1971). Low nitrogen supply causes growing axillary buds to develop as plagiotropic rhizomes, whereas application of nitrogen in various forms converts them into orthotropic tillers (McIntyre, 1967, 1976). Adequate water supply and high humidity promote lateral bud growth in *A. repens* (McIntyre, 1987) and in several other species (Hillman, 1984).

Results from other rhizomatous species confirm that higher nutrient levels stimulate the activation of lateral buds, but the effects on rhizome length are less consistent (Fig. 3). In a field experiment, Carlsson and Callaghan (1990) found that the tundra graminoid *Carex bigelowii* produced rhizomes with a greater mean length at higher fertility. However, this response was only significant for rhizomes initiated on the ventral (lower) side of the parent shoot base; rhizomes emerging from the dorsal (upper) side showed no significant response to fertilization. Schmid and Bazzaz (1992) showed that four old field perennials produced more, and shorter, new rhizomes in fertilized soil compared to unfertilized soil, but the length responses were mostly insignificant (Fig. 3). Internode lengths on rhizomes of *Hydrocotyle bonariensis* did not respond significantly to nitrogen supply (Evans, 1988, 1992), whereas petiole lengths increased. *Carex flacca* formed shorter rhizomes under a higher level of fertilization but the response was not significant; rhizomes of the grass *Brachypodium pinnatum* were significantly shorter under higher nutrient availability (de Kroon and Knops, 1990). Shoot height was also measured in these studies. All graminoid species increased in shoot height under a higher level of fertilization but to a variable degree. The especially vigorous response of *B. pinnatum* is likely to project leaves into the high light zone of the canopy and may contribute to its strong competitiveness under fertile conditions. In contrast, tiller length in *Deschampsia flexuosa* hardly responded to fertility, and tiller weight was greater at low fertility (Foggo, 1989). Foggo (1989) suggested that the formation of a limited number of large tillers may increase the competitiveness of *D. flexuosa* in the shaded, low fertility environments in which this species usually occurs.

Information about plasticity in response to nutrient levels in stoloniferous species corroborates the results described in the previous paragraph. Lateral bud development in the aquatic weed *Salvinia molesta* is almost completely suppressed when it is grown at low nutrient levels, but many second and third order branches are produced when nutrients are abundant (Mitchell and Tur,

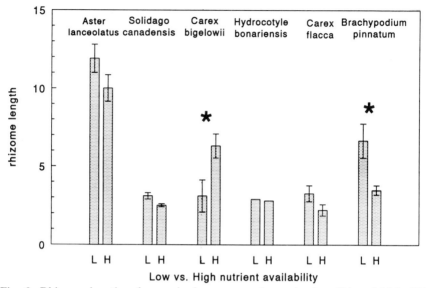

Fig. 3. Rhizome lengths of some herbaceous species under low (L) and high (H) nutrient availability. All values are means (± SE) in cm, unless indicated otherwise below. Significant treatment differences (*p* < 0·01) are indicated by asterisks. Results of *Aster lanceolatus* and *Solidago canadensis* from a single season experiment in common garden soil in which the high nutrient plots received granular NPK fertilizer and the low nutrient plots remained unfertilized (Schmid and Bazzaz, 1992). Results of *Carex bigelowii* from a 3-year field experiment in which the high nutrient plots received an NPK nutrient solution three times a year and the controls received only water (Carlsson and Callaghan, 1990). Data for ventral rhizomes only. Error bars give 95% confidence limits. Results of *Hydrocotyle bonariensis* from single season experiment in sand (outdoors) in which the high nutrient treatment received nitrogen-augmented Hoagland solution and the low nutrient treatment received no nitrogen at all (Evans, 1992). Data for severed plants that received ample light and water. No standard errors are provided. Results of *Carex flacca* and *Brachypodium pinnatum* from a greenhouse experiment in a mixture of river sand and chalk soil fertilized with an NPK nutrient solution which was ten times more concentrated for the high than for the low nutrient treatment (de Kroon and Knops, 1990). Data for plants growing under high light only. *B. pinnatum* rhizome lengths in mm.

1975; Room, 1983; Julien and Bourne, 1986). A variety of species has been shown to branch far more when grown under higher nutrient availability (Ginzo and Lovell, 1973; Slade and Hutchings, 1987a; Hutchings and Slade, 1988; Evans, 1988, 1992; Alpert, 1991). Under such conditions, *Glechoma hederacea* produces shorter stolon internodes with a higher weight per unit length (Slade and Hutchings, 1987a), but internode lengths of *Ranunculus repens* (Lovett Doust, 1987) and *Fragaria chiloensis* (Alpert, 1991) did not respond to nutrients. *Trifolium repens* also branches more freely when nutrients are abundant, but produces *longer* internodes with a higher weight per unit length (Sackville Hamilton, 1982). However, Shivji and Turkington (1989) found similar stolon internode lengths in *T. repens* both in the presence and the absence of nitrogen-fixing bacteria. In many species the total length of primary stolons and their branches is almost independent of nutrient supply, although higher-order branching can be almost entirely suppressed at low nutrient levels (McIntyre, 1965; Ginzo and Lovell, 1973; Slade and Hutchings, 1987a; Hutchings and Slade, 1988).

D. Response to Competition

A number of authors have studied morphological plasticity in situations in which density or the level of competition varied naturally or were manipulated. While the observations can be relied on, there are difficulties in interpreting the morphological responses in terms of environmental conditions. For example, lower competition is not always associated with greater availability of nutrients, and it cannot be assumed that all other environmental variables remain constant as the level of competition changes. Light availability especially is likely to be higher in patches of habitat with lower density of plants. In heterogeneous field habitats there is also the difficulty that carryover effects from parts of the plant growing in good patches to parts growing in poor patches may cause the pure morphological response of the plant to its local growing conditions to be obscured (see Slade and Hutchings, 1987c). The effects described below must thus be treated cautiously.

S. Waite (unpublished) has shown that the internodes of *Ranunculus repens* stolons were significantly shorter on mole-hills devoid of vegetation than elsewhere in a meadow habitat. This observation contrasts with the results of the greenhouse studies on this species which were reported above. Eriksson (1986) found no effect of density upon internode lengths in *Potentilla anserina*. Bishop and Davy (1985) observed more stolon branching in *Hieracium pilosella* grown at lower density. *Lycopodium annotinum* branches more in sites where nitrogen-fixing lichens are present (Svensson and Callaghan, 1988), and *Solidago canadensis* clones produce fewer but longer rhizomes when grown experimentally at greater density (Hartnett and Bazzaz, 1985). However, four old-field perennials, including *S. canadensis*, tended to produce more and shorter rhizomes when grown together with species in which rooting density was higher (Schmid and Bazzaz, 1992). It was suggested that this result may have been due to mechanical resistance of dense root mats to penetration by rhizomes. *Glechoma hederacea* stolon internodes and petioles were significantly longer when grown in competition with uncut swards of *Lolium perenne* than when growing without competition (E.A.C. Price and M.J. Hutchings, 1993, unpublished). The effect on petiole lengths was especially strong. In competition with grass swards cut to 1 cm height, however, stolon and petiole lengths were not significantly different from those on the control clones. However, the total biomass of the G. *hederacea* clones was equally reduced by the competition treatments involving uncut and cut grass swards.

E. Response to Stress

Under some environmental conditions essential resources (light, water, mineral nutrients) may be available in ample quantities but some other factor restraining growth limits their uptake by the plants. Under such circumstances, plants may invoke morphological responses which promote escape from these conditions. Resource-acquiring structures may then be non-randomly placed into microsites from which essential resources can be obtained more efficiently. Examples of such foraging responses are presented in the following paragraphs.

1. Response to Flooding

Flooding prevents adequate diffusion of gases between plants and the atmosphere, often leading to death of the plant. Elongation of immersed shoots out of the water, together with an increase in aerenchymatous tissue in roots during flooding, promotes the diffusion of oxygen from the shoots to the root tips, and is a well-documented response which can promote survival (e.g. Voesenek and Blom, 1989; Laan and Blom, 1990; Waters and Shay, 1990). The elongation response is caused by an increased synthesis of ethylene, and reduced diffusion of ethylene from within the plant to the atmosphere, which results in an increase mainly in cell expansion but in some cases also in cell division (Voesenek *et al.*, 1990). Stem elongation can also be stimulated by external application of ethylene to non-immersed plants (van der Sman *et al.*, 1991), whereas immersion of the plant in a solution of silver nitrate (an inhibitor of ethylene action) inhibits elongation (Voesenek and Blom, 1989). Species with a higher tolerance of flooding show greater responses to shoot immersion, with both developed parts and newly forming parts elongating, whereas flooding-intolerant species show small responses which are confined to newly produced parts. The responsiveness of the plant also alters as it moves through different developmental phases.

2. Response to Burial

Shoot height also increases in a saltatory fashion following burial in sand, as shown by Seliskar (1990) in *Scirpus americanus*. Whether this is due to an etiolation response in tissue formed in the dark during burial, or to an increase in ethylene concentration within the plant, is not known.

3. Response to Salinity

Salzman (1985) found that whereas salinity strongly limited biomass accumulation in *Ambrosia psilostachya*, ramet dispersal was greater under saline than non-saline conditions. She interpreted this result as an expression of a commitment to extensive rhizome spread, which increased both the rate at which plants explored new territory and the probability of encountering more favourable habitat.

4. Response to Touch

When *Arabidopsis thaliana* is stimulated by touch, it develops shorter leaves and longer stems (Braam and Davis, 1990). Calcium ions and calmodulin are involved in transduction of signals from the environment, enabling plants to sense and respond to environmental change. If the responses to touch were induced by neighbouring plants, they could result in stem etiolation, which might overtop these neighbours and reduce physical interference from them.

F. Response to Host Quality by Parasitic Plants

Those plants which are heterotrophic depend on resources which are available in discrete patches (host plants) within a background environment which provides no resources at all. Like foraging animals, parasitic plants thus make decisions about whether or not to exploit patches of resources, and about when to leave them, having extracted a certain amount of resources. Kelly (1990) examined these questions in *Cuscuta subinclusa* and showed that stems coiled more on hosts which gave a higher yield of resource per unit of coil length. Thus, the parasitic plant was capable of assessing the quality of the host prior to resource uptake. The level of investment in coiling upon the host depended on the expected reward, measured in amount of resource acquired. The coiling cues are as yet unclear. No relationship was found between coiling and host nitrogen content, and it was suggested that host flavonoid levels mediated the coiling response. However, in a later study with *Cuscuta europaea* Kelly (1992) did find that stems were more likely to accept (i.e. coil upon) hosts of higher nutritional value.

G. The Mechanism of Morphological Plasticity in Shoots

The shape of a plant both above- and below-ground is a product of the differential activity of meristems. Each actively growing apex completely or partially inhibits the development of lateral structures derived from the same or different apices. This is known as correlative inhibition. In orthotropic shoots the highest lateral buds are usually the first to grow out following decapitation of the apex; in plagiotropic shoots it is the laterals nearest the roots which develop preferentially following this treatment (Hillman, 1984). It is well-documented that major factors which influence plant morphology include the effects of plant growth substances and phytochrome on apical dominance and internode elongation. Here we briefly review what is known about these regulation mechanisms.

Matthysse and Scott (1984), Tamas (1987) and Woodward and Marshall (1988) have provided succinct accounts of the effects of plant growth substances on axis elongation and lateral bud growth on stems, tillers and roots. Auxins, cytokinins and gibberellins appear to be the most important of the plant growth substances in shaping gross morphology. Auxins are synthesized mainly in shoot meristems and expanding leaves, cytokinins in the roots, and gibberellins in both shoots and roots. Auxins impose apical dominance in shoots and inhibit the growth of tillers and lateral buds on stems, but they promote branching in the root system. Cytokinins and their analogues promote the growth and development of shoot branches and tiller buds, and appear to impose apical dominance in the root system (Johnston and Jeffcoat, 1977; Sharif and Dale, 1980; Jinks and Marshall, 1982; Tamas, 1987), but the evidence for this is debatable (Hillman, 1986; Roberts and Hooley, 1988). Gibberellins promote main stem development and reduce the development of tiller buds (Jewiss, 1972; Johnston and Jeffcoat, 1977; Isbell and Morgan, 1982; Woodward and Marshall, 1988). However, their application to root systems may result in the release of buds from apical dominance (Watson *et al.*, 1982), probably because root apex growth is inhibited (see also Tamas, 1987). Matthysse and Scott (1984) suggest that auxins and cytokinins are important in controlling rapid responses by plants to short-term, unpredictable fluctuations in growing conditions, whereas gibberellins govern responses to gradual changes, such as those taking place over the course of a season.

There are now strong arguments for auxins, and in particular for the active substance indole-acetic acid (IAA), being a signal which directly or indirectly suppresses the activity of lateral meristems (Hillman, 1986; Tamas, 1987; Roberts and Hooley, 1988; Tamas *et al.*, 1992). Cytokinins transported from roots to shoot are probably needed for the activity of lateral buds, but the expression of the activity may be under apical control (Tamas, 1987). The patterns of water and nutrient distribution throughout the plant play a role in apical dominance, but possibly *via* hormone-directed metabolite transport (Hillman, 1984; Salisbury and Marinos, 1985). Shein and Jackson (1972) suggested that the growth of laterals and main stems depends upon a complex hormone balance which could be modified by several internal and external factors including water and nutrient status, light regime and plant age. A delicate interaction may exist between auxins, gibberellins and cytokinins, producing the degree of apical dominance which is expressed in the phenotype (Jewiss, 1972; Woolley and Wareing, 1972; Woodward and Marshall, 1988).

Scarcity of nutrients and water normally result in stronger apical dominance (Gregory and Veale, 1957; Phillips, 1975; Hillman, 1984; Salisbury and Marinos, 1985) and there is a positive relationship between both soil nutrient availability (especially nitrogen supply) and water supply, and cytokinin synthesis (Aung et al., 1969; Banko and Boe, 1975; Menhenett and Wareing, 1975; McIntyre, 1976; Qureshi and McIntyre, 1979; Menzel, 1980; Lovell and Lovell, 1985). The effect of nitrogen application on the propensity of rhizome buds to grow out as orthotropic tillers rather than plagiotropic rhizomes, as reported in a previous section, may be due to a combined effect of low levels of IAA and high levels of gibberellic acid (Bendixen, 1970; see Hutchings and Mogie, 1990). However, in general there is little information about the effects of environmental conditions upon the concentrations of plant growth substances, except for hormones that seem to be less important for the control of apical dominance, such as abscisic acid (e.g. Davies and Zhang, 1991). Together with the fact that most of the work on this subject has been carried out with a restricted group of legume and grass species, this lack of information hinders an interpretation of the effects of the environment on morphology in terms of the actions of growth substances.

Changes in shoot morphology in response to light quality operate through photochemical changes in the state of phytochrome, the main sensor of shade light quality. The degree to which a stem elongates depends on the ratio between P_{fr}, the active form of phytochrome, and the total phytochrome content, P_{total}. High rates of stem extension take place when no P_{fr} has been formed, as in seedlings growing towards the soil surface immediately after germination. The low R/FR ratio of light which has passed through a plant canopy reduces the photoequilibrium, P_{fr}/P_{total}, compared with plants in the open, thus promoting stem elongation. The shade cast by plant canopies in terrestrial communities causes changes in the value of the R/FR ratio over a range within which the photoequilibrium is highly unstable. The ratio of P_{fr} to P_{total} can thus provide a very sensitive indication of the presence of shade and the degree of shade cast (Smith, 1982; Holmes, 1983). It is, however, far more stable at R/FR ratios above 1·5. The R/FR ratio increases rapidly with passage of light through water, and for this reason the photoequilibrium is unlikely to be sensitive enough to operate as a sensor of shade cast by plant leaves in aquatic habitats (Smith, 1982). Artificial manipulation of the ambient R/FR light regime can dramatically change plant morphology, as already discussed (Casal et al., 1985; Deregibus et al., 1985). Imposition of very high R/FR ratios promotes vigorous branch production by roses and tomatoes even when they are grown at densities at which severe mutual shading occurs (Novoplansky et al., 1990b). Flowering and fruiting can also be increased spectacularly by such treatment.

Phytochrome does not seem to affect internode elongation directly. It appears to act indirectly *via* gibberellin. Localized shading of stems promotes stem elongation by inducing greater rates of cell division and expansion (e.g. Garrison and Briggs, 1972 for *Helianthus*), and these responses have traditionally been regarded as an expression of altered gibberellin metabolism. Indeed, when gibberellic acid is applied to *Xanthium* internodes, rates of cell division and elongation increase and the growth rate in internode length more than doubles compared to control plants (Maksymowych *et al.*, 1984). Recent research suggests that in darkness or under low R/FR ratios phytochrome alters the sensitivity of the stem tissue to gibberellic acid, thus enhancing the ability of internodes to elongate (Garcia-Martinez *et al.*, 1987; Martinez-Garcia and Garcia-Martinez, 1992; Ross and Reid, 1992).

Photon flux density also affects shoot morphology, but the mechanisms by which the effects are exerted have received little attention. It is thought that some species sense changes in photon flux density *via* blue light photoreceptors (Holmes, 1983). The responses, which again result in increased stem extension rates, are fast, but in many cases limited to the earliest stages of growth (the hypocotyl stage). It is believed that such responses may allow the young plant to avoid small obstacles such as stones and mounds of soil (Holmes, 1983).

H. Preferential Shoot Proliferation in Favourable Patches

Most habitats are heterogeneous, consisting of patches which differ in resource availability. In low vegetation, above-ground patchiness may be imposed by the spatial arrangement of dwarf shrubs and persistent clumps of perennial herbs (e.g. Eriksson, 1986; van der Hoeven *et al.*, 1990), and modified by micro-topography and grazing (Gibson, 1988a; Svensson and Callaghan, 1988). Forest canopies are patchy because of gaps and because tree species with different architectures and light transmission properties occur together. Here we review studies which have looked at the selective placement of shoots, and of the leaves on shoots, within the patchy environment, rather than at morphological plasticity itself.

The probability of buds growing out, and of leaves and branches being produced at any node, varies with local site quality. For example, Jones and Harper (1987a,b) showed that fewer buds developed in parts of *Betula pendula* crowns with high local shoot densities and presumably low light. A similar regulatory effect was seen regardless of whether crowding was caused by shoots of the same plant or shoots of a different plant. As a consequence, trees grown in isolation develop symmetrically, but when grown close together they show a strong tendency to grow more on those sides which face away from neighbours (Franco, 1986; Young and Hubbell, 1991). Weiner *et al.* (1990) showed that branch distribution up stems differed between isolated and competing plants of *Impatiens pallida*. Isolated plants produced branches from lower on the main stem, and more second-order branches, than crowded plants. The vertical distribution of first- and second-order branches on isolated plants was the same, but competing plants produced second-order branches only from their highest first-order branches. Suppressed competing plants produced branches and leaves only at the top of their stems, where access to light would be greatest. Thus, probability of formation of branches depended on the favourability of local conditions.

The capacity to concentrate in light patches in a heterogeneous environment appears, at least in bryophytes, to be greater for species with a high relative growth rate under uniform conditions, and for species of open rather than shaded habitats (Rincon and Grime, 1989). However, During (1990) suggested that this observation may be a consequence of the differences in growth form between the species that Rincon and Grime used (acrocarps *versus* pleurocarps), rather than due to differences in growth rate or habitat.

When light patchiness is created by the placement of leaves and shoots of competing plants, the responses of species to environmental patchiness may have consequences for the competitive interactions between species. Recent experiments have shown that the success of the most dominant species of herbaceous phanerogams in gaining access to good habitat patches appears to be primarily due to their ability to rapidly develop a large mass of leaves (and roots) in those patches (Campbell *et al.*, 1991a,b). However, the percentage of leaf and shoot weight increments which individual species achieve in light-rich patches under heterogeneous experimental conditions is negatively correlated with their proportional contribution to the total biomass of a mixture of competing species. In other words, subordinate species show greater *precision* than dominant species in locating the more favourable patches (Campbell *et al.*, 1991b). It seems therefore that dominant species produce large leaf and root masses in both good and poor quality habitat patches, while subordinate species have a greater capacity to select the better patches. As this placement is achieved by morphological plasticity, it may be hypothesized that subordinate species may have more plastic shoot architectures than dominant species. To our knowledge data with which to test this hypothesis are lacking. Greater selectivity in the placement of resource-gathering structures in higher quality habitat patches may be an important attribute enabling less competitive species to avoid competitive exclusion from mixed communities.

Many species with plagiotropic stems do not shorten their internodes under more favourable conditions (Figs. 1, 3). Such species would therefore not be expected to concentrate the placement of their ramets in favourable patches as a result of morphological plasticity. Even stoloniferous species with qualitatively similar internode length responses to those of orthotropic shoots are inefficient at placing ramets in resource-rich patches, as demonstrated recently by Birch and Hutchings (1994). Clones of *Glechoma hederacea* grow-ing under heterogeneous nutrient supply produced shorter internodes in nutrient-rich patches, but the density of ramets in these patches was hardly altered by this response. This result is confirmed by model simulations in which the distribution of ramets of *G. hederacea* in a patchy environment was calculated on the basis of its morphological responses (Cain, 1994). The degree of plasticity of plagiotropic stems therefore may not be sufficient to result in a marked concentration of ramets in favourable patches of habitat, in contrast to the predictions of the models of Sutherland and Stillman (1988, 1990).

The speed with which an internode ceases to elongate has great significance for foraging. Once internode elongation has stopped, the node at its apex is fixed in position and leaves and (in plagiotropic stems) roots are produced from this site. Rapid fixation of the position of the node allows resource uptake to commence quickly. For example, Birch and Hutchings (1992) and Dong (1993) found that stolon internodes in two stoloniferous species reached their final lengths rapidly. When a ramet at the end of a fully elon-gated stolon internode is in a nutrient-rich patch, *Glechoma hederacea* exploits the nutrients by very rapid development of roots from that ramet (Birch and Hutchings, 1994); when the ramet is in a poor patch, however, its roots begin to develop later, and root growth is much less profuse. In marked contrast to the rapid completion of internode growth, a protracted period of growth of petioles, and of the leaf lamina, allows the plant to make ongoing responses to unfavourable or deteriorating light conditions long after the position of the ramet has been fixed on the soil surface (Birch and Hutchings, 1992; Dong, 1993).

I. Conclusions

The data show that all orthotropic stems and stem analogues (such as leaf sheaths) respond similarly to variations in light quantity and quality; under a more favourable light regime branching increases and internode elongation decreases. The *degree* of plasticity differs between species, however, and there is evidence that the response is greater in shade-intolerant than shade-tolerant species. This response to canopy shade provides orthotropic shoots with some capacity to overgrow competitors for light, but this will be of little value for species of short stature in habitats in which most of the shade is cast by canopy trees.

While plagiotropic and orthotropic stems respond similarly to light with respect to branching by lateral meristems, comparison of the responses of internode lengths is less straightforward. Some species with plagiotropic stems exhibit reduced internode lengths under more favourable light conditions, but lack of a response, and even an increase in internode lengths

have been reported for others. The responses to variations in nutrient availability are equally inconsistent. Even when species with plagiotropic stems do
shorten their internodes under more favourable nutrient conditions, the
degree of plasticity which has been reported appears to be too small to
concentrate ramets efficiently in the high quality patches in the habitat. An
understanding of the differences in response to environmental quality between orthotropic and plagiotropic stems is hindered by a lack of information
about the interactive effects of the environment and plant growth substances
on stem elongation.

V. THE FORAGING ACTIVITIES OF ROOTS

Root morphology can be described in terms of the density of formation of
laterals and the rate of root extension, and also in terms of the topology of the
branching system. Both types of description are used here to review the
plastic responses of roots to variation in the availability of soil resources. This
material is supplemented with information on the proximal mechanisms that
control the responses of roots. In a final section we investigate the effectiveness of roots at exploiting patches of high nutrition within the soil volume,
and the consequences for plant growth.

A. Plasticity in Root Branching and Elongation

The distinction between resource-acquiring structures and spacers is not as
clear-cut in roots as it is in stems. Edaphic resources (water, nutrients) are
taken up all along the root surface, but the younger, finer roots and the
terminal zones bearing root hairs are the most important areas for these
activities (Passioura, 1988). Root extension increases both the area of the
absorbing surface and the volume of soil explored. Local root surface area
can be increased by lateral root branching.

Unlike stems and stem homologues, roots do not have a segmented metameric construction (Steeves and Sussex, 1989; but see Barlow, 1989). In
angiosperms, lateral root primordia originate in the pericycle—the outermost
cell layer of the stele—and arise in ranks (orthostichies) which are usually
positioned opposite the protoxylem poles of the vascular system (Lloret et al.,
1988; Steeves and Sussex, 1989). The longitudinal spacing of lateral roots
usually has some degree of regularity. The number of lateral roots in each of
the ranks can vary widely. Lateral roots are usually initiated at a certain
distance from the root tip and branching does not occur on older parts of
roots (Drew et al., 1973; Lloret et al., 1988), although dormant root primordia
as well as the formation of 'adventive' roots are reported for some species
(see Barlow, 1989).

The fact that branching of roots may be stimulated under higher levels of soil nutrition was recognized as early as the late 1800s (see Wiersum, 1958). Weaver (1919) produced numerous diagrams showing that roots proliferated selectively in pockets of soil that were rich in soil resources. Drew *et al.* (1973) were among the first to experimentally study the effects of local nutrient enrichment on root morphology. Single root axes of barley (*Hordeum vulgare*) were grown into three compartments in which the concentration of nutrients could be controlled separately. High nitrate concentration in a given compartment promoted the formation of more first and second order laterals per unit of primary root length within that compartment, and greater lateral root extension (Fig. 4). These responses were significant irrespective of the level of nutrition experienced by other parts of the root system. In contrast to the growth of the lateral roots, the rate of extension growth of the main root axis was *not* affected by nitrate availability in the compartment in which it was growing. These results were essentially corroborated by later work (Drew, 1975—see Fig. 4; Drew and Saker, 1975, 1978; Eissenstat and Caldwell, 1988a; Granato and Raper, 1989; Jackson and Caldwell, 1989; Burns, 1991). Caldwell and his co-workers examined the extension growth of roots in three cold desert perennials, making no distinction between main root axes and laterals. Local enrichment of the soil by injection with a nutrient solution significantly stimulated root elongation at different times of the year (Eissenstat and Caldwell, 1988a). Species differed markedly both in the degree to which root growth was stimulated and in the time between nutrient application and response. *Agropyron desertorum* showed a fourfold increase in the relative growth rate of root length within one day, while *Artemisia tridentata* and especially *Pseudoroegneria spicata* responded less vigorously (Jackson and Caldwell, 1989). In the latter species extension growth was not affected until several weeks after nutrient application. It is likely that the root growth of *A. desertorum* also responds more strongly to water availability than the root growth of the other two species, because roots of *A. desertorum* can more rapidly invade disturbed soil patches which are not enriched in nutrients (Eissenstat and Caldwell, 1989), and because this species is a stronger competitor for water (Eissenstat and Caldwell, 1988b).

The mobility of nutrient ions in soil may differ by several orders of magnitude. Diffusion coefficients decline from nitrate to ammonium to phosphate (Nye and Tinker, 1977; Caldwell, 1988). Bray (1954), Harper (1985) and Fitter (1991) hypothesized that root morphological characteristics may depend on the diffusive properties of the nutrient for which the plant is foraging. Prolific branching producing a high root density would seem to be required for effective acquisition of immobile ions such as phosphate. For nitrate a less intensive branching pattern should suffice. However, there is little evidence to show that the morphology of roots depends on the type of nutrient which is in short supply. Drew (1975) found that the nitrate, ammonium and phosphate anions were equally effective in inducing lateral

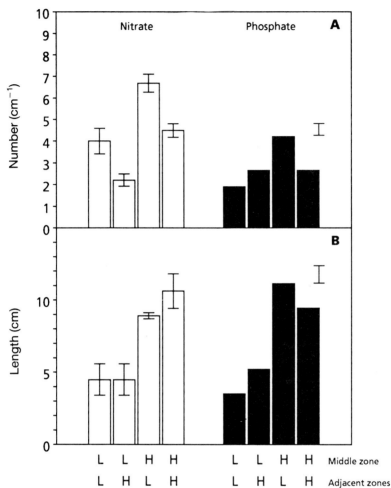

Fig. 4. Effects of nitrate and phosphate supply on (A) the numbers of lateral roots per cm of main root axis, and (B) the lengths of individual lateral roots in barley. Main root axes were divided into three zones and nutrients were supplied independently to each of these zones. Data given are those for the first-order laterals that developed in the middle zone. This zone experienced either a low (L) or a high (H) concentration of nitrate or phosphate. Adjacent rooting zones also grew in either low or high nutrient solution. In the nitrate experiment plants were grown hydroponically. In the phosphate experiment they were grown in sand. Nitrate data given as means ± SE, phosphate data as means with separate bars showing the LSD at the 5% level. After Drew *et al.* (1973) (nitrate) and Drew (1975) (phosphate).

root formation and growth (see Fig. 4), while potassium availability had no effect. Wiersum (1958) grew roots over ion adsorbing resins and reported a larger effect of nitrate than of other ions on root branching. However, he also failed to demonstrate an effect of cations. In contrast, Passioura and Wetselaar (1972) suggested that the concentration of roots of wheat in the soil corresponded more closely with the distribution of ammonium than of nitrate, while roots of birch displayed no preference for either form of nitrogen (Crabtree and Bazzaz, 1992). Jackson and Caldwell (1989) showed that an increase in root extension could be induced by N–P–K fertilization as well as by N and P alone, and that the magnitude of the response was a function of the level of enrichment. Ion specificity in the responses of roots may be lacking because in reality roots always forage for multiple resources simultaneously.

In comparison with the diameter of stems, root diameter shows little plasticity. In many species, specific root length (SRL, root length per unit root dry weight) does not change significantly as a function of nutrient availability. In species which do respond, SRL is always higher under more nutrient-poor conditions (Fitter, 1985; Robinson and Rorison, 1985, 1987, 1988; Boot and Mensink, 1990; Hetrick et al., 1991). Thus, when nutrients are scarce, thinner roots are formed, which explore the soil more efficiently (Nye, 1973) at lower cost.

B. Plasticity in Root Topology and Architecture

A different way of examining root morphology is the topological analysis of branching pattern introduced by Fitter (Fitter, 1986, 1987, 1991; Fitter et al., 1991). Roots are divided into a number of "links", where a link is defined as a root segment between either two branch junctions or nodes (internal link) or between a branch junction and an apex (external link). The branching pattern of a root system is typified by a number of parameters such as the total number of external links and the maximal and total path lengths of the external links to the base (origin) of the root system. Additional (nontopological) parameters include link length and branch angles. Using this approach, Fitter (1986) described the plasticity of roots of Trifolium pratense in response to water availability. Under drought the root system was characterized by linear growth and internal branching giving rise to a herringbone type of structure. Abundant soil moisture resulted in predominantly external branching and a more dichotomous structure. Increasing water availability increased the external link length (indicating that branch initiation occurred further behind the growing tip of the root), but interior link length was unaffected. Subsequent studies with a variety of other species showed roughly similar responses of root morphology to different levels of nutrient availability (Fitter et al., 1988; Fitter and Stickland, 1991; Fitter, 1993, 1994).

Internal and external link lengths were usually lower when nutrient availability was greater but some species showed the opposite responses (Fitter and Stickland, 1992). In general, the levels of plasticity shown by individual species were small compared to the differences in root topology between species. A number of species (including all of the grass species studied) did not respond significantly to different levels of edaphic resources. Species in the same genus frequently behaved in a similar way, suggesting that root architecture is partly determined by phylogeny (Fitter and Stickland, 1991).

Fitter *et al.* (1988) showed that species from poor soils generally had a more herringbone root structure with longer links than species from rich soils, but this conclusion was only partly corroborated by the results of a subsequent study (Fitter and Stickland, 1991). The larger soil volume occupied by the roots of species from relatively nutrient-poor habitats may be important for acquiring scarce resources. Hence, species-specific root architecture may have greater functional significance than morphological plasticity (Fitter, 1991).

Topological characterization of the entire root system appears to be less sensitive than measurements of root branching and root extension rate for reflecting responses to variation in resource supply (see previous section). Thus, topological analysis carried out by itself may obscure some of the morphological plasticity of which roots are capable.

C. Effects of Infection by Mycorrhizal Fungi

Infection by arbuscular mycorrhizal (AM) fungi may significantly alter root morphology and plasticity. Mycorrhizal hyphae play an important role in the uptake of some nutrients, and may effectively take over the role of the fine roots and root hairs. Inoculation of roots with AM fungi has been shown to reduce root branching, lateral root extension and the development of root hairs (Steeves and Sussex, 1989; Hetrick *et al.*, 1991). The morphology of infected roots responds weakly to differences in nutrition compared to uninfected roots (Hetrick *et al.*, 1991). St John *et al.* (1983a) grew inoculated plants in pots with soil patches of different quality. It was shown that total hyphal length was significantly higher in patches rich in organic matter compared with patches containing only sand. This was probably caused by increased branching of hyphae in the nutrient-rich patches. Thus, branching of mycorrhizal hyphae seems to respond to local variations in soil nutrients in a similar way to the branching of roots reported above (St John *et al.* 1983b; see also section VI on the foraging activities of fungi).

D. The Mechanism of Morphological Plasticity in Roots

The root tip exerts an inhibiting effect on the development of lateral roots, comparable with the way in which the shoot apex suppresses the development of lateral structures. Hence, root branching is stimulated by removal or death of the root tip (Lloret *et al.*, 1988; Callaway, 1990). It has been suggested that, in principle, every cell in the pericycle is capable of lateral root formation (Barlow, 1989). The regular placement of lateral roots seems to be the result of the interplay between an inhibiting effect on branching caused by the apex and a promoting effect which develops in mature regions of the root (Lloret *et al.*, 1988). In addition, existing lateral roots may have an inhibiting influence on the formation of new laterals (Steeves and Sussex, 1989).

The general effects of plant growth substances on apical dominance have been discussed in the section on the foraging activities of shoots. Much less is known about the regulation of apical dominance in the root than in the shoot. Granato and Raper (1989) suggest that, when nitrate levels in the substrate are low, inhibition of lateral root formation may be due to a reduction of nitrate reductase activity in the root tip which in turn reduces the concentration of organic nitrogen compounds. This may explain the responses of branching to availability of nitrate, but not to other nutrients. It is not clear why lower levels of nutrition usually reduce the extension growth of lateral roots but not that of the main axis. One reason may be that the primary root apex is stronger than the lateral root apices as a sink for carbohydrates and reduced nitrogen compounds (Granato and Raper, 1989).

E. Preferential Root Proliferation in Favourable Patches

In many habitats the spatial distribution of water and nutrients is profoundly heterogeneous even at a scale as small as a few centimetres (Frankland *et al.*, 1963; Hall, 1971; Gibson, 1988a; Svensson and Callaghan, 1988; Lechowicz and Bell, 1991). For example, Jackson and Caldwell (1993) recently found that nitrate and ammonium concentrations in the soil of a cold desert habitat varied more than tenfold at a scale of 50 cm, and that there was still a threefold variation at a scale of 3 cm. Soil patches of different quality may be created at these scales by abiotic factors (soil type differences, soil depth, micro-topography) as well as by biotic factors such as treefalls and stemflow in forests (Lechowicz and Bell, 1991) and persistent turfs of perennial plants in grasslands and deserts (Gibson, 1988a,b; Hook *et al.*, 1991; Jackson and Caldwell, 1993). How effectively are roots placed in the richer patches within the soil volume, and to what extent can the resources in these patches actually be acquired? What fraction of the growth achieved under a homogeneous supply of soil resources can be realized when similar amounts of resources are patchily distributed?

Caldwell *et al.* (1991a) locally enriched soil in a cold desert habitat and showed that within 3 weeks *Agropyron desertorum* had developed a higher root density in patches of soil supplied with nutrients than in control patches that received only water (Fig. 5A). This concentration of roots was probably

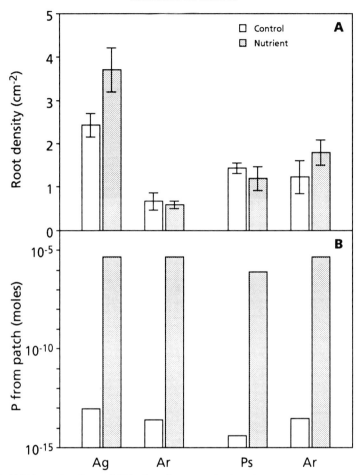

Fig. 5. (A) Root density and (B) phosphate uptake in a field experiment in which soil cores were either enriched with an NPK nutrient solution or supplied with distilled water (controls). Roots of two pairs of species were allowed to grow into the cores simultaneously, in two combinations: *Artemisia tridentata* (Ar) competed with either *Agropyron desertorum* (Ag) or with *Pseudoroegneria spicata* (Ps). The experiment lasted for 5 weeks; root data came from a harvest after 3 weeks. Root density is the number of root intersections in cut planes of soil cores. Note log scale in B. After Caldwell *et al.* (1991a,b).

the result of vigorous root extension growth in response to nutrient enrichment, as reported for this species by Jackson and Caldwell (1989). The uptake of phosphorus was orders of magnitude higher from the enriched than from control soil patches (Caldwell *et al.*, 1991b) (Fig. 5B). This was partly due to the aggregation of roots, and partly a result of higher rates of phosphorus uptake per unit of root length, in enriched patches of soil (Caldwell *et al.*,

1992). In contrast to *A. desertorum*, two other species, *Pseudoroegneria spicata* and *Artemisia tridentata*, proliferated roots equally in fertilized and control patches (Fig. 5A), but exhibited higher phosphorus uptake rates in the enriched patches. In a subsequent study (Van Auken *et al.*, 1992) it was shown that *Artemisia tridentata* and *Agropyron desertorum* were more capable than *P. spicata* of exploiting distant soil microsites enriched in phosphate. This was probably a result of the higher root extension growth of the former species.

De Jager and Posno (1979) compared the root growth of three *Plantago* species in a split-root experiment in which phosphorus was withheld from most of the root system. *P. major*, a species from relatively nutrient-rich sites, proliferated roots more rapidly in soil compartments supplied with ample phosphorus than *P. media*, a species from relatively nutrient-poor habitats. The response of *P. lanceolata* was intermediate to the other two species.

Crick and Grime (1987) grew plants of *Agrostis stolonifera* and *Scirpus sylvaticus* in arenas in which their roots could be equally distributed between several compartments. The strength of the nutrient solution provided in each compartment could be controlled independently. In addition to uniform high and low nutrient treatments to all compartments, nutrients were applied either in a stable configuration to particular compartments or as pulses applied consecutively to selected compartments. After 27 days of growth, root dry weight of *A. stolonifera* was markedly higher in compartments with a high nutrient supply for the duration of the experiment than in compartments with a low nutrient supply. In contrast, roots of *S. sylvaticus* did not accumulate preferentially in high nutrient compartments. Crick and Grime (1987) interpreted the differences in responsiveness between the species as adaptations to habitats with fertile (*A. stolonifera*) and infertile soils (*S. sylvaticus*). More recently, Grime *et al.* (1991) were unable to repeat this result with seedlings of 11 species grown in pots of sand divided into sectors with low and high nutrient supply, although there were no physical barriers between sand of different quality in this experiment. The extent to which roots developed preferentially in the richer sectors of the soil was similar for species from both fertile and infertile habitats. Fast-growing species developed a higher biomass than slow-growing species not because they placed more roots in nutrient-rich sectors, but probably because their nutrient uptake rates were higher (see also Section VIII.A and Fig. 9).

An experimental study referred to earlier, by Drew *et al.* (1973) analysed the effects of local nutrient enrichment upon root morphology in barley. It also analysed the effects of local nutrient enrichment upon the localization of root growth. It involved subjecting different fractions of the root systems of plants to either high or low nutrient supply provided in separated compartments. When one-third of the entire root system received a nutrient-rich solution, total lateral root length per unit of length of the primary axis was ten times higher, and the total root biomass six times higher, in the high nutrient compartment than their values in the low nutrient compartments. Later in the experiment, when the lateral roots had grown out, whole plant relative growth rate (RGR) under localized supply of nutrients approached the value attained by control plants growing under homogeneous nutrient supply (Drew and Saker, 1975). When phosphate was supplied to 2 cm of the main root axis—a fraction amounting to only a few per cent of its total length—whole-plant RGR was more than 80% of its value in control plants in which the whole root system received phosphate. When applied to 4 cm of the main root axis, the RGR for the whole plant was similar to that of controls. There were increased lengths of lateral roots, and increased phosphate absorption rates per unit of root length in the enriched compartment, compared with both other parts of the root system in treated plants, and the root system of control plants (Drew and Saker, 1975). Similar results have been obtained in split-root experiments with maize (de Jager, 1982) and lettuce (Burns, 1991).

Birch and Hutchings (1994) grew clones of *Glechoma hederacea* in flats in which a given amount of potting compost was either uniformly mixed with sand or confined to a circular patch occupying only 11% of the total area of the flat. Due to vigorous rooting within the high nutrient patch, total clone biomass in the heterogeneous treatment was more than twice as high as in the treatment in which the nutrients were uniformly distributed (see Fig. 8).

The results of these studies show that high nutrient patches can be rapidly occupied by species with morphologically responsive root systems. For such species whole plant growth in a patchy environment can be similar to, or even higher than in an environment in which the same supply of resources is uniformly distributed.

F. Conclusions

Roots of many species tend to aggregate where resources are abundant in the soil. The aggregation is the result of an increase in the formation and growth of lateral roots in response to local enrichment. The extension growth of the main root axis may be unaffected by the nutrient status of the patches in the soil, suggesting that the soil is continuously searched, while plasticity in lateral root formation is responsible for the exploitation of the resources in those favourable patches that are encountered. Not all species have equal plasticity in root morphology; the roots of some species do not seem able to grow selectively into favourable patches. The results of some studies suggest that more competitive species and species from soils which are relatively rich in nutrients have higher levels of root plasticity than less competitive species and species from poor soils, but conflicting results have been reported. It has been suggested that large unresponsive root systems are better at utilizing short, unpredictable nutrient pulses, such as those which occur in unproductive habitats, than are fast growing roots with high morphological plasticity (see Section VII).

VI. THE FORAGING ACTIVITIES OF FUNGI

Rayner and Franks (1987) have drawn parallels between the foraging behaviour of the mycelia of fungi and the social organization of colonies of ants. Both fungi and ants must mobilize units of structure which search for resources, from a food base upon which they are initially dependent, into new habitat. Parts of the newly explored habitat may lack the resources which are sought. Achievement of the objective of finding essential resources, and making the best use of them, necessitates coordination and cooperation between the parts of the structure which find the resources and the established food base.

There is strong evidence for coordinated foraging activities in fungi, at least in species from stable environments (Dowson et al., 1986, 1988, 1989a,b). Dowson and his co-workers placed wood blocks ("food bases") inoculated with mycelial cord-forming basidiomycetes in plastic dishes filled with sand. One or more sources of nutrition, such as wood blocks, beech leaves or pine cones, were placed in the dishes at some distance from the food base. Both verbal descriptions and photographic evidence are provided of the ability of the mycelia to selectively forage for these sources of nutrition (Fig. 6). Once such a source has been discovered by the radially expanding fungus, the mycelia connecting the food base to it are stimulated to thicken, presumably to enable transport of more materials from the resource to the rest of the fungal body. Many mycelia which were not originally growing towards the resource are seen to reorientate towards it, and those mycelia which are still unsuccessful in discovering sources of nutrition eventually regress. As a result the fungus stops growing radially from its source, showing instead preferential growth towards the resource-rich areas (Fig. 6). The degree of change in growth form was greater when more valuable sources of nutrition were discovered. Regardless of the number of sources of nutrition offered, similar behaviour was always observed (Dowson et al., 1989b), with the fungus redirecting its growth towards newly discovered sources of nutrition and proliferating upon them. The mechanisms which coordinate such foraging activities in fungi are not known.

Fig. 6. Outgrowth pattern of *Hypholoma fasciculare* from a large inoculum and its response to contact with an equal sized bait after (a) 8 days, (b) 20 days, (c) 31 days, showing the regular margin present on contact with the bait, and thickening of connective cords, (d) 51 days, showing irregular morphology at the edge of the mycelium, (e) 68 days, showing outgrowth from the bait and regression of non-connective mycelium, (f) 85 days. From Dowson et al. (1986).

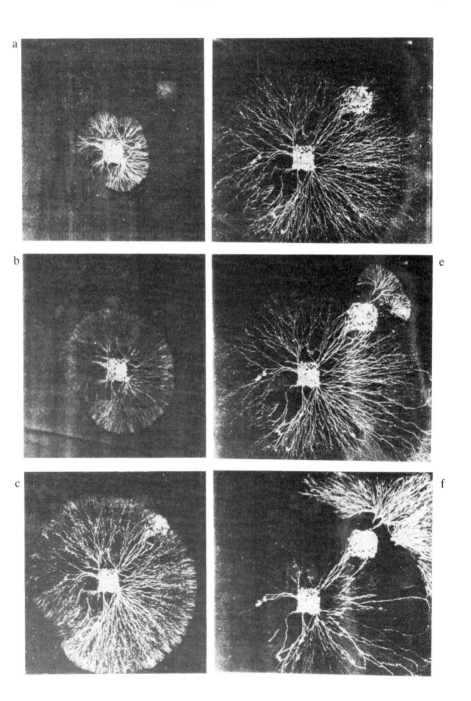

There are marked differences in the growth forms of fungi. Species which are specialized in their choice of substrate exhibit strong rhizomorphic outgrowth, with marked apical dominance, whereas those with less specialized requirements have a more diffuse growth pattern with less apical dominance (Rayner and Franks, 1987). The patterns of expansion growth also differ. Among the fairy ring fungi the species *Clitocybe nebularis* grows in a nutritionally rich habitat, exploiting recently fallen leaf litter. It has a rapid extension rate and a sparse mycelium, it vacates occupied ground rapidly, without exploiting all of the resources on offer, and it is uncompetitive in interactions with other species. Several other woodland litter decomposers can be contrasted with this, particularly *Marasmius wynnei* (Dowson *et al.*, 1989a), which has very dense, slowly-moving mycelia which utilize a high proportion of the resources on offer at any site. It would be very interesting to investigate the extent to which these species differ in their capacity to locate discrete sources of nutrition in their vicinity and the levels of resource extraction they can accomplish in environments with different patch structure.

VII. WHOLE-PLANT COORDINATION OF FORAGING

In our discussion of the foraging activities of shoots and roots we have presented information about the morphological responses of plants and plant parts to differences in resource supply, and the capacity of plants to select favourable habitat patches for the placement of resource-acquiring structures. Many resources acquired locally are transported from the site of acquisition to other locations, so that the benefits of acquisition can appear elsewhere in the structure of the plant. We now discuss the extent to which the morphological responses of the plant to its environment are influenced by the resource status of the whole plant. We also consider whether the level of morphological plasticity expressed locally is influenced by conditions experienced by other parts of the plant located in habitat of different quality. Local patterns of growth could be profoundly affected by transport of resources and growth substances between connected modules. Such integration could alleviate local resource shortage (Pitelka and Ashmun, 1985; Marshall, 1990), and this might enable the morphology of plant parts growing under low resource supply to more closely resemble that of parts growing under high resource supply. We finally discuss resource uptake and integration in clonal plants in patchy habitats in which the distributions of different resources are negatively correlated in space.

A. Effects of Integration on Local Foraging Responses

In Section V we presented evidence showing that local nutrient enrichment of the soil may promote local formation and extension of lateral roots. Local conditions determine *where* lateral root growth is promoted (Drew *et al.*, 1973; Drew and Saker, 1975), but the *magnitude* of the local response depends on the conditions experienced by the rest of the root system. An experiment by Drew (1975) illustrates this well. He subjected roots of barley plants to either a uniform or localized nutrient supply. Part of the root system given a high phosphate supply produced more and longer lateral roots when the rest of the root system was receiving low phosphate rather than high phosphate (see Fig. 4). This suggests that the local morphological response is stronger when phosphate is more limiting to the plant. In addition, lateral root formation under low local phosphate availability was higher when the rest of the root system was given high compared to low phosphate supply (Fig. 4), suggesting that the effects of locally abundant phosphate on root morphology were carried over into adjacent parts of the root system growing under phosphate deficiency. Broadly similar effects were produced when the nitrate and ammonium supply to different sections of the root system were varied (Drew, 1975). However, effects were less clear for nitrate (Drew *et al.*, 1973; see Fig. 4). Similar effects of integration in root systems were observed in fertilization experiments with three *Plantago* species (de Jager and Posno, 1979) and with pea seedlings (Gersani and Sachs, 1992), but not in a similar study with maize (de Jager, 1982). When local nutrient availability allows for rapid root initiation, the root may become the preferred sink for carbohydrates and auxins provided by the shoot, inducing enhanced root development in sections with higher nutrient supply (Gersani and Sachs, 1992). Morphological responses to local water stress may also be affected by integration. For example, Evans (1992) grew the rhizomatous herb *Hydrocotyle bonariensis* under different levels of water availability. Large root systems were produced by ramets in response to water stress, but root systems were much smaller when water-stressed ramets were connected to ramets rooting in wet soil.

The effects of integration on shoot morphology have mostly been investigated in species with plagiotropic shoots. Slade and Hutchings (1987c) grew single stolons of *Glechoma hederacea* clones from resource-poor into resource-rich conditions and *vice versa*. Under unshaded conditions, short stolon internodes were formed irrespective of the conditions experienced by other parts of the stolon. Internodes were longer under shaded conditions, but when stolons grew from unshaded conditions into shade, the first two internodes formed under shade were shorter than the subsequent ones (Fig. 7A). Thus the effects of high photon flux density on stolon morphology were carried over acropetally into adjacent parts of the stolon growing under low photon flux density. The morphology of still more distal regions of the stolon was unaffected by such integration. This observation may be explained by the fact that most translocates from the ramets growing in unshaded conditions are transported into new, higher-order stolons growing from their leaf axils,

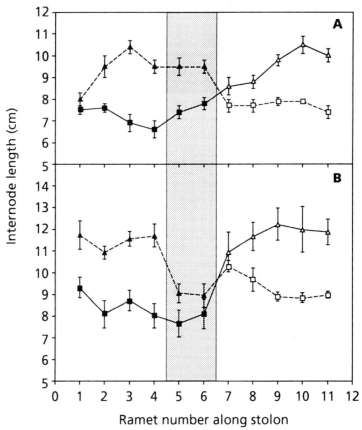

Fig. 7. Mean (± sᴇ) lengths of sequentially produced primary stolon internodes of (A) *Glechoma hederacea* and (B) *Lamiastrum galeobdolon*. In the treatment marked (■–△) clones grew from high light into low light conditions; in the treatment marked (▲–□) they grew from low light into high light conditions. The shaded band in the figure indicates the stolon internodes which were still elongating under "old" light conditions at the time when the stolon apex grew into "new" light conditions. In A, low light was neutral; in B, low light also had a low R/FR ratio. After Slade and Hutchings (1987c) for *G. hederacea* and Dong (1993) for *L. galeobdolon*.

instead of along the original stolon axis (see Price *et al.*, 1992; Price and Hutchings, 1992).

Lamiastrum galeobdolon (like *G. hederacea*, a member of the Labiatae) displayed similar localized morphological responses to light (Dong, 1993) but did not show the same carry-over effect: instead, its internodes *abruptly* increased in length when stolons grew from high light into shade (Fig. 7B). It did exhibit another effect of integration, however. When stolons grew from low into high light conditions, the last two internodes formed under low light were already as short as those formed under high photon flux density (Fig. 7B). These internodes were still elongating when the apex of the stolon penetrated the high light patch. Thus, exposing the apical region of the stolon to high light resulted in significant internode shortening even when older, expanding internodes experienced low light. In the opposite situation, when the apical region was subjected to low light and the expanding internodes to high light, the influence of high light upon morphology prevailed, and short internodes were also formed (Fig. 7B). Dong (1993) speculated that these responses were caused by hormone transport enabling the plant to sense the smallest high light patches in its environment and immediately respond to them. Responses of *L. galeobdolon* petiole lengths to light supply were also investigated. Petiole extension by ramets under low photon flux density was less when they were connected to ramets growing under high photon flux density than when the entire clone was growing under low light conditions. This effect was not seen when only the apical region of the stolon was in high light.

Integration had no effect on the length of internodes formed by the rhizomatous species *Hydrocotyle bonariensis* under lower photon flux density (Evans, 1992). However, the tendency of lateral rhizome buds to grow out under low resource levels was higher when the rhizome was connected to a sibling rhizome growing under high resource levels than when the connection was severed. The longer petioles and larger leaf blades produced under shaded conditions *increased* further in size when connected to a rhizome branch that was growing under unshaded conditions. Evans (1992) suggested that integration may thus facilitate the projection of leaf blades into the high light zone of the canopy. It is noteworthy that *L. galeobdolon*, in which the effect of integration on petiole growth was the opposite of this, is a species of woodlands, unlike *H. bonariensis*. In woodlands it may not be possible for a species with plagiotropic shoots to increase its access to light by employing morphological responses which raise the positions of leaves. Like *H. bonariensis*, *Trifolium repens* ramets in a grass sward produced longer petioles and more secondary stolons when connected to sibling ramets growing in bare soil than they did when the sibling ramets were also growing in competition with grasses (Turkington and Klein, 1991). Stolon internodes of *T. repens* also increased in length when connected to ramets growing under more favourable conditions.

In many plant species, structural branches or ramets have a high degree of autonomy (Watson and Casper, 1984; Sprugel *et al.*, 1991; Price *et al.*, 1992). This occurs when the sectorial arrangement of vascular strands in orthostichies results in an absence of vascular connections between branches or ramets, preventing exchange of carbohydrates and other resources between them. These semi- to fully autonomous structures have been named integrated physiological units (IPUs). Necessarily, the morphology of IPUs will be a pure response to local conditions which is unaffected by translocation of resources from other IPUs. Thus, in a patchy habitat, despite their being physically connected, the morphology of each IPU of a single plant could be strikingly different. For example, such sectoriality in *Glechoma hederacea* clones precludes resource transport between the sibling stolons that originate from a single ramet (Price *et al.*, 1992), and consequently each stolon, and the ramets it bears, responds independently to its local environment (Slade and Hutchings, 1987a). It has been suggested that an autonomous, non-integrated behaviour of structural branches or ramets may enhance foraging efficiency by increasing the growth of those resource-acquiring structures which are in the most favourable patches of habitat (de Kroon and Schieving, 1990; Sprugel *et al.*, 1991). In time this will increase the proportion of resource-acquiring structures occupying favourable patches. In contrast, integration between branches occupying habitat of differing quality would result in more uniform growth of all parts of the plant; the branch in poor quality habitat would be sustained, and its growth increased by translocation from the branch in good quality habitat. Provision of support could, however, limit the growth of the branch in good quality habitat.

B. Division of Labour

Most natural habitats are spatially and temporally patchy in the provision of resources. It is possible that few microsites provide all essential resources in adequate quantities. The availability of different essential resources may also be negatively correlated in space (Friedman and Alpert, 1991). For example, a high local availability of nutrients may lead to a local concentration of biomass of tall species, producing a low availability of light at ground level. When the spatial supply of different resources is negatively correlated there may be benefits in a "division of labour" in resource capture between different parts of the plant (Callaghan, 1988). The species most likely to exhibit such behaviour are clonal plants with plagiotropic stems.

An example is seen in the clonal graminoid *Carex bigelowii*, which has long-lived ramets interconnected by persistent rhizomes (Jónsdóttir and Callaghan, 1988, 1990). Different ramet generations undertake different tasks of resource acquisition. The youngest ramets are photosynthetic and they provide carbohydrates which support the forward growth of rhizomes and new ramets, and sustain the root systems of several generations of older ramets. The older generations of ramets are leafless. The combined root systems of these old generations of ramets exploit a large volume of soil for nutrients and water, and a proportion of these resources is transported to the growing points of the rhizome system, thus supplementing the limited capacity of the youngest, photosynthetic generations of ramets to acquire soil-based resources. A similar division of labour may occur in the tundra club-moss *Lycopodium annotinum* (Headley *et al.*, 1988) and the forest under-storey perennial *Podophyllum peltatum* (de Kroon *et al.*, 1991; Landa *et al.*, 1992). However, only a small percentage of the total nutrient uptake by the older roots is transported acropetally towards the growing tips (Headley *et al.*, 1988; Jónsdóttir and Callaghan, 1990), and the importance of this trans-port for the growth and survival of younger parts has yet to be established. The results of rhizome severing studies do not provide a general answer: severing old rhizome segments significantly decreased shoot survival and new rhizome growth in *C. bigelowii* (Jónsdóttir and Callaghan, 1988), but in *P. peltatum* new rhizome growth was unaffected, at least in the short term (de Kroon *et al.*, 1991).

Data on the stoloniferous herb *Fragaria chiloensis* suggest that water and photosynthates can simultaneously move in opposite directions within the same stolon. Connected ramets can therefore exchange different resources and rectify local deficiencies in a heterogeneous habitat. Alpert and Mooney (1986) showed that if a ramet growing under low light levels, but with ample water supply, was connected to a ramet under water stress but with sufficient light, both grew successfully, although isolated ramets could not survive under each of these conditions. Reciprocal exchange of water and carbon may thus enable the plant to grow in a patchy environment which would otherwise be uninhabitable. A subsequent study (Friedman and Alpert, 1991) pur-ported to show similar effects of reciprocal translocation of carbohydrates and nitrogen in *F. chiloensis*. However, no evidence was obtained to show that nitrogen was actually translocated between sibling ramets under the experi-mental conditions applied. In an experiment with two species of *Potentilla*, plants reached a higher total biomass under conditions in which either the mother ramet was growing under shaded and the daughter stolon under unshaded conditions, or *vice versa*, as compared to uniform shaded or unshaded conditions (J. Stuefer, H.J. During and H. de Kroon, 1993, unpub-lished). There were indications that ramets under unshaded conditions ex-perienced water stress, and Stuefer *et al.* suggested that the higher biomass obtained under heterogeneous conditions was the result of reciprocal translo-cation of water and carbohydrates between mother and daughter parts.

Physiological integration, and a division of labour between ramets, may also provide clonal plants with the ability to exploit below-ground resources in a resource-rich patch of habitat, while avoiding the above-ground compe-tition that could result from the biomass produced with these resources. Birch and Hutchings (1994) have shown that *Glechoma hederacea* ramets rooted vigorously in nutrient-rich patches surrounded by nutrient-poor sand, and that this resulted in a high ramet production by the clone (Fig. 8). The majority of these ramets were placed *beyond* the nutrient-rich patch and they occupied a wide area of ground. As a result, the risk of extreme ramet density was alleviated and severe intraclonal competition for light was avoided.

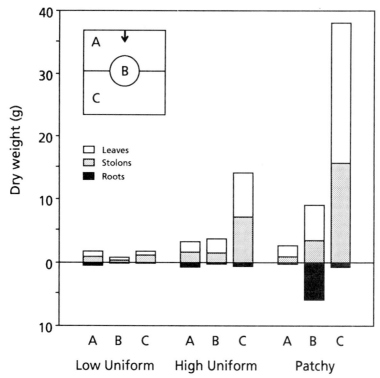

Fig. 8. Division of labour in the clonal herb *Glechoma hederacea*. Clones were grown in square boxes in one of three artificial soil environments made by distributing washed sand and potting compost throughout the box in different ways. In the "High Uniform" treatment, potting compost was mixed evenly with sand throughout the box. The "Patchy" treatment provided a heterogeneous soil environment made by using the same quantity of potting compost as in High Uniform, but concentrating half of it in a 30 cm circle in the centre of each box (11% of the box area). In the treatment "Low Uniform", compost was distributed homogeneously, as in the High Uniform treatment, but the total quantity of potting compost was halved. At harvest the biomass of the plants in each of three compartments was determined: (A) the near half of the flat with parts of the clone that developed prior to entering the patch, (B) the central circle and (C) the far half of the flat predominantly containing stolons with basipetal connections to ramets growing in the central circle. In the case of the patchy resource distribution, the majority of the roots developed in the central nutrient-rich circle while most of the above-ground biomass was formed outside the circle. After Birch and Hutchings (1993, unpublished).

C. Conclusions

Data suggest that locally favourable conditions may intensify the local placement of resource-acquiring structures, irrespective of the conditions experienced by other parts of the plant. Total resource demand by a plant does not affect the *location* of placement, but may affect the *magnitude* of the morphological response to environmental conditions. This has been demonstrated convincingly for root systems which branch more vigorously in nutrient-rich patches when nutrients are more limiting to the plant. We are not aware of any studies showing similar responses for shoot systems. When clones with interconnected ramets are subjected to heterogeneous conditions, different studies have shown that local responses of bud activation to patch quality are damped, responses of shoot height increment are either damped or intensified, and stolon internode length responses are mostly unaffected by integration. Thus, resources received by part of a plant through intraclonal transport may alter the morphology of the recipient part from the pattern which would be produced purely in response to its local resource supply. Such effects of integration have been interpreted in terms of foraging efficiency. Patches of habitat may not be simply classifiable as "favourable" to a particular degree dependent on the availability of a single resource. Each essential resource may vary independently in time and space, so that all parts of the habitat may be favourable for some resources but unfavourable for others. When the spatial supply of different resources is negatively correlated, ramets of clonal species may divide the tasks of acquiring different resources in a manner which reflects the local availability of each resource. In such situations resource sharing between ramets is beneficial to the growth of the clone.

VIII. PHYSIOLOGICAL PLASTICITY AS AN ALTERNATIVE TO MORPHOLOGICAL PLASTICITY FOR RESOURCE ACQUISITION

Foraging responses by a plant in a patchy environment may enhance resource acquisition by placing more leaves and roots in patches of the habitat with higher concentrations of resources. However, total resource gain depends not only upon the effective location of leaves and roots in such patches but also on the rate of resource uptake by those leaves and roots. Variation is possible in the rate of acquisition of essential resources as the level of resource availability changes, because of physiological plasticity.

Marked physiological plasticity is predicted to be a particularly valuable attribute in conditions in which resources become available in the form of short, unpredictable pulses above a low background level (e.g. Grime *et al.*, 1986; Hutchings, 1988). This is because the pulses may be too brief and too unpredictable in location to enable them to be exploited sufficiently fast by morphological responses. In such situations physiological plasticity may serve either as an adjunct to, or as a more valuable asset than, morphological plasticity for resource acquisition.

In this section we discuss resource uptake under low levels of resource availability, and the ability of plants to rapidly increase their uptake rates when resource levels increase. We subsequently review what is known about plasticity in nutrient uptake rates in response to nutrient pulses, and plasticity in carbon assimilation in response to sunflecks. Special attention is paid to the levels of physiological plasticity exhibited by species from habitats with differing levels of resource supply, and the extent to which whole plant growth may profit from plasticity in resource uptake rates.

A. Acquisition of Nutrients

Under conditions of high nutrient availability, species from infertile habitats display lower nutrient uptake rates than species from more fertile habitats (e.g. Bradshaw et al., 1964; Chapin, 1980; Chapin et al., 1982; Campbell and Grime, 1989; Garnier et al., 1989; Kachi and Rorison, 1990; Poorter et al., 1991). However, their absorption rates vary less than those of species from fertile habitats over a given range of nutrient concentrations, and thus, when nutrients are scarce, the nutrient uptake rates of species from infertile habitats can equal or even exceed those of species from fertile habitats, even though these rates are low. In general, nutrient uptake rates are greater when the nutrient demand of the plant is higher (Robinson, 1989), and it has been suggested that the magnitude of the uptake rate at a given level of supply is negatively related to the internal nutrient concentration of the plant (de Jager, 1984; Burns, 1991).

In habitats with inherently infertile soils, such as tundra and chalk grassland, a large proportion of the nutrients annually available to plants may be provided in short, unpredictably occurring flushes (see Crick and Grime, 1987; Campbell and Grime, 1989; Jonasson and Chapin, 1991). Physiological plasticity allowing rapid changes in the capacity of roots to absorb nutrients may ensure that these flushes are captured before they can be exploited by the microbial community in the soil. Most mobile nutrients which are not captured by higher plants are absorbed by microorganisms within a few days of release into the soil (Campbell and Grime, 1989; Jonasson and Chapin, 1991).

An elegant illustration of the ability of some plant species to respond rapidly to an unpredictable flush of nutrients is provided by the work of Jackson et al. (1990) on three species which grow in phosphate-poor soils in cold desert habitat. The soil on one side of well-established plants was supplied with a solution containing phosphate, while the soil on the other side was given an equal volume of distilled water. Roots were extracted from the sectors treated in these different ways at various times after treatment. For all three species, those parts of the root system extracted from the phosphate-enriched soil showed significant increases in the rate of phosphate uptake per gram of root compared with roots from the soil watered with distilled water.

Two species of bunchgrasses showed this effect within 3 days of treatment. Mean rates of phosphate uptake increased by up to 80% in roots from the enriched patches, with the largest increases being seen when the external phosphate concentration was highest. Later experiments (Jackson and Caldwell, 1991; Caldwell *et al.*, 1991a) confirmed these results (see Fig. 5). Enrichment of soil by nitrogen appeared to significantly increase the uptake of nitrogen and potassium, while phosphate enrichment increased the uptake of phosphate. One of the treated species, *Pseudoroegneria spicata*, showed no increased root development following soil enrichment (Jackson and Caldwell, 1989; Caldwell *et al.*, 1991b) (see Fig. 5). Thus, its response to a pulse of nutrients was entirely physiological. In this short-term experiment (3–5 weeks), the change in uptake kinetics achieved by means of physiological plasticity in *P. spicata* was more effective in acquiring phosphate from enriched as opposed to control patches than that of *A. desertorum*, which combined *both* morphological and physiological responses to increased phosphate supply. Finally, it is important to realize that even without a change in uptake kinetics, resource acquisition is markedly increased when soil resource concentrations are elevated (Caldwell *et al.*, 1992).

A higher degree of physiological plasticity may result in higher growth rates under a regime of ephemeral nutrient patches. Poorter and Lambers (1986) grew *Plantago major* in a growth chamber under continuous or fluctuating nutrient supply. The same total amount of nutrients was applied in pulses of different duration. Relative growth rate of the plants was inversely related to pulse length, but with decreasing pulse duration, a highly plastic inbred line maintained higher growth rates than a marginally plastic inbred line. The magnitude of the difference was greater when plants of the two lines were grown in competition.

A number of studies have investigated whether slow-growing species from poor soils are better at utilizing nutrient pulses than fast-growing species from richer soils. De Jager and Posno (1979) supplied *Plantago* seedlings after a period of phosphate depletion with a full-strength nutrient solution, applied either to the entire root system or to a small part of it. Phosphate uptake rate $(\mathrm{mmol\,g^{-1}\,d^{-1}})$ of *P. media*, a species from relatively nutrient-poor habitats, increased more strongly than that of *P. major*, a species of more fertile habitats, in response to a localized supply of phosphate. However, Kachi and Rorison (1990) found that *Holcus lanatus*, a grass species from nutrient-rich habitats, had similar nitrogen uptake rates to *Festuca rubra*, a grass species from nutrient-poor habitats, when their roots were subjected to high nutrient levels after a period of nutrient depletion.

Campbell and Grime (1989) compared the nutrient uptake characteristics of *Festuca ovina* with those of *Arrhenatherum elatius*, another grass species from fertile habitats. Both were grown hydroponically in conditions in which periods of low nutrient supply were interspersed with pulses during which nutrient-rich solutions were applied. Pulse duration varied from 80 seconds to 6 days, with a single pulse applied at the midpoint of each 6-day period throughout the experiment. Nitrogen absorption rates per unit of root weight were greater in *A. elatius* when long nutrient pulses were supplied, but greater in *F. ovina* for all pulse durations shorter than 10 hours. Under these short pulse durations *F. ovina* maintained higher root:shoot ratios and more viable leaf tissue than *A. elatius*. As a result, under all pulse lengths shorter than 10 hours, nitrogen capture by plants of the two species was similar, despite the far greater size of *A. elatius*, and *F. ovina* had higher relative growth rates. *A. elatius* grew faster at longer pulse lengths. However, as the concentration of the nutrient solution used was the same in all pulse treatments in this experiment, total nutrient supply to the plants increased with increasing pulse duration. Thus, the results must be interpreted with caution, because responses to pulse duration cannot be separated from responses to total nutrient supply.

Crick and Grime (1987) grew *Scirpus sylvaticus*, a species of nutrient-poor habitats, and *Agrostis stolonifera*, a species of nutrient-rich habitats, in an apparatus enabling given fractions of the root system to be supplied with high or low nutrient solutions for specified lengths of time. In one treatment (no. 4; see Fig. 9A) nutrients were supplied in short (1-day) pulses to sectors of the root system selected virtually at random, i.e. the timing and location of the nutrient pulses were effectively unpredictable to the plant. In another treatment (no. 2), the whole root system received a continuous supply of low nutrient solution. Relative growth rates of both species were altered little by imposition of either of these treatments, but both species achieved a greater rate of nitrogen uptake when provided with unpredictable pulses of nutrients (Fig. 9). However, the proportional increase in nitrogen uptake rate between the pulse treatment and the continuously low nutrient treatment was considerably greater (nearly fivefold) for *S. sylvaticus*, the species from nutrient-poor habitats, than for *A. stolonifera*, in which it was only about 50%. We would conclude from this analysis that *S. sylvaticus* is more effective than *A. stolonifera* in capturing ephemeral nutrient pulses above a low background level.

However, in two further treatments, the same quantity of nutrients as provided in the pulse treatment was supplied either in a more stable configur-

---

Fig. 9. Relative growth rate (RGR) (B) and nitrogen uptake rate per plant (C) in an experiment in which *Agrostis stolonifera* and *Scirpus sylvaticus* were grown in a root growth arena that allowed a separate control of nutrient supply in each of nine root sectors. The roots were subjected to five patterns of nutrient supply differing in spatial and temporal configuration as depicted in A. See text for further explanation. Recalculated from Crick and Grime (1987).

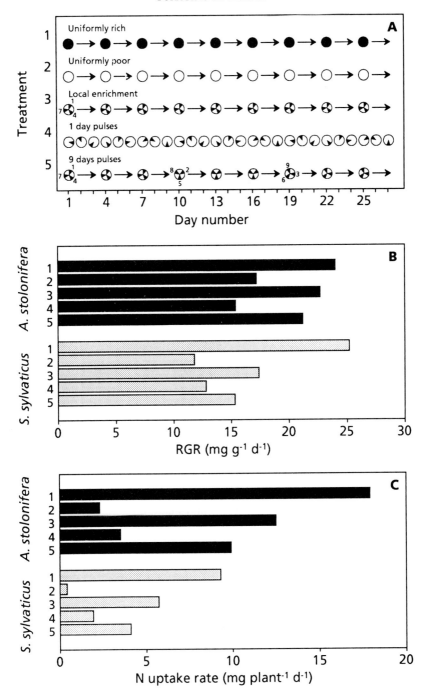

ation (9-day pulses; treatment no. 5) or in a completely stable configuration (treatment no. 3). The relative growth rates and nitrogen uptake rates of both species were greater in both of these treatments than in the 1-day pulse treatment (Fig. 9). The proportional reduction in uptake rate in the random pulse treatment (no. 4) compared with the stable configuration treatment (no. 3) was approximately the same in both species (72% in *A. stolonifera*, 67% in *S. sylvaticus*; see Fig. 9C). If the nutrient uptake efficiencies under random pulse supply (no. 4) are calculated as a percentage of that achieved when the same nutrient supply is provided in the more stable configuration of treatment no. 5, the two species still appear to show little difference; the value for *A. stolonifera* was 35%, while that for *S. sylvaticus* was 46%. This analysis is far less convincing in showing that the species from infertile habitats was more efficient in acquiring brief, unpredictable pulses of nutrients. For *S. sylvaticus*, a relatively large root mass (approximately twice as large as the root mass of *A. stolonifera*; Crick and Grime, 1987, Fig. 2) was important for the acquisition of short nutrient pulses. *A. stolonifera* accomplished nutrient acquisition from these short pulses—as in all other treatments—by having higher nutrient absorption rates (see Fig. 9).

The maintenance of a large long-lived mass of roots, and an ability to commence absorption of nutrients quickly when conditions permit, despite long periods of low nutrient availability, seem to be of great importance in enabling species to acquire nutrient pulses of short duration (Crick and Grime, 1987; Campbell and Grime, 1989; Kachi and Rorison, 1990). However, the available data do not unambiguously show that species from nutrient-poor habitats are better capable of taking up nutrients supplied in the form of ephemeral pulses than species from nutrient-rich habitats.

B. Acquisition of Carbon

Björkman (1981) succinctly lists the differences in photosynthetic characteristics between shade-tolerant ("shade") and shade-intolerant ("sun") species. Leaves of obligatory shade species typically have low dark respiration rates resulting in low light compensation points which allow them to maintain higher rates of photosynthesis under shaded conditions than sun species. At higher photon flux densities the inherently low contents of photosynthetic enzymes such as Rubisco in leaves of shade plants result in inherently low maximum rates of photosynthesis compared to sun plants. Moreover, high photon flux densities may cause damage to the photosynthetic equipment of shade plants through photoinhibition. Sun and shade species also differ in a number of leaf morphological characteristics, with shade plants developing larger leaf areas per unit of plant weight (leaf area ratios) and larger leaf areas per unit weight of leaf (specific leaf areas) in response to shading than sun plants. Shade-intolerant species may acclimate to shade and develop physiological and morphological characteristics similar to those of shade species. However, their responses to photon flux density are not of a magnitude sufficient to match the efficiency of light use in shade plants.

In tropical forests, and in temperate deciduous forest during the summer, sunflecks typically contribute 40–50% of the total daily photon flux density (PFD) (Chazdon, 1988; Chazdon and Pearcy, 1991), and on clear days this value can rise to 90%. Sunflecks may strike leaves for a very small fraction of the day. In a Mexican rain forest the total duration of sunflecks ranged from 0 to 42 minutes per day at different points on the forest floor (Chazdon, 1988). Fifty-six per cent of the sunflecks were less than 4 seconds, and 90% less than 32 seconds in duration. While the exact time at which sunflecks will occur is unpredictable, their temporal pattern through the day is far from random. In tropical forest in north-eastern Australia, 70% of sunflecks occurred within one minute of the preceding sunfleck and only 5% were preceded by low light periods longer than an hour (Pearcy, 1988). Fewer sunflecks occur in the morning and evening when the sun is low in the sky, because of the longer path length of its rays through foliage. Peak photon flux densities during sunflecks are relatively low and rarely amount to values reached outside the forest.

Sunflecks are most important for the growth of plants in forest understoreys. For example, Pearcy (1983) found that the relative growth rate of tree seedlings in Hawaiian tropical rain forest was linearly related to sunfleck activity (but see Pfitsch and Pearcy, 1992). On clear days 40–60% of the daily CO_2 uptake of plants in forest understoreys may take place during sunflecks (e.g. Pfitsch and Pearcy, 1989a), and it has been shown that net carbon gain during sunflecks of a given PFD can be substantially greater than predicted from knowing the steady-state rate of photosynthesis of a species at the same PFD (e.g. Chazdon and Pearcy, 1986a; Pons and Pearcy, 1992). Chazdon (1988), Pearcy (1990) and Chazdon and Pearcy (1991) have succinctly reviewed the mechanisms responsible for this efficient use of light pulses. The efficiency appears to be due to the maintenance of high levels of photosynthetic induction during a sequence of sunflecks, combined with substantial post-illumination CO_2 fixation. Here we briefly summarize what is known about these processes.

Leaves exposed to high radiation levels after a long period of low light require a light-priming induction period of up to 60 minutes before a steady-state photosynthetic rate is reached. During the first 6–10 minutes of exposure to high light, the photosynthetic rate is limited primarily by biochemical processes, in particular by the activity of the carboxylating enzyme Rubisco (Chazdon and Pearcy, 1986a; Pfitsch and Pearcy, 1989b; Pons et al., 1992). Stomatal conductance may become a limiting factor later on during exposure to high light. To our knowledge only one species (the fern *Poly-*

podium virginianum) is known to lack a measurable induction period in the field (Gildner and Larson, 1992).

It has been shown that the degree of induction accumulates during a sequence of rapid consecutive sunflecks. Over a sequence of five 30- or 60-second sunflecks, each separated by 2 minutes of low light, the induction state of leaves of two Australian rain-forest species increased two to three times, and rate of induction during these brief sunflecks matched that achieved under constant illumination (Chazdon and Pearcy, 1986a). The shade-tolerant species *Alocasia macrorrhiza* reached higher levels of induction during the first lightfleck than the shade-intolerant *Toona australis*, and maintained this higher level of induction throughout a sequence of lightflecks. In soybean a constant induction state could be achieved by providing 1-second lightflecks; the degree of induction reached was greater when the lightflecks were provided at a higher frequency (Pons *et al.*, 1992). Thus, the clustered temporal pattern of sunflecks in forests may be of great importance for carbon gain as leaves are partially induced by each preceding sunfleck and can therefore use some of the available light from nearly all sunflecks occurring during the day. Tinoco-Ojanguren and Pearcy (1992) showed that the shade-tolerant Mexican rain-forest species *Piper aequale* was better able to make use of such a sequence of sunflecks than the shade-intolerant *Piper auritum* because shade-acclimated plants of *P. aequale* increased more rapidly in stomatal conductance than *P. auritum* in response to sunflecks.

Induction is lost when leaves are subjected to continuous shade (e.g. Pons *et al.*, 1992). Chazdon and Pearcy (1986a) found that completely induced leaves of *A. macrorrhiza* took over an hour in continuous low light conditions to return to a completely uninduced state, but the loss of induction was far more rapid for *T. australis*. Similarly, the stomatal conductance of *P. aequale* decreased more slowly after a sunfleck than that of *P. auritum* (Tinoco-Ojanguren and Pearcy, 1992). This suggests that shade-tolerant species lose induction less rapidly than shade-intolerant species. However, *Adenocaulon bicolor*, a shade-tolerant herb from Californian redwood forests, also rapidly lost induction under shade (Pfitsch and Pearcy, 1989b), which limited its capacity to use sunflecks interspersed by periods of shade longer than 1–2 minutes.

The efficiency of sunfleck use is increased substantially by post-illumination CO_2 fixation (Chazdon, 1988; Pearcy, 1990). For example, when *A. macrorrhiza* is given a 5-second sunfleck, CO_2 assimilation during and after the sunfleck can give 60% greater net carbon gain than would occur during 5 seconds of steady-state photosynthesis at the same photon flux density (Sharkey *et al.*, 1986). This increase appears to be due to a build-up of pools of triose phosphates—photosynthetic intermediates that are synthesized immediately after the sunfleck. Sun-grown plants do not have the large pool of triose phosphates following exposure to a sunfleck, and therefore cannot match the benefit gained from short sunflecks by shade-grown plants (Sharkey *et al.*, 1986). Chazdon and Pearcy (1986b) found that at all sunfleck durations the shade-tolerant *A. macrorrhiza* had both a higher photosynthetic efficiency and a greater net carbon gain than the shade-intolerant *T. australis*. It is unclear whether this was due to maintenance of a higher state of photosynthetic induction during a sequence of sunflecks, to a higher level of post-illumination CO_2 fixation, or both.

Post-illumination CO_2 fixation may compensate for both low photosynthesis under the relatively low peak PFDs of most of the sunflecks in forest understoreys, and for the typically low state of induction of the leaves of plants in such habitats. This compensation is most apparent for short sunflecks (less than approximately 40 seconds); photosynthetic efficiency rapidly decreases with increasing sunfleck duration, for a number of reasons. For example, leaves of C_3 plants release a burst of CO_2 upon darkening as a result of respiring a residual pool of photorespiratory metabolites, but no such burst is apparent after lightflecks shorter than 20 seconds (Pearcy, 1990). In addition, if the PFD of a sunfleck is very high, the efficiency with which it can be used may be reduced because of photoinhibition, particularly in those shade-tolerant species in which photosynthesis is saturated at low light levels (Chazdon, 1988). This will be especially important during long (more than about 10 minutes) sunflecks, but there is no evidence to date to show that photoinhibition significantly limits carbon gain under natural regimes of short sunflecks (Pearcy, 1990). High photon flux densities may also limit photosynthesis through thermal damage and water stress resulting in stomatal closure, but again such effects will only become apparent during extended sunfleck duration (Chazdon, 1988; Pearcy, 1990; but see Pfitsch and Pearcy, 1992).

It is of interest to consider whether shade-tolerant species can grow faster than shade-intolerant species in a sunfleck-dominated light regime. Rincon and Grime (1989) grew bryophyte species from open and shaded habitats under a number of artificial light regimes. In addition to uniform high and low light, plants were exposed to regimes of either low light with lightflecks provided in a spatially and temporally stable configuration, or spatially unpredictable lightflecks of 20 minutes' duration. Four species from open habitats grew equally well when the same total amount of light was provided as either stable patches or as unpredictable lightflecks, but one of the two species from shaded habitats (*Thamnobryum alopecurum*) grew significantly faster in the unpredictable lightfleck regime than in the stable lightfleck regime. This was also the only species with a significantly higher relative growth rate under unpredictable lightflecks than under uniformly low light. These results suggest that at least one shade-tolerant species grows faster than shade-intolerant species under a temporally unstable lightfleck regime. However, it is unclear whether these bryophytes would respond similarly in a forest understorey habitat in which most of the sunflecks are likely to be much shorter than those applied by Rincon and Grime (1989), because, as discussed earlier, lightfleck photosynthesis is most efficient when the pulses of light occur in clusters.

We conclude that the photosynthetic machinery is well equipped to utilize brief sunflecks that are clustered in time. Such a temporal pattern of sunflecks is typical in many forest understorey habitats. There appears to be much variation in photosynthetic characteristics between species. Although some data suggest that shade-tolerant species can use sunflecks more efficiently than shade-intolerant species, the evidence is inconsistent.

C. Conclusions

In many habitats, the usual level of supply of resources is low. Most resources become available in the form of ephemeral pulses which are unpredictable in time and space. Plants possess mechanisms for efficient acquisition of the resources in such pulses. Nutrient uptake capacity of roots rapidly increases when they are exposed to high nutrient levels following periods of nutrient scarcity, and levels of photosynthesis immediately increase in response to sunflecks interrupting periods of low light. These responses are most effective in terms of the amounts of resources acquired and the growth that is realized with them. A number of studies have investigated whether species from habitats characterized by transient resource pulses are able to utilize such brief pulses more effectively than species from habitats in which resource availability is usually higher and more stable. Support for this hypothesis is equivocal, but few comparative studies have been carried out to date, and few species have been investigated. Levels of physiological plasticity are not always higher in species from habitats in which resource supply is ephemeral; the capacity to maintain viable resource-acquiring structures during periods of resource shortage may be at least as important as physiological plasticity for the capture of temporally unpredictable resource pulses.

IX. GENERAL TRENDS IN THE FORAGING ACTIVITIES OF PLANTS AND THEIR INTERPRETATION

In this review we have explored and documented the foraging behaviour of plants, and concentrates on the ability of plants to actively promote the exploitation of patches of habitat with high levels of resource availability. In this section we present a summary of the hypotheses about foraging behaviour which have been proposed in the literature. Secondly, we survey the patterns which have emerged. Thirdly, we evaluate the extent to which the hypotheses are supported by the patterns. The section is concluded with a list of questions in need of solutions.

A. Hypotheses

An essential consequence of the modular construction of plants is that leaves and root tips (resource-acquiring structures, or "feeding sites") are located at the ends of branches ("spacers") which project them into habitat space (Bell, 1984). The morphology of shoots and roots can be described using a few simple parameters such as spacer length (the distance between consecutive potential branching points), the branching propensity and the branching angle. It has been hypothesized, implicitly more than explicitly, that the heterogeneous distribution of resources in most habitats has acted as one of the major selective forces upon the evolution of morphological plasticity (e.g. Harper, 1985; Hutchings, 1988; Schmid, 1990; Sutherland, 1990; Oborny, 1991). According to this view, morphological plasticity has evolved in such a way as to project resource-acquiring structures into the more favourable patches within the environment and to promote the avoidance or vacation of habitat patches of low quality (Cook, 1983; Hutchings and Bradbury, 1986; Sutherland and Stillman, 1988). Such behaviour is expected to increase resource acquisition from the habitat and thus to enhance plant fitness. Analogous organs, such as orthotropic stems, plagiotropic stems and root branches, have been subjected to similar environmental constraints and therefore will have developed convergent patterns of morphological plasticity (Grime et al., 1986; Hutchings, 1988). It is thus predicted that (i) shaded orthotropic shoots will produce long unbranched stems which effectively grow towards the high light zone at the top of a vegetation canopy (Grime, 1966; Lovell and Lovell, 1985), (ii) clonal plants with plagiotropic stems search horizontal space and place ramets non-randomly within their environment by shortening their spacers and increasing their branching intensity when favourable microhabitats are encountered (Slade and Hutchings, 1987a,b; Hutchings and Slade, 1988; Hutchings and Mogie, 1990), (iii) long root axes search the soil volume and when they grow into a patch of high resource supply, lateral root formation is increased and the resources are exploited (Drew et al., 1973; Fitter, 1991). All of these plastic modifications are expected to be local responses to patch quality. They should ideally be independent of the conditions experienced by the rest of the plant, as widespread integration may decrease foraging efficiency (de Kroon and Schieving, 1990).

Different foraging syndromes may have evolved in habitats which differ in patch structure. It is unlikely that morphological plasticity will enhance resource acquisition if patches are either much smaller or much larger than the distance between adjacent resource-acquiring structures (Sutherland and

Stillman, 1988). In addition, morphological responses may take place too slowly to allow effective exploitation of patches of high resource availability which only last for a short period of time. These considerations have led a number of authors to suggest that foraging for resource-rich patches is not a profitable behaviour in habitats with inherently low resource availability, in which resources become available in the form of ephemeral pulses (e.g. Grime et al., 1986; Sibly and Grime, 1986; Hutchings, 1988; de Kroon and Schieving, 1990). Such habitats might be found, for example, in tundra communities (nutrient-poor) and in forest understoreys (light-poor). It is hypothesized that in such habitats more "conservative" methods have been selected for gaining resources, including resource-acquiring structures with a large total surface area placed on branches with low levels of morphological plasticity. In order to acquire the short-lived resource pulses, species in such habitats should be able to increase resource uptake and assimilation rapidly when resource levels increase—i.e. they should have high levels of physiological plasticity (Grime et al., 1986; Campbell and Grime, 1989; Robinson, 1989). It is thus predicted that species of forest understoreys will possess a photosynthetic mechanism which is capable of effective carbon acquisition under a sunfleck regime, and that species from tundra will be highly plastic in nutrient uptake.

It is important to accept that several alternative foraging solutions for the acquisition of essential resources may be employed in a single habitat. For example, in relatively nutrient-rich habitats, some fast-growing species may dominate the vegetation ("space-consolidators" sensu de Kroon and Schieving, 1990) and maintain dominance for a considerable period of time. Selective placement of orthotropic shoots in high light patches may (de Kroon and Knops, 1990) or may not (Campbell et al., 1991a) be important for the maintenance of the dominant position of such species. "Subordinate" species (sensu Grime, 1987), with a lower growth potential, may coexist in the same habitats by foraging efficiently in horizontal and/or vertical space (Campbell et al., 1991a). If these subordinate species are clonal, they are likely to possess a guerilla growth form compared to a more phalanx growth form in clonal dominants (Lovell and Lovell, 1985). Such relatively guerilla species would be likely to exhibit a pattern of morphological plasticity characterized by a reduction of spacer length and an increase in branching intensity under more favourable conditions. This has been referred to as the foraging model in earlier publications (e.g. Sutherland and Stillman, 1988; de Kroon and Schieving, 1990; Oborny, 1991). Here we take a broader view and recognize other resource-acquiring syndromes as alternative examples of foraging behaviour. In particular, the foraging behaviours of orthotropic stems, plagiotropic stems and roots should be regarded as separate phenomena as they may have evolved independently in response to the particular types of environmental patchiness experienced by each of these types of organs.

B. Patterns

To what extent are these predictions borne out by the data reviewed in Sections III to VII? It has been shown in many species that branching indeed intensifies when resource levels increase. This is a very general response, observed in both shoots and roots and in response to a variety of resources. Branching is not only affected by local conditions; when resources are imported into one part of a plant from other parts, branching in the recipient part is usually also increased.

In orthotropic shoots and in roots, a reduction of spacer length under high resource availability is also a very general observation. Usually this is a local response which is relatively independent of the conditions experienced by other parts of the plant. Together with increased branching, the reduction in spacer length results in an increase in placement of resource-acquiring structures in favourable patches. The morphological responses of shoots have been found to be particularly sensitive to changes in the ratio of R/FR light, such as are caused by canopy shading. Stem analogues such as leaf sheaths respond just as vigorously as true stems. Lateral roots formed in nutrient-rich patches have relatively high growth rates, which increase local root surface area. While the extent of lateral root formation is highly plastic, the total length of the main root axis has been shown to be unresponsive to resource availability. Plasticity in root morphology is markedly reduced when roots are associated with mycorrhizal fungi. These fungi may adopt the foraging function of the roots. Fungi growing independently have also been shown to be able to search for, and converge efficiently upon favourable patches within the soil volume.

In contrast to both orthotropic stems and roots, the internode lengths of plagiotropic stems exhibit a variety of responses to resource availability. Under more favourable conditions, the formation of both shorter and longer internodes has been reported in different species; other species fail to show significant responses. For species which do shorten their internode lengths under high resource availability, the magnitude of the response seems to be much smaller than that shown by orthotropic shoots. As a result, plagiotropic stems are less capable of placing ramets selectively in favourable patches within a heterogeneous habitat. The total length of the axis of the plagiotropic shoot is again relatively independent of ambient conditions, even when internode lengths are plastic.

While the foraging patterns of orthotropic shoots and roots described above are qualitatively similar for many species, the magnitude of the responses differs widely; in some species the morphological responses to environmental quality are insignificant. Are these differences in foraging behaviour in accordance with the predictions which have been made for species from different habitats, as summarized above? There is some evidence that orthotropic shoots in shade-intolerant species are indeed more plastic than those of shade-tolerant species. It is less clear whether shade-tolerant species are better capable of utilizing sunflecks than shade-intolerant species. Sunfleck photosynthesis has been well studied, but few species have been compared to date.

The correlations between the levels of morphological and physiological plasticity of roots and the habitat of species are also weak. This may again be due in part to the fact that few comparative studies have yet been carried out. To date, it has not been conclusively shown that species from fertile soils are morphologically more plastic than species from infertile soils. The capacity to take up nutrients that are supplied in the form of ephemeral pulses can also be similar for species of both types of habitat. In general, the relative importance of morphological *versus* physiological plasticity for the exploitation of resource pulses of any duration has not yet been established. The capacity to maintain a large, viable root system under prolonged periods of nutrient depletion appears to be important for the efficient utilization of ephemeral nutrient pulses. This trait appears to be more characteristic of species from infertile habitats than of species from fertile habitats.

C. Evaluation

An appropriate null-model of foraging is that resource availability affects only the growth of the plant, without associated changes in modular construction occurring which would enhance the placement of resource-acquiring structures in localities with high resource availability. Increased growth implies the construction of larger branches, with greater branch length and diameter, and/or the production of a larger number of branches as a consequence of increased meristem activity. As growth proceeds, the probability of branching at a particular node in the plant's structure will usually increase.

The empirical patterns reviewed above show two almost universal responses which accord with the null-model: branching probability (meristem activity) at each node and branch diameter increase in response to increasing resource availability, although in some cases the changes are insignificant. However, the responses of branch (spacer) length convincingly falsify the null-hypothesis. Many examples exist of species which shorten their shoot and root spacer lengths under higher levels of resource availability. Stem analogues such as leaf sheaths respond as effectively as true stems. This observation is in support of a functional interpretation for the plasticity they exhibit. Changes in spacer length are local responses to ambient resource levels, and resources imported from other parts of the plant usually have little effect on the response. This also suggests that the responses are significant for foraging, rather than expressions of growth. As a result of shorter spacers and increased branching intensity, the placement of resource-acquiring structures is concentrated in more favourable patches of the environment, and, together with generally higher uptake rates, resource acquisition and growth may increase markedly. This set of responses is in accordance with the foraging hypothesis. Thus, while an example of enhanced growth, the response of branching may also be important for future resource acquisition.

Surprisingly, the responses of plagiotropic stems are usually qualitatively different from, and quantitatively smaller than, those of orthotropic stems. Qualitatively different responses to light availability are unexpected, given the homology of these organs (Sachs, 1988). This is especially true for above-ground stolons. However, subterranean plagiotropic stems (rhizomes) are architecturally less similar to orthotropic stems and phylogenetically more ancient (Mogie and Hutchings, 1990). Thus, they may not exhibit foraging responses because such responses were probably not ancestral traits. The primary function of rhizomes may be storage. If this is the case, the longer rhizomes which are formed under higher resource availability in some species should be interpreted as expressions of resource accumulation and growth.

A corollary to the quantitatively small responses of plagiotropic spacer lengths is that the habitat is continuously searched. When favourable patches are encountered, the foraging responses of the shoots and roots of individual ramets enable efficient exploitation of the resources to take place without compromising the rate of exploration of horizontal space. It is worth recalling (Section V.A) that the length of main root axes has also been shown to be unresponsive to resource availability, enabling the root system to search the soil volume continuously. Orthotropic shoots do not exhibit this behaviour but always shorten under higher levels of photon flux density. Possession of an elongation response in orthotropic shoots can be profitable because the growth of these shoots is nearly always directed towards the high light zone in the top of the canopy. Roots and plagiotropic stems experience a patch structure which is less predictable and this may warrant the continuous exploration of new microhabitats for resources.

The data collected so far do not unambiguously support the premise that species from different habitats possess markedly different patterns of morphological and physiological plasticity. Species differ in the extent to which they are morphologically and physiologically plastic, but correlations with habitat type are at best incomplete. A notable exception is the extension response to light availability in orthotropic shoots. This is generally far greater in shade-intolerant than shade-tolerant species, which suggests that vigorous stem height extension has been selected in shade-intolerant species as a response to above-ground competition for light. For the acquisition of ephemeral resource pulses in resource-poor habitats, the ability to survive periods of resource depletion appears to be at least as important as the possession of higher levels of physiological plasticity. This ability is enhanced by economic use of resources and slow growth. These characteristics are typical for species from resource-poor habitats (Berendse and Aerts, 1987; Lambers and Poorter, 1992; Aerts and van der Peijl, 1993).

D. Unsolved Questions

The juxtaposition of hypotheses and patterns draws attention to significant gaps in our knowledge about the foraging behaviour of plants. Here we identify some of the gaps in order to suggest topics for future research.

1. Generality of Responses

In order to reveal general trends in foraging behaviour, data are needed from a large number of species. Because comprehensive sets of data are lacking, it is as yet unclear whether there is a tendency for species from habitats which contrast in resource availability to differ in the levels of morphological and physiological plasticity which they possess. Comparative studies in which traits are screened in a large number of species (see for example Grime *et al.*, 1981; Poorter *et al.*, 1990; Westoby *et al.*, 1990) are advocated as a remedy. There is a special need for between-species comparison of those plastic characteristics which have only been examined in a few species, but for which the regulation mechanism has been well elucidated. One such example involves the physiology of light fleck photosynthesis, which has been extensively studied, but about which information has been collected from a very limited number of species. Another subject about which much is known is the mechanism underlying the etiolation of stems in response to a change in light quality. In most experiments in which morphological plasticity has been investigated in species with plagiotropic stems, however, consideration has only been given to the analysis of responses to variation in light *quantity*, rather than quality. Rectification of this situation would be informative.

2. Patch Structure, Patch Quality and the Profitability of Different Types of Behaviour

In recent years an increasing number of papers has been published describing the patch structure of different habitats. The importance of these studies for the interpretation of foraging responses cannot be emphasized too strongly. The profitability of any level of morphological plasticity will depend on the spatial and temporal variability in resource supply within the habitat (Harper, 1985; Sutherland and Stillman, 1988), but virtually nothing is known about the nature of this relationship. Many studies on morphological plasticity, including some of our own (e.g. Slade and Hutchings, 1987a,b; de Kroon and Knops, 1990) implicitly assume that the grain of environmental heterogeneity will match average spacer length so that plasticity will lead to an advantage to the plant in terms of its resource acquisition. However, this premise is rarely put to the test. In combination with ongoing efforts to quantify patch structure, we believe that studies of plasticity should increasingly be directed towards quantification of the advantages of the particular behaviours which are displayed by species in given environmental settings, compared with other possible behaviours.

3. Comparative Plasticity of Different Types of Spacers

One question which emerges from the concept of foraging as a general phenomenon, is whether different types of spacers employ similar solutions when contending with a given problem of environmental heterogeneity. Few studies have compared the foraging characteristics of spacers such as orthotropic stems, plagiotropic stems, stem analogues (such as petioles and leaf sheaths), and roots. It can be hypothesized that the degree of morphological plasticity shown by a spacer is attuned to the pattern of resource heterogeneity which it encounters. Alternatively, it has been suggested that high levels of morphological plasticity are expressions of strong competitive ability in a species, and hence that a positive correlation is expected between the levels of plasticity exhibited by roots and shoots (Campbell *et al.*, 1991a,b). Tests of such hypotheses call for studies in which levels of plasticity are compared between organs in a variety of environmental conditions, and in which the overall benefits of a given behaviour are measured in terms of competitive ability and resource acquisition. It should also be accepted that the generally low levels of plasticity observed in plagiotropic stems can perhaps only be understood if the plasticity of the roots and shoots of the same species is also considered. However, the relative benefits conferred by different degrees of plasticity in each of these organs have never been assessed for any plant species.

4. Morphological versus Physiological Plasticity

An acute unsolved problem is determination of the relative importance of morphological *versus* physiological plasticity in a given environmental setting. One way of tackling this problem empirically is to subject closely related species (or genotypes or varieties of single species) which differ in these modes of plasticity to environments with a specified patch structure, and to measure their rates of acquisition of essential resources. Studies of this type on phosphate acquisition with species of bunchgrasses (Caldwell *et al.*, 1991a,b); Van Auken *et al.*, 1992) have provided evidence which suggests that physiological plasticity may be more important for resource acquisition in patchy environments than previously thought. Further studies should be carried out on this topic. In addition there is a need for similar experiments in which patch duration is varied. Other suitable combinations of species or subspecies for comparison can easily be identified (for published examples see, e.g., Poorter and Lambers, 1986; Callaway, 1990; Tinoco-Ojanguren and Pearcy, 1992). Ultimately, simulation studies will be a necessary tool to compare the merits of morphological *versus* physiological plasticity (Caldwell *et al.*, 1992; see Section X).

5. Foraging Intensity and Whole-plant Growth

Are the patterns of shoot and root foraging modified in similar ways if the growth of the whole plant is resource limited? It has been shown that both root branching and root elongation increase more vigorously in response to higher local levels of nutrient availability if the nutrient demand of the whole plant is higher. We can also enquire whether shoot branching and leaf production will increase more markedly in high light patches if the growth of the plant is more carbon limited than when it is not. Such responses would be expected because higher levels of branching under the same regimes of photosynthetically active radiation may increase assimilate production (see Novoplansky *et al.*, 1990b), but to our knowledge no experiments have been carried out to examine this phenomenon.

6. Foraging and Resource Integration in Clonal Plants

If part of an integrated clonal plant growing under low resource levels imports resources from another part of the clone growing under more favourable conditions, one would expect that morphological responses to local resource scarcity will be damped as a result of the elevation of internal resource concentrations. Surprisingly, it appears that this is not always the case. Instead, integration may even intensify the morphological responses to local growing conditions (see Alpert, 1991; Evans, 1992). The physiological mechanism behind this reaction and the functional significance of integration for local foraging responses deserve further study. One hypothesis to explain this result is that augmentation of the local response could stimulate a behaviour which would promote escape from locally unfavourable conditions. Integration could thus increase foraging efficiency in a patchy habitat, in contrast to suggestions made by de Kroon and Schieving (1990). The importance of local foraging and integration for resource acquisition may markedly increase if habitat patches are characterized not by some level of favourability, but instead by an adequate supply of some resources and an inadequate supply of others. The consequences of plasticity and integration in such a multiple-resource habitat patch structure are only now beginning to be considered.

7. Mechanisms and Constraints

Some of the mechanisms underlying the foraging responses of plants to variable environmental conditions are still elusive. For example, there is still no clear understanding of the way in which environmental conditions affect apical dominance through changes in the levels of, or the sensitivity to, plant growth substances. The importance of such regulation mechanisms underlying plasticity should be emphasized. The developmental and physiological traits behind the responses that we observe are the traits which are heritable. Some of these traits may be readily modified by natural selection while others may be embedded in a developmental program which constrains their evolution (Sachs, 1988). Indeed, some of the similarities observed in morphological plasticity between species may be due to phylogenetic constraints (de Kroon and van Groenendael, 1990; Fitter and Stickland, 1991), while other morphological characteristics may vary widely within a genus (Mogie and Hutchings, 1990). Another constraint acting on the evolution of plasticity may be a (physiological or genetic) trade-off between morphological and physiological plasticity, causing species with marked morphological *and* physiological plasticity to be rare, as suggested by Fitter and Stickland (1991). Ultimately, information on traits underlying plasticity, and on the inheritance of these traits, will be indispensable for an understanding of the ecological and evolutionary significance of phenotypic plasticity.

X. TOWARDS A FORAGING RESEARCH PROGRAMME

This review has presented information about plasticity in the modular construction of plants, and the consequences of this plasticity for the placement of resource-acquiring structures within a heterogeneous environment. Our ultimate goal is to understand the advantages of different resource-acquiring syndromes and their ecological and evolutionary consequences. An investigation of morphological plasticity will be only one of the elements, albeit an essential one, in a research programme which is directed towards the goal of evaluating foraging behaviour. A complete and accurate evaluation of foraging behaviour requires an assessment of its economics, i.e. a quantification of its costs and benefits. Such an analysis must be based on the premise that the behaviour which results in the highest net resource extraction from the environment will have a selective advantage. Here we outline the requirements for an analysis of the costs and benefits of foraging in plants, as a perspective for future research.

A. Cost–Benefit Analysis

In accordance with foraging models (Stephens and Krebs, 1986) and current views about the costs of plant structures (Bloom *et al.*, 1985; Chapin, 1989),

we define the costs of a certain behaviour as the long-term resource investment in spacers at the level of the whole plant. The benefits are defined as the
long-term resource extraction. An appropriate unit of measurement is the
amount of resources invested or acquired per unit of plant biomass and per
unit of time. Following Hunt *et al.* (1990), the efficiency of foraging can be
expressed as the amount of resources captured as a proportion of the amount
of resources that the environment supplies. The growth that the plant can
achieve will depend, in turn, on the efficiency of resource utilization, i.e. the
rate of biomass production per unit of resource captured. Note that it is not
the strategy with the highest efficiency, but the strategy with the highest
resource gain per unit of time which is most advantageous in a given environmental setting (cf. Stephens and Krebs, 1986).

Figure 10 illustrates the different components of a cost–benefit analysis of
foraging, and their interrelationships. The costs of investment in plant structures can be divided between the costs of construction and costs of maintenance. There will also be *negative* costs associated with the resorption of
resources from dying plant parts. Such costs are the subject of an extensive
body of literature (e.g. Chapin, 1989; Jonasson, 1989; Carlsson and
Callaghan, 1990; Pugnaire and Chapin, 1992). Costs of maintenance may be
significant for carbon due to respiration, but low for mineral nutrients, while a
large proportion of the invested nutrients but hardly any carbon may be
recovered when plant structures senesce (Chapin, 1989). These processes
have been extensively documented for leaves but much less so for spacers
such as orthotropic and plagiotropic stems. A notable exception is provided
by work on tundra species in which resorption of nutrients from senescing
plagiotropic spacers was shown to be between 50 and 90% (Callaghan, 1980;
Headley *et al.*, 1985; Carlsson and Callaghan, 1990). Investment costs of
spacers in terms of the allocation of resources per plant per year may thus be
relatively small, especially for nutrients. Although methods for calculating
such costs are readily available, to our knowledge no such cost analyses have
been carried out to date.

An important, but largely unexplored element in the cost–benefit analysis
of foraging, is the relationship between morphological plasticity and investment costs. Resource investment in spacers is likely to increase with spacer
length, but this relationship will not be proportional if longer spacers are also
thinner, as for shoots growing under shade (Hutchings and Mogie, 1990). For
example, when growing under low photon flux density, *Glechoma hederacea*
produces stolon internodes which are 50% longer but which have a 40%
lower weight than the internodes of clones growing at high photon flux
density (Slade and Hutchings, 1987b). Consequently, under unfavourable
light conditions longer stems may be produced with little if any costs in terms
of reduced biomass allocation to leaves and roots. An illustration of this has
recently been seen in *Amaranthus quitensis* (Ballaré *et al.*, 1991b).

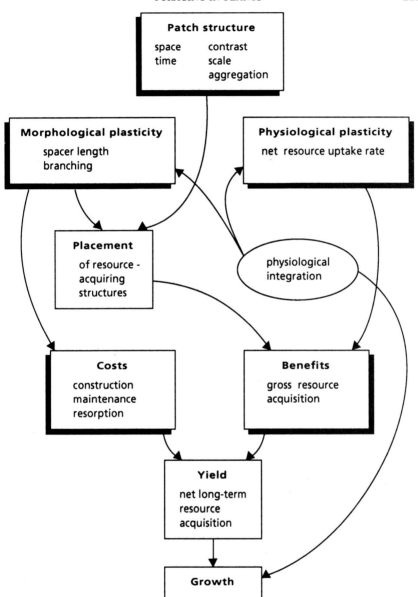

Fig. 10. A diagram of the components of a plant foraging research programme, and the interrelationships between these components. Central elements are the levels of physiological and morphological plasticity and how they influence the costs and benefits of foraging, mediated by the patch structure of the habitat.

220   M.J. HUTCHINGS AND H. DE KROON

If the production of longer spacers is combined with reduced branching intensity, the relationship between morphological plasticity and resource investment becomes more complex. De Kroon and Schieving (1991) formulated a simplified branching model of a clonal plant and calculated the allocation to spacers as a function of resource availability, based on plastic changes in the length and specific weight of spacers and the weight of ramets. Simulations showed that opposing responses in branching intensity and spacer length may compensate for each other in such a way that resource investment in spacers was equal under both low and high resource availability. Surprisingly, this compensation appeared to be incomplete when intermediate levels of resources were available, so that resource investment in spacers was predicted to be larger under these conditions. One virtue of formalizing the relationship between plasticity and investment is that it highlights factors which are important determinants of the relationship, but which are easily overlooked. The model of de Kroon and Schieving (1991; and 1990 using a preliminary version of the model) suggests that the developmental time of spacers *versus* ramets—i.e. the time each of these structures needs to reach its maximum weight—may be critical for the amount of resources allocated to spacers. Such developmental times are only now beginning to be regarded as important attributes for ecological study (e.g. the plastochron index; Birch and Hutchings, 1992; Dong, 1993).

As emphasized above, the patch structure of the habitat is a critical factor governing the benefit gained from morphological plasticity. The pattern of habitat heterogeneity, together with the degree of morphological plasticity which a species exhibits, will determine the extent to which resource-acquiring structures will be placed in the more resource-rich patches of the environment. Ultimately, both morphological and physiological plasticity will contribute to the amount of resources captured by a plant (Fig. 10). The costs of physiological plasticity (the energetic costs of resource uptake and assimilation), and of resource transport, also have to be included in any cost–benefit analysis, although they are not shown separately in Fig. 10 for the sake of simplicity. Physiological integration may modify the magnitude of both the morphological and physiological responses and hence will change the quantity of resource accumulated. By enabling translocation of resources from ramets with low growth rate to ramets with high growth rate, integration could also increase the resource utilization efficiency and growth of a clonal plant (Caraco and Kelly, 1991; see Salzman and Parker, 1985; Slade and Hutchings, 1987d).

B. Foraging Models

Significant progress in plant foraging research will be possible if the different elements depicted in Fig. 10, and their relationships, are modelled and interrelated. The development of such foraging models may already be within reach because most of the components involved have been formalized.

The patch structure of a habitat can be characterized by the contrast in quality between patches (the relative differences in their resource supply) as well as by their scale and distribution in space and time (Addicot *et al.*, 1987; Kotliar and Wiens, 1990). A measure of environmental predictability could also be usefully devised to allow comparisons between habitats (Oborny, 1994). For clonal plants, there is a variety of models which simulate morphological plasticity and the placement of ramets in patchy environments (Ford, 1987; Sutherland and Stillman, 1988, 1990; Oborny, 1994; see also Rubin, 1987), while other spatially explicit models (Bell, 1986; Callaghan *et al.*, 1990; Cain *et al.*, 1991; Klimes, 1992; Cain, 1994) could easily be adapted for this use. There are as yet fewer models of this type for orthotropic shoots and roots (Aono and Kunii, 1984; Bell, 1986; Fitter, 1987). From a knowledge of the patch structure of the habitat and the amount of morphological plasticity available in a species, spatial models will enable prediction of the local conditions that resource-acquiring structures will experience. Models of nutrient uptake (Nye and Tinker, 1977; Barber, 1984; Robinson, 1989; Caldwell *et al.*, 1992) or of photosynthesis (Gross *et al.*, 1991) will then allow calculation of the quantity of resources acquired by the species (i.e. the "benefits" in Fig. 10), taking into account the level of physiological plasticity that the species exhibits. Lastly, investment costs of spacers may be calculated based on estimates of the amount of resources involved in construction, maintenance and resorption (see Chapin, 1989). By using allocation models such as that of de Kroon and Schieving (1991), the effects of morphological plasticity on these costs can also be incorporated.

It is beyond the scope of this review to dwell on the goals of foraging models. Here we confine ourselves to identifying two complementary aims. First, using the scheme of Fig. 10, theoretical models can be constructed to generate hypotheses about probable foraging behaviour in a habitat with a generalized patch structure. Those constraints on the possibilities for foraging behaviour, which are dictated by physiological and genetic regulation mechanisms, should be taken into account in such modelling. Secondly, models may be formulated in which each of the elements of the foraging process is calibrated on the basis of empirical data. With both of these approaches, each of the components is interrelated and the net benefits calculated in terms of net long-term resource extraction. The relative importance of, for instance, morphological *versus* physiological plasticity could be assessed by means of sensitivity analyses, and foraging efficiencies could be computed. Specific foraging syndromes could be contrasted with null-models in which plants place their resource-acquiring structures at random within the habitat, or according to fixed growth rules. It should be possible to identify the environmental parameters with the strongest impact on foraging behaviour. Ultimately, modelling approaches will prove to be imperative in solving the questions raised in the previous section.

We believe that plant and animal foraging can be fruitfully related by comparing the economics of behaviour in relation to the patch structure of the habitat. As a consequence of such comparison, the use of the term foraging in plant ecology will surpass the level of a metaphor (see Oborny, 1991). At the same time, despite the same underlying economic principles, some animal foraging models (such as the Marginal Value Theorem) will only be applicable to plants after considerable adjustment. Many of the assumptions of these models, such as decisions being taken prior to resource uptake, a finite amount of resources per patch (and hence the exhaustion of the patch through the activities of a foraging organism), the discrete nature of patches and resources (prey), and the separation in time of search and exploitation (Stephens and Krebs, 1986), may apply directly to plants only in exceptional cases (Kelly, 1990). It will be a challenge to formulate such adjustments to generate models which are applicable to both plants and animals, and to seek a unified foraging theory to encompass both.

> Elsewhere immense research into the nature, habits and constitution of the triffid went on. Earnest experimenters set out to determine in the interests of science how far and for how long it could walk; whether it could be said to have a front, or could it march in any direction with equal clumsiness; what proportion of its time it must spend with its roots in the ground; what reactions it showed to the presence of various chemicals in the soil; and a vast quantity of other questions, both useful and useless.
>
> (from John Wyndham, *The Day of the Triffids*, 1951)

ACKNOWLEDGEMENTS

We are grateful to Henk Konings, Thijs Pons and Hendrik Poorter for scrutinizing some of the sections of this review and for providing valuable references. Comments on the manuscript made by Michael Cain, Alastair Fitter, Philip Grime and Beata Oborny were also gratefully appreciated. Ming Dong and Josef Stuefer made illuminating analyses of our conceptualization of foraging behaviour in plants. We would also like to thank Colin Birch, Monica Geber, Jan van Groenendael, Vincent Pelling, Liz Price, Andy Slade and Maxine Watson for thought-provoking discussion throughout the years. MJH gratefully acknowledges grant and studentship support from the Natural Environment Research Council, UK. HdK gratefully acknowledges financial support from the Niels Stensen Foundation and the Royal Netherlands Academy of Sciences.

REFERENCES

Addicot, J.F., Aho, J.M., Antolin, M.F., Padilla, D.K., Richardson, J.S. and Soluk, D.A. (1987). Ecological neighbourhoods: scaling environmental patterns. *Oikos* **49**, 340–346.

Aerts, R. and van der Peijl, M.J. (1993). A simple model to explain the dominance of low-productive perennials in nutrient-poor habitats. *Oikos* **66**, 144–147.

Alpert, P. (1991). Nitrogen sharing among ramets increases clonal growth in *Fragaria chiloensis*. *Ecology* **72**, 69–80.

Alpert, P. and Mooney, H.A. (1986). Resource sharing among ramets in the clonal herb, *Fragaria chiloensis*. *Oecologia* **70**, 227–233.

Aono, M. and Kunii, T.L. (1984). Botanical tree image generation. *IEEE Comp. Graph. Appl.* **1984**, 10–34.

Aung, L.H., de Hertogh, A.A. and Staby, G. (1969). Temperature regulation of endogenous gibberellin activity and development in *Tulipa gesneriana* L. *Plant Physiol.* **44**, 403–406.

Ballaré, C.L., Sanchez, R.A., Scopel, A.L., Casal, J.J. and Ghersa, C.M. (1987). Early detection of neighbour plants by phytochrome perception of spectral changes in reflected sunlight. *Plant, Cell Environ.* **10**, 551–557.

Ballaré, C.L., Sanchez, R.A., Scopel, A.L. and Ghersa, C.M. (1988). Morphological responses of *Datura ferox* seedlings to the presence of neighbours: their relationships with canopy microclimate. *Oecologia* **76**, 288–293.

Ballaré, C.L., Scopel, A.L. and Sanchez, R.A. (1990). Far-red radiation reflected from adjacent leaves: an early signal of competition in plant canopies. *Science* **247**, 329–332.

Ballaré, C.L., Scopel, A.L. and Sanchez, R.A. (1991a). Photocontrol of stem elongation in plant neighbourhoods: effects of photon fluence rate under natural conditions of radiation. *Plant, Cell Environ.* **14**, 57–65.

Ballaré, C.L., Scopel, A.L. and Sanchez, R.A. (1991b). On the opportunity cost of the photosynthate invested in stem elongation reactions mediated by phytochrome. *Oecologia* **86**, 561–567.

Banko, R.J. and Boe, A.A. (1975). Effects of pH, temperature, nutrition, ethephon, and chlormequat on endogenous cytokinin levels of *Coleus blumei* Benth. *J. Am. Soc. Hort. Sci.* **100**, 168–172.

Barber, S.A. (1984). *Soil Nutrient Bioavailability*. Wiley, New York.

Barlow, P.W. (1989). Meristems, metamers and modules and the development of shoot and root systems. *Bot. J. Linn. Soc.* **100**, 255–279.

Bates, J.W. (1988). The effect of shoot spacing on the growth and branch development of the moss *Rhytidiadelphus triquetrus*. *New Phytol.* **109**, 499–504.

Bazzaz, F.A. and Harper, J.L. (1977). Demographic analysis of growth of *Linum usitatissimum*. *New Phytol.* **78**, 193–208.

Bell, A.D. (1984). Dynamic morphology: a contribution to plant population ecology. In: *Perspectives on Plant Population Ecology* (Ed. by R. Dirzo and J. Sarukhan), pp. 48–65. Sinauer, Sunderland.

Bell, A.D. (1986). The simulation of branching patterns in modular organisms. *Philos. Trans. R. Soc. Lond.* **B313**, 143–159.

Bendixen, L.E. (1970). Altering growth form to precondition yellow nutsedge for control. *Weed Sci.* **18**, 599–603.

Berendse, F. and Aerts, R. (1987). Nitrogen-use-efficiency: a biologically meaningful definition? *Funct. Ecol.* **1**, 293–296.

Birch, C.P.D. and Hutchings, M.J. (1992). Analysis of ramet development in the stoloniferous herb *Glechoma hederacea* L. using a plastochron index. *Oikos* **63**, 387–394.

Birch, C.P.D. and Hutchings, M.J. (1994). Studies of growth in the clonal herb *Glechoma hederacea*. III. Exploitation of patchily distributed soil resources. *J. Ecol.* **82** (in press).

Bishop, G.F. and Davy, A.J. (1985). Density and the commitment of apical meristems to clonal growth and reproduction in *Hieracium pilosella*. *Oecologia* **66**, 417–422.

Björkman, O. (1981). Responses to different quantum flux densities. In: *Physiological Plant Ecology I. Responses to the Physical Environment* (Ed. by O.L. Lange, P.S. Nobel, C.B. Osmond and H. Ziegler), pp. 57–107. Encyclopedia of Plant Physiology 12A. Springer, Berlin.

Bloom, A.J., Chapin, F.S. and Mooney, H.A. (1985). Resource limitation in plants—an economic analogy. *Ann. Rev. Ecol. Syst.* **16**, 363–392.

Boot, R.G.A. and Mensink, M. (1990). Size and morphology of root systems of perennial grasses from contrasting habitats as affected by nitrogen supply. *Plant Soil* **129**, 291–299.

Braam, J. and Davis, R.W. (1990). Rain-, wind-, and touch-induced expression of calmodulin and calmodulin-related genes in *Arabidopsis*. *Cell* **60**, 357–364.

Bradshaw, A.D. (1965). Evolutionary significance of phenotypic plasticity in plants. *Adv. Genet.* **13**, 115–155.

Bradshaw, A.D., Chadwick, M.J., Jowett, D. and Snaydon, R.W. (1964). Experimental investigations into the mineral nutrition of several grass species. IV. Nitrogen level. *J. Ecol.* **52**, 665–676.

Bray, R.H. (1954). A nutrient mobility concept of soil-plant relationships. *Soil Sci.* **78**, 9–22.

Burns, I.G. (1991). Short- and long-term effects of a change in the spatial distribution of nitrate in the root zone on N uptake, growth and root development of young lettuce plants. *Plant, Cell Environ.* **14**, 21–33.

Cain, M.L., Pacala, S.W. and Silander, J.A. (1991). Stochastic simulation of clonal growth in the tall goldenrod, *Solidago altissima*. *Oecologia* **88**, 477–485.

Cain, M.L. (1994). Consequences of foraging in clonal plant species. *Ecology* (in press).

Caldwell, M.M. (1988). Plant root systems and competition. In: *Proceedings of the XIV International Botanical Congress* (Ed. by W. Greuter and B. Zimmer), pp. 385–404. Koeltz, Königstein.

Caldwell, M.M., Manwaring, J.H. and Durham, S.L. (1991a). The microscale distribution of neighbouring plant roots in fertile soil microsites. *Funct. Ecol.* **5**, 765–772.

Caldwell, M.M., Manwaring, J.H. and Jackson, R.B. (1991b). Exploitation of phosphate from fertile soil microsites by three Great Basin perennials when in competition. *Funct. Ecol.* **5**, 757–764.

Caldwell, M.M., Dudley, L.M. and Lilieholm, B. (1992). Soil solution phosphate, root uptake kinetics and nutrient acquisition: implications for a patchy soil environment. *Oecologia* **89**, 305–309.

Callaghan, T.V. (1980). Age related patterns of nutrient allocation in *Lycopodium annotinum* from Swedish Lapland. *Oikos* **35**, 373–386.

Callaghan, T.V. (1988). Physiological and demographic implications of modular construction in cold environments. In: *Plant Population Ecology* (Ed. by A.J. Davy, M.J. Hutchings and A.R. Watkinson), pp. 111–135. *Symp. Brit. Ecol. Soc.* **28**. Blackwell Scientific Publications, Oxford.

Callaghan, T.V., Svensson, B.M., Bowman, H., Lindley, D.K. and Carlsson, B.A. (1990). Models of clonal plant growth based on population dynamics and architecture. *Oikos* **57**, 257–269.

Callaway, R.M. (1990). Effects of soil water distribution on the lateral root development of three species of Californian oaks. *Am. J. Bot.* **77**, 1469–1475.

Campbell, B.D. and Grime, J.P. (1989). A comparative study of plant responsiveness to the duration of episodes of mineral nutrient enrichment. *New Phytol.* **112**, 261–267.

Campbell, B.D., Grime, J.P. and Mackey, J.M.L. (1991a). A trade-off between scale and precision in resource foraging. *Oecologia* **87**, 532–538.

Campbell, B.D., Grime, J.P. Mackey, J.M.L. and Jalili, A. (1991b). The quest for a mechanistic understanding of resource competition in plant communities: the role of experiments. *Funct. Ecol.* **5**, 241–253.

Caraco, T. and Kelly, C.K. (1991). On the adaptive value of physiological integration in clonal plants. *Ecology* **72**, 81–93.

Carlsson, B.A. and Callaghan, T.V. (1990). Programmed tiller differentiation, intraclonal density regulation and nutrient dynamics in *Carex bigelowii. Oikos* **58**, 219–230.

Casal, J.J., Deregibus, V.A. and Sanchez, R.A. (1985). Variations in tiller dynamics and morphology in *Lolium multiflorum* Lam. vegetative and reproductive plants as affected by differences in red/far-red irradiation. *Ann. Bot.* **56**, 553–559.

Casal, J.J., Sanchez, R.A. and Deregibus, V.A. (1987). The effect of light quality on shoot extension growth in three species of grass. *Ann. Bot.* **59**, 1–7.

Casal, J.J., Sanchez, R.A. and Gibson, D. (1990). The significance of changes in the red/far-red ratio, associated with either neighbour plants or twilight, for tillering in *Lolium multiflorum* Lam. *New Phytol.* **116**, 565–572.

Chapin, F.S. (1980). The mineral nutrition of wild plants. *Ann. Rev. Ecol. Syst.* **11**, 233–260.

Chapin, F.S. (1989). The costs of tundra plant structures: evaluation of concepts and currencies. *Am. Nat.* **133**, 1–19.

Chapin, F.S., Follett, J.M. and O'Connor, K.F. (1982). Growth, phosphate absorption, and phosphorus chemical fractions in two *Chionochloa* species. *J. Ecol.* **70**, 305–321.

Chapin, F.S., Schulze, E.-D. and Mooney, H.A. (1990). The ecology and economics of storage in plants. *Ann. Rev. Ecol. Syst.* **21**, 423–447.

Chazdon, R.L. (1988). Sunflecks and their importance to forest understorey plants. *Adv. Ecol. Res.* **18**, 1–63.

Chazdon, R.L. and Pearcy, R.W. (1986a). Photosynthetic responses to light variation in rainforest species. I. Induction under constant and fluctuating light conditions. *Oecologia* **69**, 517–523.

Chazdon, R.L. and Pearcy, R.W. (1986b). Photosynthetic responses to light variation in rainforest species. II. Carbon gain and photosynthetic efficiency during lightflecks. *Oecologia* **69**, 524–531.

Chazdon, R.L. and Pearcy, R.W. (1991). The importance of sunflecks for forest understorey plants. *BioScience* **41**, 760–766.

Child, R., Morgan, D.C. and Smith, H. (1981). Morphogenesis and simulated shade-light quality. In: *Plants and the Daylight Spectrum* (Ed. by H. Smith), pp. 409–420. Academic Press, London.

Cook, R.E. (1983). Clonal plant populations. *Am. Sci.* **71**, 244–253.

Corré, W.J. (1983). Growth and morphogenesis of sun and shade plants. II. The influence of light quality. *Acta Bot. Neerl.* **32**, 185–202.

Crabtree, R.C. and Bazzaz, F.A. (1992). Seedlings of black birch (*Betula lenta* L.) as foragers for nitrogen. *New Phytol.* **122**, 617–625.

Crick, J.C. and Grime, J.P. (1987). Morphological plasticity and mineral nutrient capture in two herbaceous species of contrasted ecology. *New Phytol.* **107**, 403–414.

Davies, W.J. and Zhang, J. (1991). Root signals and the regulation of growth and development of plants in drying soil. *Ann. Rev. Plant Physiol. Plant Mol. Biol.* **42**, 55–76.

Deregibus, V.A., Sanchez, R.A. and Casal, J.J. (1983). Effects of light quality on tiller production in *Lolium* spp. *Plant Physiol.* **72**, 900–902.

Deregibus, V.A., Sanchez, R.A., Casal, J.J. and Trlica, M.J. (1985). Tillering responses to enrichment of red light beneath the canopy in a humid natural grassland. *J. Appl. Ecol.* **22**, 199–206.

Dong, M. (1993) Morphological plasticity of the clonal herb *Lamiastrum galeobdolon* (L.) Ehrend. & Polatschek in response to partial shading. *New Phytol.* **124**, 291–300.

Dowson, C.G., Rayner, A.D.M. and Boddy, L. (1986). Outgrowth patterns of mycelial cord-forming basidiomycetes from and between woody resource units in soil. *J. Gen. Microbiol.* **132**, 203–211.

Dowson, C.G., Rayner, A.D.M. and Boddy, L. (1988). Foraging patterns of *Phallus impudicus, Phanerochaete laevis* and *Steccherinum fimbriatum* between discontinuous resource units in soil. *FEMS Microbiol. Ecol.* **53**, 291–298.

Dowson, C.G., Rayner, A.D.M. and Boddy, L. (1989a). Spatial dynamics and interactions of the wood fairy ring fungus *Clitocybe nebularis*. *New Phytol.* **111**, 699–705.

Dowson, C.G., Springham, P., Rayner, A.D.M. and Boddy, L. (1989b). Resource relationships of foraging mycelial systems of *Phanerochaete velutina* and *Hypholoma fasciculare* in soil. *New Phytol.* **111**, 501–509.

Drew, M.C. (1975). Comparison of the effects of a localized supply of phosphate, nitrate, ammonium and potassium on the growth of the seminal root system, and the shoot, in barley. *New Phytol.* **75**, 479–490.

Drew, M.C. and Saker, L.R. (1975). Nutrient supply and the growth of the seminal root system in barley. II. Localized, compensatory increases in lateral root growth and rates of nitrate uptake when nitrate supply is restricted to only part of the root system. *J. Exp. Bot.* **26**, 79–90.

Drew, M.C. and Saker, L.R. (1978). Nutrient supply and the growth of the seminal root system in barley. III. Compensatory increases in growth of lateral roots, and in rates of phosphate uptake, in response to a localized supply of phosphate. *J. Exp. Bot.* **29**, 435–451.

Drew, M.C., Saker, L.R. & Ashley, T.W. (1973). Nutrient supply and the growth of the seminal root system in barley. I. The effect of nitrate concentration on the growth of axes and laterals. *J. Exp. Bot.* **24**, 1189–1202.

During, H.J. (1990). Clonal growth patterns among bryophytes. In: *Clonal Growth in Plants: Regulation and Function* (Ed. by J. van Groenendael and H. de Kroon), pp. 153–176. SPB Academic Publishers, The Hague.

Eissenstat, D.M. and Caldwell, M.M. (1988a). Seasonal timing of root growth in favorable microsites. *Ecology* **69**, 870–873.

Eissenstat, D.M. and Caldwell, M.M. (1988b). Competitive ability is linked to rates of water extraction. A field study of two arid tussock grasses. *Oecologia* **75**, 1–7.

Eissenstat, D.M. and Caldwell, M.M. (1989). Invasive root growth into disturbed soil of two tussock grasses that differ in competitive effectiveness. *Funct. Ecol.* **3**, 345–353.

Ellison, A.M. (1987). Density-dependent dynamics of *Salicornia europaea* monocultures. *Ecology* **68**, 737–741.

Ellison, A.M. and Niklas, K.J. (1988). Branching patterns of *Salicornia europaea* (Chenopodiaceae) at different successional stages: a comparison of theoretical and real plants. *Am. J. Bot.* **75**, 501–512.

Eriksson, O. (1986). Mobility and space capture in the stoloniferous plant *Potentilla anserina*. *Oikos* **46**, 82–87.

Evans, J.P. (1988). Nitrogen translocation in a clonal dune perennial, *Hydrocotyle bonariensis*. *Oecologia* **77**, 64–68.

Evans, J.P. (1992). The effect of local resource availability and clonal integration on ramet functional morphology in *Hydrocotyle bonariensis*. *Oecologia* **89**, 265–276.

Fitter, A.H. (1985). Functional significance of root morphology and root system architecture. In: *Ecological Interactions in Soil. Plants, Microbes and Animals* (Ed by A.H. Fitter, D. Atkinson and D.J. Read), pp. 87–106. Blackwell, Oxford.

Fitter, A.H. (1986). The topology and geometry of plant root systems: influence of watering rate on root system topology in *Trifolium pratense*. *Ann. Bot.* **58**, 91–101.

Fitter, A.H. (1987). An architectural approach to the comparative ecology of plant root systems. *New Phytol.* **106 (Suppl)**, 61–77.

Fitter, A.H. (1991). The ecological significance of root system architecture: an economic approach. In: *Plant Root Growth. An Ecological Perspective* (Ed. by D. Atkinson), pp. 229–243. Blackwell, Oxford.

Fitter, A.H. (1994). Architecture and biomass allocation as components of the plastic response of root systems to soil heterogeneity. In: *Exploitation of Environmental Heterogeneity by Plants: Ecophysiological Processes Above and Below Ground* (Ed. by M.M. Caldwell and R.W. Pearcy). Academic Press, San Diego (in press).

Fitter, A.H. and Ashmore, C.J. (1974). Responses of two *Veronica* species to a simulated woodland light climate. *New Phytol.* **73**, 997–1001.

Fitter, A.H., Nichols, R. and Harvey, M.L. (1988). Root system architecture in relation to life history and nutrient supply. *Funct. Ecol.* **2**, 345–351.

Fitter, A.H. and Stickland, T.R. (1991). Architectural analysis of plant root systems. 2. Influence of nutrient supply on architecture in contrasting plant species. *New Phytol.* **118**, 383–389.

Fitter, A.H. and Stickland, T.R. (1992). Architectural analysis of plant root systems. 3. Studies on plants under field conditions. *New Phytol.* **121**, 243–248.

Fitter, A.H., Stickland, T.R., Harvey, M.L. and Wilson, G.W. (1991). Architectural analysis of plant root systems. 1. Architectural correlates of exploitation efficiency. *New Phytol.* **118**, 375–382.

Foggo, M.N. (1989). Vegetative responses of *Deschampsia flexuosa* (L.) Trin. (Poaceae) seedlings to nitrogen supply and photosynthetically active radiation. *Funct. Ecol.* **3**, 337–343.

Ford, E.D. and Diggle, P.J. (1981). Competition for light in a plant monoculture modelled as a spatial stochastic process. *Ann. Bot.* **48**, 481–500.

Ford, H. (1987). Investigating the ecological and evolutionary significance of plant growth form using stochastic simulation. *Ann. Bot.* **59**, 487–494.

Franco, M. (1986). The influence of neighbours on the growth of modular organisms with an example from trees. *Philos. Trans. R. Soc., Lond.* **B313**, 209–225.

Frankland, B. and Letendre, R.J. (1978). Phytochrome and effects of shading on growth of woodland plants. *Photochem. & Photobiol.* **27**, 223–230.

Frankland, J.C., Ovington, J.D. and Macrae, C. (1963). Spatial and seasonal variations in soil, litter and ground vegetation in some Lake District woodlands. *J. Ecol.* **51**, 97–112.

Friedman, D. and Alpert, P. (1991). Reciprocal transport between ramets increases growth of *Fragaria chiloensis* when light and nitrogen occur in separate patches but only if patches are rich. *Oecologia* **86**, 76–80.

Garcia-Martinez, J.L., Keith, B., Bonner, B.A., Stafford, A. and Rappaport, L. (1987). Phytochrome regulation of the response to exogenous gibberellins by epicotyls of *Vigna sinensis*. *Plant Physiol.* **85**, 212–216.

Garnier, E., Koch, G.W., Roy, J. and Mooney, H.A. (1989). Responses of wild plants to nitrate availability. Relationships between growth rate and nitrate uptake parameters, a case study with two *Bromus* species, and a survey. *Oecologia* **79**, 542–550.

Garrison, R. and Briggs, W.R. (1972). Internodal growth in localized darkness. *Bot. Gaz.* **133**, 270–276.

Gersani, M. and Sachs, T. (1992). Developmental correlations between roots in heterogeneous environments. *Plant, Cell Environ.* **15**, 463–469.

Gibson, D.J. (1988a). The relationship of sheep grazing and soil heterogeneity to plant spatial patterns in dune grassland. *J. Ecol.* **76**, 233–252.

Gibson, D.J. (1988b). The maintenance of plant and soil heterogeneity in dune grassland. *J. Ecol.* **76**, 497–508.

Gildner, B.S. and Larson, D.W. (1992). Photosynthetic response to sunflecks in the desiccation-tolerant fern *Polypodium virginianum*. *Oecologia* **89**, 390–396.

Ginzo, H.D. and Lovell, P.H. (1973). Aspects of the comparative physiology of *Ranunculus bulbosus* L. and *Ranunculus repens* L. I. Response to nitrogen. *Ann. Bot.* **37**, 753–764.

Givnish, T.J. (1982). On the adaptive significance of leaf height in forest herbs. *Am. Nat.* **120**, 353–381.

Givnish, T.J. (1986). Biomechanical constraints on crown geometry in forest herbs. In: *On the Economy of Plant Form and Function* (Ed. by T.J. Givnish), pp. 525–583. Cambridge University Press, Cambridge.

Granato, T.C. and Raper, C.D. (1989). Proliferation of maize (*Zea mays* L.) roots in response to localized supply of nitrate. *J. Exp. Bot.* **40**, 263–275.

Gregory, F.G. and Veale, J.A. (1957). A reassessment of the problem of apical dominance. *Symp. Soc. Exp. Biol.* **11**, 1–20.

Grime, J.P. (1966). Shade avoidance and shade tolerance in flowering plants. In: *Light as an Ecological Factor* (Ed. by R. Bainbridge, G.C. Evans and O. Rackham), pp. 187–207. *Symp. Soc. Br. Ecol. Soc.* **6**. Blackwell Scientific Publications, Oxford.

Grime, J.P. (1979). *Plant Strategies and Vegetation Processes*. John Wiley & Sons, Chichester.

Grime, J.P. (1987). Dominant and subordinate components of plant communities: implications for succession, stability and diversity. In: *Colonization, Succession and Stability* (Ed. by A.J. Gray, M.J. Crawley and P.J. Edwards), pp. 412–428. Blackwell, Oxford.

Grime, J.P., Mason, G., Curtis, A.V., Rodman, J., Band, S.R., Mowforth, M.A.G., Neal, A.M. and Shaw, S. (1981). A comparative study of germination characteristics in a local flora. *J. Ecol.* **69**, 1017–1059.

Grime, J.P., Crick, J.C. and Rincon, J.E. (1986). The ecological significance of plasticity. In: *Plasticity in Plants* (Ed. by D.H. Jennings and A.J. Trewavas), pp. 5–29. Biologists Limited, Cambridge.

Grime, J.P., Campbell, B.D., Mackey, J.M.L. and Crick, J.C. (1991). Root plasticity, nitrogen capture and competitive ability. In: *Plant Root Growth. An Ecological Perspective* (Ed. by D. Atkinson), pp. 381–397. Blackwell, Oxford.

Gross, L.J., Kirschbaum, M.U.F. and Pearcy, R.W. (1991). A dynamic model of photosynthesis in varying light taking account of stomatal conductance, C_3-cycle intermediates, photorespiration and Rubisco activation. *Plant, Cell Environ.* **14**, 881–893.

Hall, J.B. (1971). Pattern in a chalk grassland community. *J. Ecol.* **59**, 749–762.

Hara, T., Kimura, M. and Kikuzawa, K. (1991). Growth patterns of tree height and stem diameter in populations of *Abies veitchii, A. mariesii* and *Betula ermanii*. *J. Ecol.* **79**, 1085–1098.

Harper, J.L. (1985). Modules, branches and the capture of resources. In: *Population Biology and Evolution of Clonal Organisms* (Ed. by J.B.C. Jackson, L.W. Buss and R.E. Cook), pp. 1–33. Yale University Press, New Haven.

Hartnett, D.C. and Bazzaz, F.A. (1985). The regulation of leaf, ramet and genet densities in experimental populations of the rhizomatous perennial *Solidago canadensis*. *J. Ecol.* **73**, 429–443.

Headley, A.D., Callaghan, T.V. and Lee, J.A. (1985). The phosphorus economy of the evergreen tundra plant, *Lycopodium annotinum*. *Oikos* **45**, 233–245.

Headley, A.D., Callaghan, T.V. and Lee, J.A. (1988). Phosphate and nitrate movement in the clonal plants *Lycopodium annotinum* L. and *Diphasiastrum complanatum* (L.) Holub. *New Phytol.* **110**, 487–495.

Hetrick, B.A.D., Wilson, G.W.T. and Leslie, J.F. (1991). Root architecture of warm- and cool-season grasses: relationship to mycorrhizal dependence. *Can. J. Bot.* **69**, 112–118.

Hillman, J.R. (1984). Apical dominance. In: *Advanced Plant Physiology* (Ed. by M.B. Wilkins), pp. 127–148. Pitman, Bath.

Hillman, J.R. (1986). Apical dominance and correlations by hormones. In: *Plant Growth Substances 1985* (Ed. by M. Bopp), pp. 341–349. Springer, Berlin.

Holmes, M.G. (1976). Spectral energy distribution in the natural environment and its implications for phytochrome function. In: *Light and Plant Development* (Ed. by H. Smith), pp. 407–476. Butterworths, London.

Holmes, M.G. (1981). Spectral distribution of radiation within plant canopies. In: *Plants and the Daylight Spectrum* (Ed. by H. Smith), pp. 147–158. Academic Press, London.

Holmes, M.G. (1983). Perception of shade. *Philos. Trans. R. Soc., Lond.* **B303**, 503–521.

Holmes, M.G. and Smith, H. (1975). The function of phytochrome in plants growing in the natural environment. *Nature* **254**, 512–514.

Hook, P.B., Burke, I.C. and Lauenroth, W.K. (1991). Heterogeneity of soil and plant N and C associated with individual plants and openings in North American short-grass steppe. *Plant Soil* **138**, 247–256.

Hunt, R., Warren Wilson, J. and Hand, D.W. (1990). Integrated analysis of resource capture and utilization. *Ann. Bot.* **65**, 643–648.

Hutchings, M.J. (1976). Spectral transmission and the aerial profile in mature stands of *Mercurialis perennis* L. *Ann. Bot.* **40**, 1207–1216.

Hutchings, M.J. (1986). The structure of plant populations. In: *Plant Ecology* (Ed. by M.J. Crawley), pp. 97–136. Blackwell Scientific Publications, Oxford.

Hutchings, M.J. (1988). Differential foraging for resources and structural plasticity in plants. *Trends Ecol. Evol.* **3**, 200–204.

Hutchings, M.J. and Bradbury, I.K. (1986). Ecological perspectives on clonal perennial herbs. *BioScience* **36**, 178–182.

Hutchings, M.J. and Mogie, M. (1990). The spatial structure of clonal plants: control and consequences. In: *Clonal Growth in Plants: Regulation and Function* (Ed. by J. van Groenendael and H. de Kroon), pp. 57–76. SPB Academic Publishers, The Hague.

Hutchings, M.J. and Slade, A.J. (1988). Morphological plasticity, foraging and integration in clonal perennial herbs. In: *Plant Population Ecology* (Ed. by A.J. Davy, M.J. Hutchings and A.R. Watkinson), pp. 83–109. *Symp. Br. Ecol. Soc.* **28**. Blackwell Scientific Publications, Oxford.

Hutchings, M.J. and Turkington, R. (1993). Plasticity of branching patterns in the clonal herbs *Trifolium repens* and *Glechoma hederacea*. *Proceedings of the Vth International Symposium of Plant Biosystematics* (Ed. by P.S. Hoch) (in press).

Isbell, V.R. and Morgan, P.W. (1982). Manipulation of apical dominance in sorghum with growth regulators. *Crop Sci.* **22**, 30–35.

Jackson, R.B. and Caldwell, M.M. (1989). The timing and degree of root proliferation in fertile-soil microsites for three cold-desert perennials. *Oecologia* **81**, 149–153.

Jackson, R.B. and Caldwell, M.M. (1991). Kinetic responses of *Pseudoroegneria* roots to localized soil enrichment. *Plant Soil* **138**, 231–238.

Jackson, R.B. and Caldwell, M.M. (1993). The scale of nutrient heterogeneity around individual plants and its quantification with geostatistics. *Ecology* **74**, 612–614.

Jackson, R.B., Manwaring, J.H. and Caldwell, M.M. (1990). Rapid physiological adjustment of roots to localized soil enrichment. *Nature* **344**, 58–60.

de Jager, A. (1982). Effects of localized supply of H_2PO_4, NO_3, SO_4, Ca and K on the production and distribution of dry matter in young maize plants. *Neth. J. Agric. Sci.* **30**, 193–203.

de Jager, A. (1984). Effects of a localized supply of H_2PO_4, NO_3, Ca and K on the concentration of that nutrient in the plant and the rate of uptake by roots in young maize plants in solution culture. *Neth. J. Agric. Sci.* **32**, 43–56.

de Jager, A. and Posno, M. (1979). A comparison of the reaction to a localized supply of phosphate in *Plantago major, Plantago lanceolata* and *Plantago media*. *Acta Bot. Neerl.* **28**, 479–489.

Jerling, L. (1988). Clone dynamics, population dynamics and vegetation pattern in *Glaux maritima* on a Baltic sea shore meadow. *Vegetatio* **74**, 171–185.

Jewiss, O.R. (1972). Tillering in grasses—its significance and control. *J. Br. Grassl. Soc.* **27**, 65–82.

Jinks, R.L. and Marshall, C. (1982). Hormonal regulation of tiller bud development and internode elongation in *Agrostis stolonifera* L. In: *Chemical Manipulation of Crop Growth and Development* (Ed. by J.S. McLaren), pp. 525–542. Butterworths, London.

Johnston, G.F.S. and Jeffcoat, B. (1977). Effects of some growth regulators on tiller bud elongation in cereals. *New Phytol.* **79**, 239–245.

Jonasson, S. (1989). Implications of leaf longevity, leaf nutrient re-absorption and translocation for the resource economy of five evergreen plant species. *Oikos* **56**, 121–131.

Jonasson, S. and Chapin, F.S. (1991). Seasonal uptake and allocation of phosphorus in *Eriophorum vaginatum* L. measured by labelling with ^{32}P. *New Phytol.* **118**, 349–357.

Jones, M. (1985). Modular demography and form in silver birch. In: *Studies in Plant Demography: A Festschrift for John L. Harper* (Ed. by J. White), pp. 223–238. Academic Press, London.

Jones, M. and Harper, J.L. (1987a). The influence of neighbours on the growth of trees. I. The demography of buds in *Betula pendula*. *Proc. R. Soc. Lond.* **B232**, 1–18.

Jones, M. and Harper, J.L. (1987b). The influence of neighbours on the growth of trees. II. The fate of buds on long and short shoots in *Betula pendula*. *Proc. R. Soc. Lond.* **B232**, 19–33.

Jónsdóttir, I.S. and Callaghan, T.V. (1988). Interrelationships between different generations of interconnected tillers of *Carex bigelowii*. *Oikos* **52**, 120–128.

Jónsdóttir, I.S. and Callaghan, T.V. (1990). Intraclonal translocation of ammonium and nitrate nitrogen in *Carex bigelowii* Torr. ex Schwein. using ^{15}N and nitrate reductase assays. *New Phytol.* **114**, 419–428.

Julien, M.H. and Bourne, A.S. (1986). Compensatory branching and changes in nitrogen content in the aquatic weed *Salvinia molesta* in response to disbudding. *Oecologia* **70**, 250–257.

Kachi, N. and Rorison, I.H. (1990). Effects of nutrient depletion on growth of *Holcus lanatus* L. and *Festuca ovina* L. and on the ability of their roots to absorb nitrogen at warm and cool temperatures. *New Phytol.* **115**, 531–537.

Kelly, C.K. (1990). Plant foraging: a marginal value model and coiling response in *Cuscuta subinclusa*. *Ecology* **71**, 1916–1925.

Kelly, C.K. (1992). Resource choice in *Cuscuta europaea*. *Proc. Natl Acad. Sci.* **89**, 12194–12197.

Klimes, L. (1992). The clone architecture of *Rumex alpinus* (Polygonaceae). *Oikos* **63**, 402–409.

Kotliar, N.B. and Wiens, J.A. (1990). Multiple scales of patchiness and patch structure: a hierarchical framework for the study of heterogeneity. *Oikos* **59**, 253–260.

de Kroon, H. and Knops, J. (1990). Habitat exploration through morphological plasticity in two chalk grassland perennials. *Oikos* **59**, 39–49.

de Kroon, H. and Schieving, F. (1990). Resource partitioning in relation to clonal growth strategy. In: *Clonal Growth in Plants: Regulation and Function* (Ed. by J. van Groenendael and H. de Kroon), pp. 113–130. SPB Academic Publishing, The Hague.

de Kroon, H. and Schieving, F. (1991). Resource allocation patterns as a function of clonal morphology: a general model applied to a foraging clonal plant. *J. Ecol.* **79**, 519–530.

de Kroon, H. and van Groenendael, J. (1990). Regulation and function of clonal growth in plants: an evaluation. In: *Clonal Growth in Plants: Regulation and Function* (Ed. by J. van Groenendael and H. de Kroon), pp. 177–186. SPB Academic Publishing, The Hague.

de Kroon, H., Whigham, D.F. and Watson, M.A. (1991). Developmental ecology of mayapple: effects of rhizome severing, fertilization and timing of shoot senescence. *Funct. Ecol.* **5**, 360–368.

Laan, P. and Blom, C.W.P.M. (1990). Growth and survival responses of *Rumex* species to flooded and submerged conditions: the importance of shoot elongation, underwater photosynthesis and reserve carbohydrates. *J. Exp. Bot.* **41**, 775–783.

Lambers, H. and Poorter, H. (1992). Inherent variation in growth rate between higher plants: a search for physiological causes and ecological consequences. *Adv. Ecol. Res.* **23**, 187–261.

Landa, K., Benner, B., Watson, M.A. and Gartner, J. (1992). Physiological integration for carbon in mayapple (*Podophyllum peltatum*), a clonal perennial herb. *Oikos* **63**, 348–356.

Lechowicz, M.J. and Bell, G. (1991). The ecology and genetics of fitness in forest plants. II. Microspatial heterogeneity of the edaphic environment. *J. Ecol.* **79**, 687–696.

Lloret, P.G., Casero, P.J., Navascués, J. and Pulgarín, A. (1988). The effects of removal of the root tip on lateral root distribution in adventitious roots of onion. *New Phytol.* **110**, 143–149.

Lovell, P.H. and Lovell, P.J. (1985). The importance of plant form as a determining factor in competition and habitat exploitation. In: *Studies on Plant Demography: A Festschrift for John L. Harper* (Ed. by J. White), pp. 209–221. Academic Press, London.

Lovett Doust, L. (1987). Population dynamics and local specialization in a clonal perennial (*Ranunculus repens*). III. Responses to light and nutrient supply. *J. Ecol.* **75**, 555–568.

Maillette, L. (1985). Modular demography and growth patterns of two annual weeds (*Chenopodium album* L. and *Spergula arvensis* L.) in relation to flowering. In: *Studies on Plant Demography: A Festschrift for John L. Harper* (Ed. by J. White), pp. 239–256. Academic Press, London.

Maillette, L. (1986). Canopy development, leaf demography and growth dynamics of wheat and three weed species growing in pure and mixed stands. *J. Appl. Ecol.* **23**, 929–944.

Maksymowych, R., Elsner, C. and Maksymowych, A.B. (1984). Internode elongation in *Xanthium* plants treated with gibberellic acid presented in terms of relative elemental rates. *Am. J. Bot.* **71**, 239–244.

Marshall, C. (1990). Source-sink relations of interconnected ramets. In: *Clonal Growth in Plants—Regulation and Function* (Ed. by J. van Groenendael and H. de Kroon), pp. 23–41. SPB Academic Publishing, the Hague.

Martinez-Garcia, J.F. and Garcia-Martinez, J.L. (1992). Interaction of gibberellins and phytochrome in the control of cowpea epicotyl elongation. *Physiol. Plant.* **86**, 236–244.

Matthysse, A.G. and Scott, T.K. (1984). Functions of hormones at the whole plant level of organization. In: *Encyclopaedia of Plant Physiology, New Series* (Ed. by T.K. Scott), **10**, 219–243, Springer, Berlin.

McIntyre, G.I. (1965). Some effects of nitrogen supply on the growth and development of *Agropyron repens* L. Beauv. *Weed Res.* **5**, 1–12.

McIntyre, G.I. (1967). Environmental control of bud and rhizome development in the seedling of *Agropyron repens* L. Beauv. *Can. J. Bot.* **45**, 1315–1326.

McIntyre, G.I. (1976). Apical dominance in the rhizome of *Agropyron repens*: the influence of water stress on bud activity. *Can. J. Bot.* **54**, 2747–2754.

McIntyre, G.I. (1987). Studies on the growth and development of *Agropyron repens*: interacting effects of humidity, calcium and nitrogen on growth of the rhizome apex and lateral buds. *Can. J. Bot.* **65**, 1427–1432.

Menges, E.S. (1987). Biomass allocation and geometry of the clonal herb *Laportea canadensis*: adaptive responses to the environment or allometric constraints? *Am. J. Bot.* **74**, 551–563.

Menhenett, R. and Wareing, P.F. (1975). Possible involvement of growth substances in the response of tomato plants (*Lycopersicon esculentum* Mill.). *J. Hort. Sci.* **50**, 381–397.

Menzel, C.M. (1980). Tuberization in potato at high temperatures: responses to gibberellin and growth inhibitors. *Ann. Bot.* **46**, 259–265.

Mitchell, D.S. and Tur, N.M. (1975). The rate of growth of *Salvinia molesta* (*S. auriculata* Auct.) in laboratory and natural conditions. *J. Appl. Ecol.* **12**, 213–225.

Mitchell, P.L. and Woodward, F.I. (1988). Responses of three woodland herbs to reduced photosynthetically active radiation and low red to far-red radiation in shade. *J. Ecol.* **76**, 807–825.

Mogie, M. and Hutchings, M.J. (1990). Phylogeny, ontogeny and clonal growth in vascular plants. In: *Clonal Growth in Plants: Regulation and Function* (Ed. by J. van Groenendael and H. de Kroon), pp. 3–22. SPB Academic Publishing, The Hague.

Morgan, D.C. (1981). Shadelight quality effects on plant growth. In: *Plants and the Daylight Spectrum* (Ed. by H. Smith), pp. 205–221. Academic Press, London.

Morgan, D.C. and Smith, H. (1979). A systematic relationship between phytochrome-induced development and species habitat, for plants grown in simulated natural radiation. *Planta* **145**, 253–258.

Navas, M.L. and Garnier, E. (1990). Demography and growth forms of the clonal perennial *Rubia peregrina* in managed and unmanaged habitats. *J. Ecol.* **78**, 691–712.

Novoplansky, A. (1991). Developmental responses of *Portulaca* seedlings to conflicting spectral signals. *Oecologia* **88**, 138–140.

Novoplansky, A., Cohen, D. and Sachs, T. (1990a). How *Portulaca* seedlings avoid their neighbours. *Oecologia* **82**, 490–493.

Novoplansky, A., Sachs, T., Cohen, D., Bar, R., Bodenheimer, J. and Reisfeld, R. (1990b). Increasing plant productivity by changing the solar spectrum. *Solar Energy Mat.* **21**, 17–23.

Nye, P.H. (1973). The relation between the radius of a root and its nutrient absorbing power. *J. Exp. Bot.* **24**, 783–786.

Nye, P.H. and Tinker, P.B. (1977). *Solute Movement in the Soil–Root System*. University of California Press, Berkeley.

Oborny, B. (1991). Criticisms on optimal foraging in plants: a review. *Abstr. Bot.* **15**, 67–76.

Oborny, B. (1994). Growth rules in plants and environmental predictability: a simulation study. *J. Ecol.* (in press).

Ogden, J. (1970). Plant population structure and productivity. *Proc. N.Z. Ecol. Soc.* **17**, 1–9.

Passioura, J.B. (1988). Water transport in and to roots. *Ann. Rev. Plant Physiol. Plant Mol. Biol.* **39**, 245–265.

Passioura, J.B. and Wetselaar, R. (1972). Consequences of banding nitrogen fertilizers in soil. II. Effects on the growth of wheat roots. *Plant Soil* **36**, 461–473.

Pearcy, R.W. (1983). The light environment and growth of C_3 and C_4 species in the understory of a Hawaiian forest. *Oecologia* **58**, 26–32.

Pearcy, R.W. (1988). Photosynthetic utilisation of lightflecks by understory plants. *Aust. J. Plant Physiol.* **15**, 223–238.

Pearcy, R.W. (1990). Sunflecks and photosynthesis in plant canopies. *Ann. Rev. Plant Physiol. Plant Mol. Biol.* **41**, 421–453.

Peñalosa, J. (1983). Shoot dynamics and adaptive morphology of *Ipomoea phillomega* (Vell.) House (Convolvulaceae), a tropical rainforest liana. *Ann. Bot.* **52**, 737–754.

Pfitsch, W.A. and Pearcy, R.W. (1989a). Daily carbon gain by *Adenocaulon bicolor* (Asteraceae), a redwood forest understorey herb, in relation to its light environment. *Oecologia* **80**, 465–470.

Pfitsch, W.A. and Pearcy, R.W. (1989b). Steady-state and dynamic photosynthetic response of *Adenocaulon bicolor* (Asteraceae) in its redwood forest habitat. *Oecologia* **80**, 471–476.

Pfitsch, W.A. and Pearcy, R.W. (1992). Growth and reproductive allocation of *Ade-*

234 M.J. HUTCHINGS AND H. DE KROON

nocaulon bicolor following experimental removal of sunflecks. *Ecology* **73**, 2109–2117.

Phillips, I.D.J. (1975). Apical dominance. *Ann. Rev. Plant Physiol.* **26**, 341–367.

Pitelka, L.F. and Ashmun, J.W. (1985). Physiology and integration of ramets in clonal plants. In: *Population Biology and Evolution of Clonal Organisms* (Ed. by J.B.C. Jackson, L.W. Buss and R.E. Cook), pp. 399–435. Yale University Press, New Haven.

Pons, T.L. and Pearcy, R.W. (1992). Photosynthesis in flashing light in soybean leaves grown in different conditions. I. Lightfleck utilization efficiency. *Plant, Cell Environ.* **15**, 577–584.

Pons, T.L., Pearcy, R.W. and Seemann, J.R. (1992). Photosynthesis in flashing light in soybean leaves grown in different conditions. I. Photosynthetic induction state and regulation of ribulose-1,5-biphosphate carboxylase activity. *Plant, Cell Environ.* **15**, 569–576.

Poorter, H. and Lambers, H. (1986). Growth and competitive ability of a highly plastic and a marginally plastic genotype of *Plantago major* in a fluctuating environment. *Physiol. Plant.* **67**, 217–222.

Poorter, H., Remkes, C. and Lambers, H. (1990). Carbon and nitrogen economy of 24 wild species differing in relative growth rate. *Plant Physiol.* **94**, 621–627.

Poorter, H., van der Werf, A., Atkin, O.K. and Lambers, H. (1991). Respiratory energy requirements of roots vary with the potential growth rate of a species. *Physiol. Plant.* **83**, 469–475.

Price, E.A.C. and Hutchings, M.J. (1992). The causes and developmental effects of integration and independence between different parts of *Glechoma hederacea* clones. *Oikos* **63**, 376–386.

Price, E.A.C., Marshall, C. and Hutchings, M.J. (1992). Studies of growth in the clonal herb *Glechoma hederacea*. I. Patterns of physiological integration. *J. Ecol.* **80**, 25–38.

Pugnaire, F.I. and Chapin, F.S. (1992). Environmental and physiological factors governing nutrient resorption efficiency in barley. *Oecologia* **90**, 120–126.

Putz, F.E. and Holbrook, N.M. (1991). Biomechanical studies of vines. In: *Biology of Vines* (Ed. by F.E. Putz and H.A. Mooney), pp. 73–97. Cambridge University Press, Cambridge.

Quail, P.H. (1983). Rapid action of phytochrome in photomorphogenesis. In: *Encyclopedia of Plant Physiology, New Series* 16A (Ed. by W. Shropshire and H. Mohr), **16A**, 178–212. Springer-Verlag, Berlin.

Qureshi, F.A. and McIntyre, G.I. (1979). Apical dominance in the rhizome of *Agropyron repens*: the influence of nitrogen and humidity on the translocation of ^{14}C-labelled assimilates. *Can. J. Bot.* **57**, 1229–1235.

Ray, T.S. (1992). Foraging behaviour in tropical herbaceous climbers (Araceae). *J. Ecol.* **80**, 189–203.

Rayner, A.D.M. and Franks, N.R. (1987). Evolutionary and ecological parallels between ants and fungi. *Trends Ecol. Evol.* **2**, 127–133.

Rice, S.A. and Bazzaz, F.A. (1989). Quantification of plasticity of plant traits in response to light intensity: comparing phenotypes at a common weight. *Oecologia* **78**, 502–507.

Rincon, E. and Grime, J.P. (1989). Plasticity and light interception by six bryophytes of contrasted ecology. *J. Ecol.* **77**, 439–446.

Roberts, J.A. and Hooley, R. (1988). *Plant Growth Regulators*. Blackie, Glasgow.

Robinson, D. (1989). Can the nutrient demand of a plant be sustained by an increase in local inflow rate? *J. Theor. Biol.* **138**, 551–554.

Robinson, D. and Rorison, I.H. (1985). A quantitative analysis of the relationships between root distribution and nitrogen uptake from soil by two grass species. *J. Soil Sci.* **36**, 71–85.

Robinson, D. and Rorison, I.H. (1987). Root hairs and plant growth at low nitrogen availabilities. *New Phytol.* **107**, 681–693.

Robinson, D. and Rorison, I.H. (1988). Plasticity in grass species in relation to nitrogen supply. *Funct. Ecol.* **2**, 249–257.

Room, P.M. (1983). "Falling apart" as a life style: the rhizome architecture and population growth of *Salvinia molesta*. *J. Ecol.* **71**, 349–365.

Ross, J.J. and Reid, J.B. (1992). Ontogenetic and environmental effects on G_1 levels and the implications for the control of internode length. In: *Progress in Plant Growth Regulation* (Ed. C.M. Karssen, L.C. van Loon and D. Vreugdenhil), pp. 180–187. Kluwer, Dordrecht.

Rubin, J.A. (1987). Growth and refuge location in continuous, modular organisms: experimental and computer simulation studies. *Oecologia* **72**, 46–51.

Sachs, T. (1988). Ontogeny and phylogeny: phytohormones as indicators of labile changes. In: *Plant Evolutionary Biology* (Ed. by L.D. Gottlieb and S.K. Jain), pp. 157–176. Chapman and Hall, London.

Sackville Hamilton, N.R. (1982). Variation and adaptation in wild populations of white clover (*Trifolium repens* L.) in East Anglia. PhD Thesis, University of Cambridge.

St John, T.V., Coleman, D.C. and Reid, C.P.P. (1983a). Association of vesicular-arbuscular mycorrhizal hyphae with soil organic particles. *Ecology* **64**, 957–959.

St John, T.V., Coleman, D.C. and Reid, C.P.P. (1983b). Growth and spatial distribution of nutrient-absorbing organs: selective placement of soil heterogeneity. *Plant Soil* **71**, 487–493.

Salisbury, F.B. and Marinos, N.G. (1985). The ecological role of plant growth substances. In: *Encyclopedia of Plant Physiology, New Series* (Ed. by R.P. Pharis and D.M. Reid), **11**, 707–766. Springer-Verlag, Berlin.

Salzman, A.G. (1985). Habitat selection in a clonal plant. *Science* **228**, 603–604.

Salzman, A.G. and Parker, M.A. (1985). Neighbors ameliorate local salinity stress for a rhizomatous plant in a heterogeneous environment. *Oecologia* **65**, 273–277.

Schmid, B. (1990). Some ecological and evolutionary consequences of modular organization and clonal growth in plants. *Evol. Trends Plants* **4**, 25–34.

Schmid, B. and Bazzaz, F.A. (1992). Growth responses of rhizomatous plants to fertilizer application and interference. *Oikos* **65**, 13–24.

Schafer, E. and Haupt, W. (1983). Blue-light effects in phytochrome-mediated responses. In: *Encyclopedia of Plant Physiology, New Series* (Ed. by W. Shropshire and H. Mohr), **16B**, 723–744. Springer-Verlag, Berlin.

Seliskar, D.M. (1990). The role of waterlogging and sand accretion in modulating the morphology of the dune slack plant *Scirpus americanus*. *Can. J. Bot.* **68**, 1780–1787.

Sharif, R. and Dale, J.E. (1980). Growth regulating substances and the growth of tiller buds in barley; effects of cytokinins. *J. Exp. Bot.* **31**, 921–930.

Sharkey, T.D., Seemann, J.R. and Pearcy, R.W. (1986). Contribution of metabolites of photosynthesis to postillumination CO_2 assimilation in response to lightflecks. *Plant Physiol.* **82**, 1063–1068.

Shein, T. and Jackson, D.I. (1972). Interaction between hormones, light and nutrition on extension of lateral buds in *Phaseolus vulgaris* L. *Ann. Bot.* **36**, 791–800.

Shivji, A. and Turkington, R. (1989). The influence of *Rhizobium trifolii* on growth characteristics of *Trifolium repens*: integration of local environments by clonal genets of *T. repens*. *Can. J. Bot.* **67**, 1080–1084.

Sibly, R.M. and Grime, J.P. (1986). Strategies of resource capture by plants—evidence for adversity selection. *J. Theor. Biol.* **118**, 247–250.

Silvertown, J. and Gordon, D.M. (1989). A framework for plant behavior. *Ann. Rev. Ecol. Syst.* **20**, 349–366.

Skálová, H. and Krahulec, F. (1992). The response of three *Festuca rubra* clones to changes in light quality and plant density. *Funct. Ecol.* **6**, 282–290.

Slade, A.J. and Hutchings, M.J. (1987a). The effects of nutrient availability on foraging in the clonal herb *Glechoma hederacea*. *J. Ecol.* **75**, 95–112.

Slade, A.J. and Hutchings, M.J. (1987b). The effects of light intensity on foraging in the clonal herb *Glechoma hederacea*. *J. Ecol.* **75**, 639–650.

Slade, A.J. and Hutchings, M.J. (1987c). Clonal integration and plasticity in foraging behaviour in *Glechoma hederacea*. *J. Ecol.* **75**, 1023–1036.

Slade, A.J. and Hutchings, M.J. (1987d). An analysis of the costs and benefits of physiological integration between ramets in the clonal perennial herb *Glechoma hederacea*. *Oecologia* **73**, 425–431.

Smith, B.H. (1983). Demography of *Floerkea proserpinacoides*, a forest-floor annual. I. Density-dependent growth and mortality. *J. Ecol.* **71**, 391–404.

Smith, H. (1982). Light quality, photoperception and plant strategy. *Ann. Rev. Plant Physiol.* **33**, 481–518.

Smith, H. and Holmes, M.G. (1977). The function of phytochrome in the natural environment. III. Measurement and calculation of phytochrome photoequilibrium. *Photochem. Photobiol.* **25**, 547–550.

Solangaarachchi, S.M. and Harper, J.L. (1987). The effect of canopy filtered light on the growth of white clover *Trifolium repens*. *Oecologia* **72**, 372–376.

Sprugel, D.G., Hinckley, T.M. and Schaap, W. (1991). The theory and practice of branch autonomy. *Ann. Rev. Ecol. Syst.* **22**, 309–334.

Steeves, T.A. and Sussex, I.M. (1989). *Patterns in Plant Development.* Second edition. Cambridge University Press, Cambridge.

Stephens, D.W. and Krebs, J.R. (1986). *Foraging Theory.* Princeton University Press, Princeton.

Strong, D.R. and Ray, T.S. Jr (1975). Host tree location behaviour of a tropical vine (*Monstera gigantea*) by skototropism. *Science* **190**, 804–806.

Sutherland, W.J. (1990). The response of plants to patchy environments. In: *Living in a Patchy Environment* (Ed. by B. Shorrocks and I.R. Swingland), pp. 45–61. Oxford University Press, Oxford.

Sutherland, W.J. and Stillman, R.A. (1988). The foraging tactics of plants. *Oikos* **52**, 239–244.

Sutherland, W.J. and Stillman, R.A. (1990). Clonal growth: insights from models. In: *Clonal Growth in Plants: Regulation and Function* (Ed. by J. van Groenendael and H. de Kroon), pp. 95–111. SPB Academic Publishing, The Hague.

Svensson, B.M. and Callaghan, T.V. (1988). Small-scale vegetation pattern related to the growth of *Lycopodium annotinum* and variations in its micro-environment. *Vegetatio* **76**, 167–178.

Tamas, I.A. (1987). Hormonal regulation of apical dominance. In: *Plant Hormones and their Role in Plant Growth and Development* (Ed. by P.J. Davies), pp. 393–410. Kluwer, Dordrecht.

Tamas, I.A., Langridge, W.H.R., Abel, S.D., Crawford, S.W., Randall, J.D., Schell, J. and Szalay, A.A. (1992). Hormonal control of apical dominance. Studies in tobacco transformed with bacterial luciferase and *Agrobacterium rol* genes. In:

Progress in Plant Growth Regulation (Ed. by C.M. Karssen, L.C. van Loon and D. Vreugdenhil), pp. 418–430. Kluwer, Dordrecht.

Thomas, B. (1981). Specific effects of blue light on plant growth and development. In: *Plants and the Daylight Spectrum* (Ed. by H. Smith), pp. 443–459. Academic Press, London.

Thompson, L. (1993). The influence of natural canopy density on the growth of white clover, *Trifolium repens*. *Oikos* **67**, 321–324.

Thompson, L. and Harper, J.L. (1988). The effect of grasses on the quality of transmitted radiation and its influence on the growth of white clover *Trifolium repens*. *Oecologia* **75**, 343–347.

Tinoco-Ojanguren, C. and Pearcy, R.W. (1992). Dynamic stomatal behavior and its role in carbon gain during lightflecks of a gap phase and an understory *Piper* species acclimated to high and low light. *Oecologia* **92**, 222–228.

Turkington, R. and Klein, E. (1991). Integration among ramets of *Trifolium repens*. *Can. J. Bot.* **69**, 226–228.

Van Auken, O.W., Manwaring, J.H. and Caldwell, M.M. (1992). Effectiveness of phosphate acquisition by juvenile cold-desert perennials from different patterns of fertile-soil microsites. *Oecologia* **91**, 1–6.

van der Hoeven, E.C., de Kroon, H. and During, H.J. (1990). Fine-scale spatial distribution of leaves and shoots of two chalk grassland perennials. *Vegetatio* **86**, 151–160.

van der Sman, A.J.M., Voesenek, L.A.C.J., Blom, C.W.P.M., Harren, F.J.M. and Reuss, J. (1991). The role of ethylene in shoot elongation with respect to survival and seed output of flooded *Rumex maritimus* L. plants. *Funct. Ecol.* **5**, 304–313.

Voesenek, L.A.C.J. and Blom, C.W.P.M. (1989). Growth responses of *Rumex* species in relation to submergence and ethylene. *Plant Cell Environ.* **12**, 433–439.

Voesenek, L.A.C.J., Perik, P.J.M., Blom, C.W.P.M. and Sassen, M.M.A. (1990). Petiole elongation in *Rumex* species during submergence and ethylene exposure: the relative contributions of cell division and cell expansion. *J. Plant Growth Regul.* **9**, 13–17.

Wareing, P.F. (1964). The developmental physiology of rhizomatous and creeping plants. *Proc. 7th Br. Weed Control Conf.* pp. 1020–1030.

Waters, I. and Shay, J.M. (1990). A field study of the morphometric response of *Typha glauca* shoots to a water depth gradient. *Can. J. Bot.* **68**, 2339–2343.

Watson, M.A. (1984). Developmental constraints: effect on population growth and patterns of resource allocation in a clonal plant. *Am. Nat.* **123**, 411–426.

Watson, M.A. and Casper, B.B. (1984). Morphogenetic constraints on patterns of carbon distribution in plants. *Ann. Rev. Ecol. Syst.* **15**, 233–258.

Watson, M.A., Carrier, J.C. and Cook, G.L. (1982). Effects of exogenously supplied gibberellic acid (G_3) on patterns of water hyacinth development. *Aquat. Bot.* **13**, 57–68.

Weaver, J.E. (1919). *The Ecological Relations of Roots*. Carnegie Institute, Washington, DC.

Weiner, J., Berntson, G.M. and Thomas, S.C. (1990). Competition and growth form in a woodland annual. *J. Ecol.* **78**, 459–469.

Westoby, M., Rice, B. and Howell, J. (1990). Seed size and plant growth form as factors in dispersal spectra. *Ecology* **71**, 1307–1315.

White, J. (1979). The plant as a metapopulation. *Ann. Rev. Ecol. Syst.* **10**, 109–145.

Wiersum, L.K. (1958). Density of root branching as affected by substrate and separate ions. *Acta Bot. Neerl.* **7**, 174–190.

Williams, E.D. (1971). Effects of light intensity, photoperiod and nitrogen on the growth of seedlings of *Agropyron repens* (L.) Beauv. and *Agrostis gigantea* Roth. *Weed Res.* **11**, 159–170.

Woodward, E.J. and Marshall, C. (1988). Effects of plant growth regulators and nutrient supply on tiller bud outgrowth in barley (*Hordeum distichum* L.). *Ann. Bot.* **61**, 347–354.

Woolley, D.J. and Wareing, P.F. (1972). The interaction between growth promoters in apical dominance. II. Environmental effects on endogenous cytokinin and gibberellin levels in *Solanum andigena*. *New Phytol.* **71**, 1015–1025.

Young, T.P. and Hubbell, S.P. (1991). Crown asymmetry, treefalls and repeat disturbance of tropical forest gaps. *Ecology* **72**, 1464–1471.

Fire Frequency Models, Methods and Interpretations*

E.A. JOHNSON and S.L. GUTSELL

I. SUMMARY

The study of wildfires has proved to be an effective system in which to work out the models and methods for studying disturbance frequency. Fire frequency is an estimate of the probability distribution of survival or mortality

---

* This review is dedicated to the memory of M.L. Heinselman. Bud Heinselman pioneered the first fire frequency study and it was his 1969 University of Minnesota seminar on his Boundary Waters Canoe Area study that started one of us (EAJ) thinking about disturbance frequency.

ADVANCES IN ECOLOGICAL RESEARCH VOL. 25
ISBN 0–12–013925–1

from fire. It is a way of following the survival or mortality from fire for each of the units in a landscape.

Fire frequency studies are generally carried out in order to estimate one of two related distributions: time-since-fire (survivorship) and fire interval (mortality). The time-since-fire or survivorship distribution, $A(t)$, measures the cumulative proportion of the entire landscape surviving longer than time t, while the fire interval or mortality distribution, $f(t)$, is the probability of fire occurring in the landscape in the interval t to $t + \Delta t$ per unit time. Another function, the hazard of burning, $\lambda(t)$, ties together time-since-fire and fire interval distributions. It is the per capita age-specific mortality from fire, sometimes called mortality force or age-specific death rate.

Two models have been used in fire frequency studies to characterize the time-since-fire and fire interval distributions: the negative exponential and the Weibull. These two models define different hazards of burning, $\lambda(t)$. The hazard of burning in the negative exponential model is constant with time, that is, the age of the forest does not influence the probability of burning. The Weibull, on the other hand, has either a constant hazard or an increasing hazard of burning, implying that older forests have a greater probability of burning.

In order to determine the time-since-fire distribution for a given study area, a complete inventory of the entire landscape or a random (unbiased) sample of the area covered by different time-since-fire ages is required. The frequency at which different ages occur in a sample is an estimate of the area covered by these ages across the entire landscape. To determine the fire interval distribution for a study area a random sample is also required to obtain, for each sample, a record of as many past fires as possible. Each sample will therefore have its own record of fire intervals which are then pooled together with all other samples.

When sampling for fire occurrence it is essential first to learn to recognize fire evidence, to date fires and to collect and prepare tree ring data. Fire scars are the most often used evidence of past fire; however, other forms of evidence may also be used, including: dating trees which are known to have been killed by fire, using dendrochronology; examining tree ring growth release in trees that have survived a fire; and ageing trees believed to have been recruited immediately after a fire. Fire scars remain the most reliable method of dating past fires. The only accurate way to determine the correct date of a fire recorded on a fire scar is to cross-date the tree against a tree ring (width) master chronology. Cross-dating is best done using a complete disc from the base of a tree. This method is preferred over coring since the ring circuit near a scar is often distorted.

Many past fire frequency studies have not adhered to a strict sampling scheme; instead they have simply involved the casual sampling of individual fire-scarred trees with multiple fire records. These "fire chronology" studies estimate the mean fire interval as the average of all fire intervals during a designated time and area. All fire scarred trees do *not* have an equal chance of being chosen and hence it is not a random sample. Since all estimations of statistical distributions or their parameters assume random sampling, the distribution obtained is useless since the bias is unknown and unknowable.

II. INTRODUCTION

Spatial and temporal heterogeneity play a central role in understanding population, community and ecosystem processes (e.g. Weins, 1976; Whittaker and Levin, 1977). Disturbance has been recognized as one of the important generators of this heterogeneity at all scales (White, 1979; Pickett and White, 1984; Sousa, 1984; Johnson, 1992). The frequency of disturbance events has been the focus of current investigations in population dynamics (Gilpin, 1987), abundance and rarity (Hanski, 1985), species diversity (Connell, 1978; Denslow, 1987), community change (Shugart, 1984), and ecosystem processes (Sprugel, 1976; Bormann and Likens, 1979). The study of wildfires in particular has proven to be an effective system in which to work out the models and methods for studying disturbance frequency.

Fire frequency is an estimate of the probability distribution of survival or mortality from fire in a population of non-overlapping landscape units. Consequently, fire frequency over the landscape is a way of following the survival or mortality from fire for each of the units in the landscape.

Originally, fire frequency studies were simply fire history studies, that is, informal accounts of fire occurrence (Spurr, 1954) as recorded on fire-scarred trees. Heinselman (1973) revised this approach for a study in the Boundary Waters Canoe Area in Minnesota (USA) by mapping the age mosaic of times since fire* (e.g. Fig. 1). He then used the area covered by each time-since-fire year, expressed as a cumulative proportion of the study area (reasoning in Section III), to estimate the survivorship from wildfire in the landscape. Unfortunately, few investigators have followed Heinselman in constructing a "time-since-fire" map for large areas. The tendency has been to map very small areas and/or to *informally* collect fire scar data over a landscape (Arno and Sneck, 1977). In both cases the estimation of fire frequency is not reliable because the sample area is too small and/or a statistically acceptable sampling design is lacking.

---

* Heinselman (1973) used the term "stand origin" map instead of "time-since-fire". We prefer the latter term because it is more descriptive. "Stand origin" can take on many meanings depending on the investigator's idea of stand dynamics. For example, it has often implied the vague term "stand (canopy) replacing fires" which does not specify the fire behaviour. The fire in a "stand replacing fire" could be a passive or active crown fire (cf. Van Wagner, 1977). "Time-since-fire" here means only that a fire occurred in the area. No assumption is made about its behaviour.

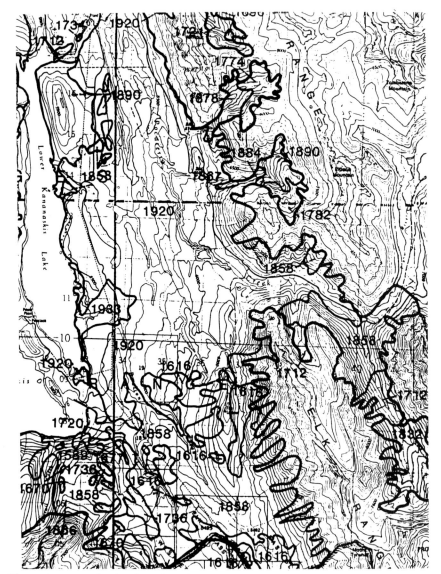

Fig. 1. An example of a time-since-fire map. The date in each patch gives the year of the last fire.

In the late 1970s, Van Wagner (1978) and Johnson (1979) formalized ways to think about fire frequency in two models. These models provided explicit statistical distributions and hypothesized mechanisms for the resulting spatial and temporal pattern of fire occurrence. The models were of particular value in giving some order to the proliferation of fire history concepts (see Romme, 1980 versus Johnson and Van Wagner, 1985). They allowed more rigorous hypotheses and empirical tests of spatial and temporal variations in fire frequency and their relationships to fire weather, fuel complexes, and topography. Since the present models of fire frequency are relatively simple, they will certainly be replaced in the future by models which better capture the salient points of fire frequency as we gain a better empirical understanding of the processes involved. One of the important points of this paper is that a lack of good empirical data of fire frequency is presently preventing this progress.

Many investigators (e.g. Heinselman, 1973; Van Wagner, 1978; Tande, 1979; Yarie, 1981; McCune, 1983) have recognized that most empirical fire frequency distributions are multimodal, indicating changes in the average fire frequency. Johnson and Larsen (1991) provided a simple graphical method for partitioning these mixed distributions into their homogenous components. Clark (1990) introduced Cox's (1972) method of proportional hazard analysis that allows the determination of the environmental processes which lead to differences in survivorship from fire in homogenous fire frequency distributions.

Most fire frequency studies have been carried out in ecosystems where fires leave both fire scars and distinct boundaries between different aged burns. The vegetation types that are most common in these ecosystems include chaparral (sclerophyllous vegetation), the boreal forest, subalpine and montane forests, northern hardwood-pine forests and pine savannas. Most of these vegetation types have average fire frequencies close to the senescence age of the canopy and a fuel structure that results in frequent crown fires because of high surface flame intensity, low basal canopy height, high canopy bulk density and lower canopy foliage moisture (see Van Wagner, 1977; Johnson, 1992).

Our objectives in this paper are to present, in an accessible manner, the fundamental mathematical basis, the sampling design, and the analytical and field methods required in fire frequency studies. As is normal in any developing area of scientific research, fire frequency studies have suffered from their share of confusion and difficulties. In particular they suffer from: (i) a lack of a clear understanding of the meaning of survivorship and mortality in fire frequency; (ii) a need for a statistical sampling design; and (iii) a need for an understanding of fire behaviour as related to fire frequency.

This review can be read in one of two ways. The first involves the traditional approach in ecology of understanding the mathematical models, the sampling design and analysis followed by the methods of giving empirical content to the models. If this approach is preferred, simply read the review in the order presented. The second is for the reader who would like an understanding of empirical fire frequency techniques, with a minimal discussion of the mathematical models. If an empirical understanding of the techniques is all that is required, we suggest that you omit Section III, parts B–F. We must stress, however, that in the long run a good understanding of the models is essential. Without this understanding the investigator will have difficulty understanding the disturbance mechanisms hypothesized in the models and the reason for the sampling design or techniques.

III. MODELS OF FIRE FREQUENCY

A. Introduction to Fire Frequency Models

In this section we will present a very brief overview of the essential distributions required for an understanding of fire frequency. A more mathematical treatment can be found in the remaining parts of this section which may be skipped in the initial reading.

As stated in Section II, fire frequency is an estimate of the probability distribution of survival or mortality from fire in a population of non-overlapping landscape units. Consequently, fire frequency over the landscape is a way of following the survival or mortality from fire for each of the units in that landscape.

Generally, fire frequency studies are carried out in order to estimate one of two related distributions: time-since-fire (survivorship) and fire interval (mortality) (see Fig. 2). The time-since-fire distribution or survivorship, $A(t)$, is the cumulative proportion of the landscape surviving longer than time t (e.g. Fig. 1). It is the probability of going without fire longer than time t. The fire interval or mortality distribution, $f(t)$, is a probability density function of the intervals between fires from a random sample of landscape units (e.g. Fig. 3). It is the probability of a fire occurring in the interval t to $t + \Delta t$ per unit time.

The hazard of burning, $\lambda(t)$ (Fig. 2), ties together the time-since-fire and the fire interval distributions. It is the per capita age-specific mortality from fire, sometimes called mortality force, age-specific death rate or instantaneous death rate. It is the probability of fire occurring in an interval, assuming survival up to the beginning of the interval. Consequently, the definition is $\lambda(t) = f(t)/A(t)$, where $f(t)$ gives the age-specific mortality and $1/A(t)$ gives the per capita rate of survival (remember $A(t)$ is the proportion of the landscape surviving). The hazard of burning, $\lambda(t)$, is not to be confused with "fire hazard" as it is used in forestry. Fire hazard is an assessment of the *potential* fire behaviour (intensity, rate of spread, fuel consumption) based *only* on the fuel load, fuel architecture and phenology of the fuel but not on the fuel moisture content.

Two models have been used in fire frequency, the negative exponential and the Weibull (see Fig. 2 for a comparison). Each model has an $A(t)$, $f(t)$ and $\lambda(t)$ distribution, and further, has parameters which characterize the distributions. The negative exponential is characterized by a single parameter which, in statistics, is called the scale parameter. The Weibull is characterized by a scale parameter as well as a shape parameter.

Up to this point we have talked about homogeneous distributions; however, most empirical fire frequencies are mixed distributions. Mixed distributions arise when different parts of the landscape have different fire frequencies (spatially heterogeneous) and/or the fire frequency has changed at some time in the past (temporally heterogeneous).

B. Fundamental Distributions

Now we present a more detailed explanation of the essential distribution in fire frequency studies. Unfortunately for most readers, fire frequency studies require an understanding of statistics not usually presented in books on introductory statistics. Instead, books on survival models (e.g. Lawless, 1982) or reliability (e.g. Nelson, 1982) will provide this background.

The random variable T denotes the time at which fire or burning occurs. This random variable can be characterized by three equivalent distributions: time-since-fire distribution or survivorship $A(t)$, cumulative mortality distribution $F(t)$, and fire interval or mortality distribution $f(t)$ (Fig. 2).

The time-since-fire or survivorship distribution function $A(t)$ is the probability of having gone without fire (survived) longer than time t, that is,

$$A(t) = Pr(T > t) \quad t \geq 0 \tag{1}$$

When $t = 0$, $A(0) = 1$ and $\lim_{t \to \infty} A(t) = 0$. Notice that $A(t)$ is a cumulative distribution (upper case letters will always denote cumulative distributions and lower case letters, density distributions).

The cumulative mortality distribution function $F(t)$ is the probability that a fire will have occurred before or at time t, that is,

$$F(t) = Pr(T \leq t) \tag{2}$$

where $F(0) = 0$ and $F(\infty) = 1$. The cumulative mortality distribution $F(t)$ is related to $A(t)$ by the equality $F(t) = 1 - A(t)$.

The fire interval or mortality distribution function, $f(t)$, is the probability of having fires in the interval t to $t + \Delta t$

$$f(t) = \frac{dF(t)}{dt} \quad t \geq 0 \tag{3}$$

or

$$f(t) = -\frac{dA(t)}{dt}$$

NEGATIVE EXPONENTIAL   **WEIBULL**

Time Since Fire Distribution (Survivorship)

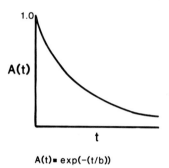

$$A(t) = \exp(-(t/b))$$

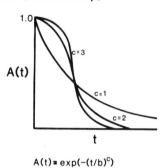

$$A(t) = \exp(-(t/b)^c)$$

Cumulative Mortality Distribution

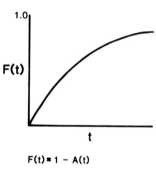

$$F(t) = 1 - A(t)$$
$$= 1 - \exp(-(t/b))$$

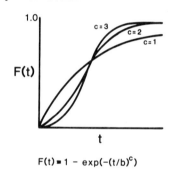

$$F(t) = 1 - \exp(-(t/b)^c)$$

Fire Interval Distribution

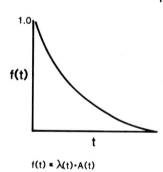

$$f(t) = \lambda(t) \cdot A(t)$$
$$= (1/b) \cdot \exp(-t/b)$$

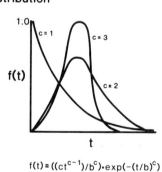

$$f(t) = ((ct^{c-1})/b^c) \cdot \exp(-(t/b)^c)$$

Hazard of Burning Function

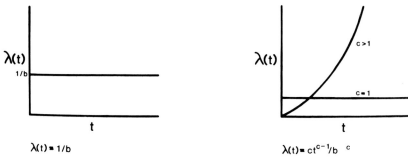

$\lambda(t) = 1/b$ $\lambda(t) = ct^{c-1}/b^c$

Fig. 2. The three fundamental distribution functions and the hazard of burning function for the negative exponential and Weibull fire frequency models: the time-since-fire (survivorship) distribution starts with landscape unburned and describes the proportion remaining (surviving) at different ages. Note that time is time from last fire to present, called running time. The cumulative mortality distribution (discussed in Section III.B) is the probability that a fire will have occurred before or at time t. The fire interval distribution is the probability of a fire interval of a given length occurring. Note that time here is fire interval, not running time. The hazard of burning, $\lambda(t)$, gives the proportion of landscape burning in each time period (age class). Consequently, it gives the rate of decrease of the time-since-fire distribution.

Remember that $\int_0^\infty f(u)du = 1$. Because of this relationship, $f(t)$ is called a probability density function and thus

$$F(t) = \int_0^t f(u)du$$

and

$$A(t) = \int_t^\infty f(u)du$$

The letter u is used above simply to avoid confusion. $F(t)$ is a function of the upper limit t of integration and not of the dummy letter u in $f(u)du$. The probability density function, $f(t)$, shows the relationship between the time-since-fire distribution, $A(t)$, and a function which defines the hazard of burning, $\lambda(t)$. The hazard of burning is defined as

$$\lambda(t) = \frac{f(t)}{A(t)} \qquad (4)$$

or perhaps more easily understood by rearranging (4), which results in

$$f(t) = \lambda(t) \cdot A(t) \qquad (5)$$

From equation (5), $f(t)$ is the probability of having a fire in the interval t to

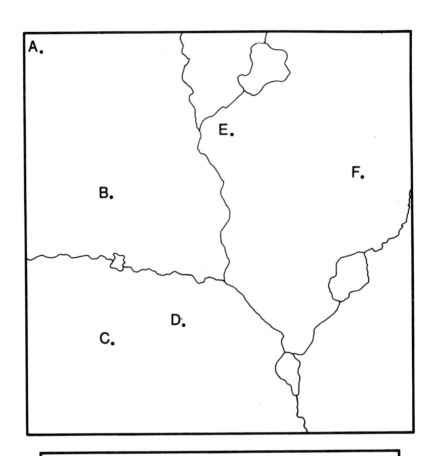

FIRE INTERVAL SAMPLES

A		B		C		D		E		F	
INTERVAL (years)	DATE	INTERVAL (years)	DATE	INTERVAL (years)	DATE	INTERVAL (years)	DATE	INTERVAL (years)	DATE	INTERVAL (years)	DATE
18	1914	24	1888	4	1886	4	1886	19	1888	4	1908
2	1896	52	1864	18	1882	18	1882	50	1869	13	1904
6	1894	10	1812	35	1864	44	1864	11	1819	6	1891
2	1888	33	1802	15	1829	15	1820	40	1808	16	1885
15	1886	40	1769	9	1814	38	1805	13	1768	32	1869
7	1871	17	1729	80	1805	5	1767	27	1755	9	1837
8	1864	12	1712	13	1775	31	1762	28	1728	91	1828
54	1856	39	1700		1762		1731	39	1700	37	1737
29	1802		1661						1661	47	1700
70	1773									22	1635
25	1703										1631
17	1678										
10	1661										
	1651										

$t + \Delta t$, $A(t)$ is the chance of surviving to the beginning of that interval and $\lambda(t)dt$ is the chance of burning in the interval. From the definition of $f(t)$,

$$\lambda(t) = \left(-\frac{1}{A(t)}\right)\left(\frac{dA(t)}{dt}\right) = -\frac{d}{dt}\ln A(t) \tag{6a}$$

and

$$\int_{0}^{t} \lambda(u)du = -\ln A(t) \tag{6b}$$

This last equation is very useful in comparing a fire frequency model's hazard to the empirically determined hazard (see next section).

All fire frequency studies must be reducible to these fundamental distributions. Studies often confuse the cumulative time-since-fire distribution, $A(t)$, with the probability density fire interval distribution, $f(t)$. This is particularly easy when both distributions have the same shape as is the case in the negative exponential (Fig. 2). The simplest way to understand which distribution is being dealt with is to remember that the time-since-fire distribution is the chance of going without fire (surviving) longer than time t while the fire interval distribution is the chance of having a fire in the interval t to $t + \Delta t$. Consequently, the first distribution gives the survivorship and the second gives the mortality rate, a distinction which should be familiar to ecologists conversant in population dynamics or life tables (Begon and Mortimer, 1986).

A moment's reflection will reveal that the time of occurrence of fires and the intervals between them must be related. This relationship is given in the renewal function

$$R(t) = F(t) + \int_{0}^{t} R(t-z)\,dF(z) \tag{7}$$

The $R(t)$ gives the expected number of times a unit will burn up to and including t. For the negative exponential, the renewal function is the Poisson distribution. The $R(t)$ for the Weibull is somewhat more complicated and a discussion can be found in Smith and Leadbetter (1963) and Clark (1989).

C. Negative Exponential and Weibull Models

The above discussion did not specify any particular functions for the distributions. Here we will assign two particular functions, the negative exponen-

---

Fig. 3. An example of fire interval data collected from a simple random sample in the study area.

250 E.A. JOHNSON AND S.L. GUTSELL

tial and the Weibull, to the distributions $A(t)$, $F(t)$ and $f(t)$. Figure 2 shows these distributions and their parameters. Notice that the negative exponential is a special case of the Weibull when its parameter $c = 1$. The parameter c, called the shape parameter, is dimensionless. As its name implies, it defines the changing shape of the Weibull model (see Fig. 2). The parameter b, called the scale parameter, has the dimensions of time.

The negative exponential and Weibull functions represent fire frequency models because they define different hazards of burning (Fig. 2). The hazard of burning of the negative exponential is constant with time, that is, the age of the forest does not influence the probability of burning. The Weibull, on the other hand, has either an increasing hazard of burning when its shape parameter $c > 1$ or a constant hazard when $c = 1$. The increasing hazard ($c > 1$) implies that older forests have a greater chance of burning. We assume that $0 \leq c < 1$ is not reasonable, that is, a decrease in hazard rate with age does not occur. Future studies may indicate otherwise.

The increasing hazard only describes a relationship between survival time and chance of burning; it does not define the nature (cause) of this relationship. The same is true of the constant hazard in the negative exponential. One must be careful not to use emotive words to imply that either "fuel accumulation" or "random ignition" are the cause of these hazards. These causes must be tested directly with empirical data.

The difference between the hazard of burning, $\lambda(t)$, in the negative exponential and the Weibull models and the importance of the hazard shape in suggesting a burning mechanism make empirical estimates of $\lambda(t)$ necessary. Equation 6b provides one means for testing the assumed form. Figure 4 shows the cumulative hazard for the negative exponential model and the empirical cumulative hazard for one time period of Heinselman's (1973) study.

D. Estimating the Parameters and Comparing Time-since-fire Distributions

We will consider two approaches to estimating the parameters of the negative exponential and Weibull models: graphical and analytical. In fire frequency studies, the best way to begin is to use the graphic approach: plot the data on a graph and then estimate the parameter. For negative exponential, semilog paper is used, and Weibull probability paper is used for the Weibull model.

In the negative exponential, the single parameter b can be estimated graphically on the $A(t)$ distribution by locating 36·8% on the ordinate and reading from the curve to the abscissa. This is because in the negative exponential age distribution, $A(t)$, 63·2% of all fire-initiated units will be younger than the sample mean (Van Wagner, 1978)(see Section VI for further explanation). The Maximum Likelihood parameter estimate is the simple mean

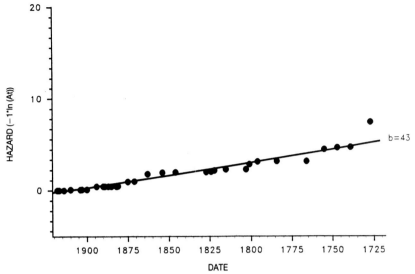

Fig. 4. The cumulative hazard for the negative exponential model. The dots are the empirical hazard of burning for the period 1727–1918 for the Boundary Waters Canoe Area (Heinselman, 1973).

$$\hat{b} = \frac{\sum\limits_{i=1}^{r} a_i x_i}{N} \tag{8}$$

where x_i is the time-since-fire age, a_i is the area in age x_i, r is the number of time-since-fire ages in the study (i.e. number of age classes) and N is the area of the total study. As can be seen in this equation, it is important to remember that percentage area $(a_i x_i/N)$ is the estimate of frequency.

The Weibull parameters can be graphically estimated from special scales on Weibull probability paper, see e.g. Nelson (1982). The Maximum Likelihood estimate (Cohen, 1965) requires an iterative process that can easily be solved on personal computers. The Maximum Likelihood of the shape parameter (c) is

$$\left[\sum_{i=1}^{r} (a_i x_i)^{\hat{c}} \ln t_i \right] \Big/ \left[\sum_{i=1}^{r} (a_i x_i)^{\hat{c}} \right] = \frac{1}{N} \left[\sum_{i=1}^{r} (a_i x_i) \right] - \frac{1}{\hat{c}} \tag{9a}$$

and the scale parameter (b) is:

$$\hat{b} = \left[\frac{1}{N} \sum_{i=1}^{r} (a_i x_i)^{\hat{c}} \right]^{1/\hat{c}} \tag{9b}$$

The starting estimate of c is obtained from the estimate on Weibull probability paper. After parameters are estimated, the goodness of fit must be tested. D'Agostino and Stephens (1986) give a detailed consideration of the many approaches and goodness of fit tests.

Often the easiest method to estimate parameters and goodness of fit is to use regressions. In the case of the negative exponential, the time-since-fire $A(t)$ equation can be linearized and the parameter b estimated by the slope of the regression line. For the Weibull, many non-linear regression programs can be used to estimate the scale and shape parameters. Initial starting estimates for the non-linear programs can be supplied by graphic methods. Remember that, because the time-since-fire distribution $A(t)$ is being estimated, the intercept of the regression is always one. Consequently, the regression must be forced through an intercept of 1·0.

Two possible methods for comparing the time-since-fire distributions of two negative exponentials and two Weibulls are, respectively, Cox's F-test for exponential distribution (Cox, 1953) and the two-sample test of Thoman and Bain (1969). These tests and others like them are necessary for determining if fire frequency distributions are different. This is the basis of any argument that vegetation types, habitat types, landforms, etc. have different fire frequencies (see Section III.F).

E. Fire History Concepts

Besides the scale and shape parameters discussed above, a good number of fire history concepts have come into use (see Romme, 1980). Many of these concepts are vague, are difficult to relate to the fundamental distributions, and use variables which are not the parameters of distributions or logically described from distributions. Often the concepts do not specify the function and, further, do not establish the non-parametric basis of the concept.

Here we give some fire history concepts which we believe to be consistent and well defined (cf. Johnson and Van Wagner, 1985). Each concept comes as a pair, one the inverse of the other. Since by definition frequency is the inverse of return period, the fire cycle (average fire interval) is the inverse of annual per cent burn (fire frequency). Also, each concept is expressed on a per unit basis (element) or as a proportion of the whole population (universe). In practical terms the element is a sample unit while the universe is the entire study area of non-overlapping contiguous sample units.

The *fire cycle* (*FC*) (universe) is the time required to burn an area equal in size to the study area. Note that each part of the study area does not have to burn once, just an *area* equal in size to the study area. The *average fire interval* (*FI*) (element) is the expected return time per stand

$$FI = FC = \int\limits_{0}^{\infty} A(t)\mathrm{d}t \tag{10}$$

which for the Weibull is

$$= b\Gamma(1/c + 1) \tag{11}$$

where Γ is a gamma function. Values of the gamma function can be found in tables of mathematical functions. For the negative exponential, the fire cycle or average fire interval is b.

The *annual percent burn* (APB) is the proportion of the universe that burns per unit time and the *fire frequency* (FF) is the probability of an element burning per unit time

$$APB = FF = \frac{1}{b \; \Gamma \left(\frac{1}{c} + 1\right)} \qquad \text{Weibull} \tag{12}$$

$$= \frac{1}{b} \qquad \text{Negative exponential} \tag{13}$$

Both annual percent burn and fire frequency are the inverse of fire cycle and average fire interval. Fire frequency can be used to convert the time-since-fire distribution into the *Age distribution of stands across the landscape*, *A*on, (personal communication, name due to C.G. Lorimer) as follows*

$$A^*(t) = FF \exp\left[-\left(\frac{t}{b}\right)^c\right] \qquad \text{Weibull} \tag{14}$$

$$= FF \exp\left[-\left(\frac{t}{b}\right)\right] \qquad \text{Negative exponential} \tag{15}$$

The *average prospective lifetime* (APL) of an element is the average time between fires

$$APL = \frac{\int\limits_0^\infty t\,A(t)\mathrm{d}t}{\int\limits_0^\infty t\,A(t)\mathrm{d}t} = b\,\frac{\Gamma(2/c)}{\Gamma(1/c)} \qquad \text{Weibull} \tag{16}$$

$$= b \qquad \text{Negative exponential} \tag{17}$$

The average prospective lifetime is the centroid of the Weibull distribution and thus could also be called the average age of the forest.

F. Spatial and Temporal Scales of Fire Frequency

The spatial and temporal scales at which fire frequency processes operate are still poorly understood. However, it is useful to try to formulate a scheme for these scales if for no other purpose than to provide a framework for further discussion and analysis. Figure 5 shows such a tentative scale diagram. Ideally, each scale should have associated with it the processes which cause it. This tentative diagram indicates the importance of scale in identifying processes causing differences in mixed and homogeneous distributions. Homogeneous distributions assume that on average the hazard of burning is constant (constant may be somewhat ambiguous, here it means that the parameters in the hazard function do not change in value) at some spatial and temporal scale (see Johnson and Van Wagner, 1985 for assumptions). If the size of a study area is small (spatially and/or temporally), changes in hazard may not be seen because the changes could be confused with variation caused by either the larger scale recurrence of years with many fires or the smaller scale area covered by large individual fires.

The distributions discussed in the preceding pages have been homogeneous. Most empirical fire frequency studies (e.g. Heinselman, 1973; Yarie, 1981; McCune, 1983; Clark, 1990; Masters, 1990; Johnson and Larsen, 1991; Johnson, 1992) however, are mixed distributions consisting of more than one homogenous distribution, each with statistically different parameters

$$A(t) = p_1 A_1(t) + p_2 A_2(t) \qquad (18)$$

where p_i is the proportion included in ith part of the mixed distribution. These mixed distributions can be easily recognized by a change in slope in the $A(t)$ distribution (e.g. Fig. 6). Mixed distributions can arise in two general ways: (i) spatially distinct landscapes with different fire frequencies; and (ii) temporal changes in fire frequency. Some situations may have both spatial and temporal differences. Several methods, both algebraic and graphic, have been developed to partition mixed distributions into homogeneous distributions (see e.g. Lawless, 1982). Some of these will be discussed in Section IV.

The few studies which have attempted to establish the causes of mixed fire frequency distributions (Clark, 1990; Masters, 1990; Johnson and Larsen, 1991) have found them to be related to large-scale processes which affect the hazard of burning. North American boreal forest fire frequency studies (reviewed in Johnson, 1992) have found that the cooler-moister climate of the Little Ice Age from the beginning of the 1700s to the end of the 1800s seems to have caused changes in the hazard of burning.

Once homogeneous fire frequency distributions are identified, what smaller scale patterns remain that are *not* due to variation in years with many fires or large individual fires? Traditionally, small-scale differences have been thought to result from processes which control fire occurrence and spread into local areas, e.g. water barriers, exposed rock ridges, valley entrance orientation, fuel moisture, fuel types. Since by the definition of homogeneous distribution the hazards are not "different", how are these changes to be

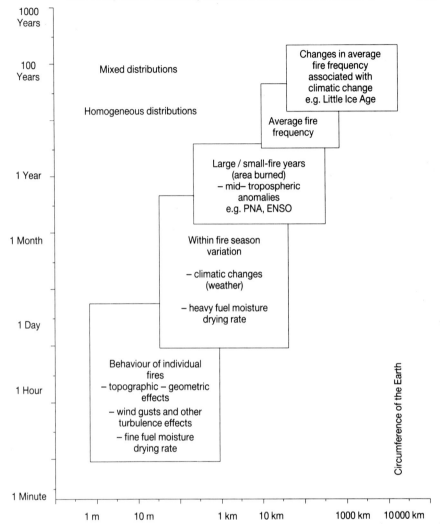

Fig. 5. Hypothetical time and space scales of important processes in fire frequency distributions.

incorporated? One approach has been to assume that the concomitant variables which cause local differences in fire occurrence affect the hazard rate in a multiplicative way (Cox, 1972). This argument can be written

$$\lambda(t; z) = \lambda_0(t) \exp(\beta' z) \qquad (19)$$

where $\lambda(t; z)$ is a hazard of burning dependent not only on time but on a vector of independent variables (z) which control local fire occurrence. $\lambda_0(t)$ is

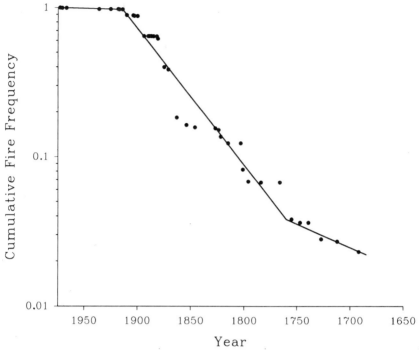

Year

Fig. 6. The time-since-fire distribution of the Boundary Waters Canoe Area (Heinsel-man, 1973). An example of a mixed fire frequency distribution and temporal partition-ing into a homogeneous fire frequency distribution (see text for explanation).

some baseline hazard rate for a landscape. This need not have any real meaning beyond being an arbitrary standard. β' is a vector of unknown regression parameters. The time-since-fire distribution conditional on z is then

$$A(t; z) = A_0(t)^{\exp \beta' z} \qquad (20)$$

where

$$A_0(t) = \exp \left(- \int_0^t \lambda_0 u \, du \right)$$

and the cumulative mortality distribution $f(t)$ conditional on z is

$$f(t; z) = \lambda_0(t) \exp (\beta'^z) \cdot \exp \left[- \exp(\beta'^z) \int_0^t \lambda_0 u \, du \right]. \qquad (21)$$

The concomitant variables z are hypotheses that certain landscape units have specific traits which allow them to have different survivorship. Simply, all we have said is that the hazard rate is *caused* by time and certain independent variables z. The variables z must be chosen so they are the mechanisms which result in the hazard of burning. Clark (1990) has been the only one, to our knowledge, who has used Cox's model to study the effects of slope and aspect on the hazard rate. He uses slope and aspect as indicators of variables that affect fire occurrence.

Another situation which may arise is when two or more independent hazards of burning compete with each other with only one occurring at any time. The hazard function under this assumption is

$$\lambda(t) = \lambda_1(t) + \lambda_2(t) \tag{22}$$

and the time-since-fire distribution is

$$A(t) = A_1(t) \cdot A_2(t) \tag{23}$$

This competing hazard model (Prentice *et al.*, 1978) would appear to be what is proposed in a study area which has frequent ground fires intermittent with few crown fires (Cooper, 1960; Weaver, 1974). However, in this case, the two competing hazards may not be independent but the occurrence of frequent ground fires could influence the occurrence of crown fires (cf. argument in Muraro, 1971). Note, however, that although opinions are strongly held about the existence of ground fire–crown fire frequencies (Houston, 1973; Brown, 1975; Arno, 1976; Tande, 1979), no published fire frequency studies to date have data to test a competing hazard model.

IV. SAMPLING DESIGN AND ANALYSIS

The basic data of fire frequency studies can be thought of as a series of maps, each of which represents the landscape mosaic of time-since-fire taken at different times in the past. As Fig. 7 shows, if these maps could be constructed at 20-year intervals into the past, the difference between the maps would be the area burned in the intervening 20 years and each map would give the time since the last fire for all parts of the study area at the time the map was made. Only the most recent fire(s) can be completely traced on each map since older fires have been progressively reduced in size as subsequent fires "overburned" them. The area in different time-since-fire ages must then reveal the progressive "overburning" of past fires by more recent fires. It is this relationship between the areas in different ages which is used to give the time-since-fire, $A(t)$, distribution. The rate of fire occurrence per unit area of the map is related to the *rate of decrease of area* covered by different time-since-fire ages. Notice that if the reoccurrence of fires is long and/or fires are always small, then the study area should have more areas of older forests. The other way of looking at the maps in Fig. 7 is to follow a sample unit through time, i.e. from map to map. The data obtained are the number of years between successive fires in each unit, i.e. the fire intervals.

In this section, we will expand on: (i) time-since-fire mapping; (ii) truncated time-since-fire mapping; (iii) time-since-fire sampling; (iv) fire interval

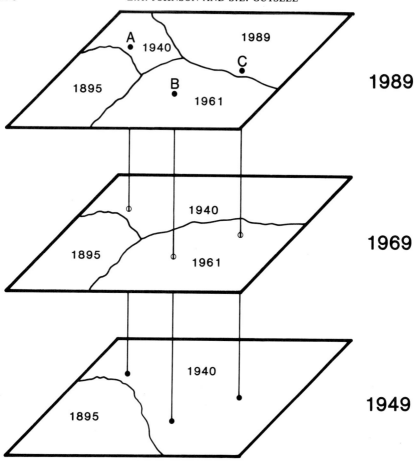

Fig. 7. A time-since-fire map at different times into the past showing the overburning of older fires by more recent fires. Points A, B and C represent random samples taken for determining fire intervals.

sampling; (v) size of landscape required for study and; (vi) how these different types of data are used to estimate fire frequency distributions and their parameters. Figure 8 gives a possible flow chart of the steps to be taken in designing and analysing the collection of fire frequency distribution data.

A. Time-since-fire Map and its Analysis

A time-since-fire map simply records the occurrence of the most recent fire in each unit of the landscape. Ideally, we would like to identify more details of the fire behaviour (fire intensity, duff consumption) in each unit area. If this were possible, then several different time-since-fire maps could be con-

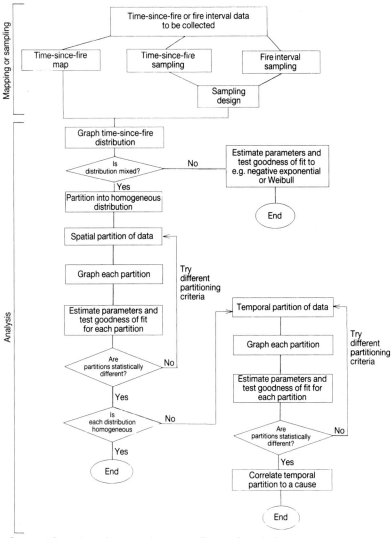

Fig. 8. General flow chart for mapping, sampling and analysis.

structed, for example, one for low intensity fires, one for high intensity fires, and also, in this case, a competing hazard model (see Section III.F). Except in very unusual cases where direct fire behaviour observations are available, these detailed maps cannot be *reconstructed* because of our poor understanding of the kinds of fire behaviour which give rise to certain fire effects, e.g. fire scars, tree survival, etc. (see Section V).

The usual method of constructing a time-since-fire map is first to use aerial photographs to obtain a preliminary identification of different aged patches (different fire dates), next to collect data in the field to determine the fire boundaries and dates of these patches, then to find patches not identified on the aerial photos, and finally to compile the aerial photo and field information into a time-since-fire map.

Aerial photographs are useful in determining some fire boundaries. Detection of these boundaries, however, depends on the difference in ages of the trees on either side of the boundary. For example, large age differences may show a distinct boundary while small age differences may not show an identifiable boundary. Also, stands that have originated from different fires cannot be differentiated by height when trees have reached their mature height. Successional sequences should not be used because they do not give actual ages and are often based on a chronosequence argument with assumptions that cannot be independently validated. Studies based on a chronosequence assume that the spatial vegetational patterns among sites represent temporal vegetational patterns at a fixed site. The underlying assumption is that there are no ecologically significant differences among sites in the spatial array except age (Jackson et al., 1988). Forest inventories which use tree height and composition to determine age should also be avoided for the reasons already given. It is also important to recognize that in detecting fire boundaries, differences in site quality, photographic scale (1:15000 is desired), quality and type of film as well as interpretational experience each play a role.

To date fires properly, all available and reliable fire evidence should be used. Different fire dates in a patch believed to be from one fire *cannot* be averaged to give a date. Dating will be discussed later in Section V.

The final time-since-fire map is constructed by transferring the fire dates and refined fire boundaries from field samples and aerial photographs to the appropriate topographic map. In many cases, fire boundaries may be related to recognizable landmarks visible on both aerial photos and the map which will improve the accuracy of the photo-to-map transfer. Most of these processes and what follows can be adapted to Geographic Information Systems (personal communication, S.R.J. Bridge and E.A. Johnson).

To estimate the individual frequencies, $A(t)$, for any age t, first determine the area burned in each age (Table 1, column 2). Next, determine the percentage of the total area that each age covers (Table 1, column 3) and finally, accumulate the percentage totals starting with 100% at the year the study was done (Table 1, column 4). Column 4 is the estimate of $A(t)$. Accumulating values from present back in time gives the progressive reduction resulting from "overburning" of past fires and hence survivorship. If we had started with the oldest date, we would have estimated $F(t)$ since $F(t) = 1 - A(t)$. Notice that column 3 is *not* an estimate of $f(t)$, but is the percentage of the total study area in age t. Consequently, it is the first step in determining the proportion of the area which has survived without fire for time t (column

Table 1
Virgin forest areas of the Boundary Waters Canoe Area, Minnesota, by stand origin
years, March 1973 (land areas only). Data from Heinselman (1973)

Fire year	Area in 1973 (acres)	Per cent of total area	Cumulative per cent of total area
1971	2032	0·5	99·5
1967	128	—	99·5
1936	7968	1·9	97·6
1925	400	0·1	97·5
1918	576	0·1	97·4
1917	1856	0·4	97
1914	32	—	97
1910	34 000	8·2	88·8
1904	1952	0·5	88·3
1903	2368	0·6	87·7
1900	512	0·1	87·6
1894	96 944	23·3	64·3
1890	32	—	64·3
1889	256	0·1	64·2
1887–8	176	—	64·2
1885–7	384	0·1	64·1
1882	288	0·1	64
1881	9968	2·4	61·6
1875	90 614	21·8	39·8
1871	5856	1·4	38·4
1863–4	83 600	20·1	18·3
1854	8112	2·0	16·3
1846	2656	0·6	15·7
1827	912	0·2	15·5
1824	1616	0·4	15·1
1822	6128	1·5	13·6
1815	5200	1·3	12·3
1803	176	—	12·3
1801	17 072	4·1	8·2
1796	5840	1·4	6·8
1784	432	0·1	6·7
1766	48	—	6·7
1755–9	12 240	2·9	3·8
1747	768	0·2	3·6
1739	160	—	3·6
1727	3408	0·8	2·8
1712	240	0·1	2·7
1692	1472	0·4	2·3
1681	8560	2·1	0·2
1648	64	—	0·2
1610	720	0·2	0·0
1595	16	—	
Total	415 782	100·0	

4). If column 3 *was* $f(t)$, it would conform to the definition given in equation (3), namely, that the value was the probability of having a fire in the interval t to $t + \Delta t$. A good number of fire frequency studies have made this mistake.

Now with an estimate of the $A(t)$ distribution, we may plot the values of $A(t)$ and t. This plotting can be done on regular (linear) graph paper; however, since we know that the data may fit either a negative exponential or Weibull model, we can use appropriate graph paper which will give a straight line if the fit to one of these models is good. Remember, negative exponential models are plotted on semilog graph paper and Weibull models are plotted on Weibull probability paper. On Weibull paper the abscissa is $\ln(t)$ and the ordinate is $\ln \ln A(t)$. These graphic methods allow a visible check of the goodness of fit of the data to the model, which enables any deviations to be recognized. Figure 6 is a plot of the $A(t)$ distribution for the data in Table 1 on semilog paper. This plot clearly shows a mixed distribution, with one break in the early 1900s and another break in the mid 1700s. Remember that mixed distributions arise when the study area is either spatially or temporally a mixture of different fire frequencies.

Next we will discuss a graphic method for partitioning mixed distributions. We believe graphic techniques are preferred, at least initially, since they show clearly how well the partitioning process is proceeding. However, this does not mean that other techniques may not, in the future, prove more efficient and useful. Partitioning can only be considered appropriate when it is shown to be supported by independent data and is consistent with fire behaviour. Just noting that a change in fire frequency occurred at the same time as something else which could have caused it does not constitute strong proof. In graphic partitioning it is best to first test for spatial divisions then for temporal divisions. Although a simple sequence is given here, in practice it is often iterative, requiring that certain steps be repeated several times and earlier steps be returned to and repeated (cf. Fig. 8).

The first step in the graphic method of partitioning data is to see if the time-since-fire map can be subdivided into spatially contiguous subareas. If so, each of these subareas will produce a homogeneous distribution. This process of subdivision must be continued until each subarea is homogeneous such that: (i) each individual homogeneous distribution gives a good statistical fit, and (ii) each subarea can be shown to have statistically different parameters. It is possible that no spatial partitioning will work or will only be partly successful so that the next step must be temporal partitioning. The temporal partitioning of a mixed distribution can be done using the graphic method of Kao (1959). The time-since-fire distribution is plotted on semilog paper. Starting at each end of the plot, a tangent line is drawn. These two lines represent the two new distributions. By tracing from the intersection of these two lines to the right, the percentage of samples in each distribution can be read. By multiplying this by the total number of samples, the number in each distribution is determined. Each of these new distributions is then plotted as a cumulative percentage of their total to give the new distribution. Finally, graphically homogeneous distributions are produced. Since temporal partitioning does not use independent data to effect the division (as was the case in spatial partitioning), it is necessary to establish which environmental variables are actually associated with the hazard change. Clearly, time cannot be the cause.

B. Truncated Time-since-fire Map and its Analysis

Fire records often include maps giving the fire's final size (see Fig. 9). If these maps have been kept for a long enough period of time so that all of the study area had been burned by fire, it is possible to construct a time-since-fire map from these individual fire maps. However, if only part of the study had burned during the time fire maps were kept, a truncated time-since-fire map and distribution can be constructed.

To the best of our knowledge, there are no examples of published studies with fire maps which have been kept long enough to produce a complete time-since-fire map. There are, however, some examples where a truncated time-since-fire map can be constructed. For example, the Los Padres National Forest in California (USA) has been keeping fire maps from 1912 to the present (here we use only to 1985) during which time 63% of the area burned. To construct the truncated time-since-fire map for Los Padres, each fire is put onto one map (e.g. Fig. 9A) and the area covered by the most recent fire outlined as the time-since-fire age (Fig. 9B). Next, the area of the total study is measured, both areas with fires in the period since 1912 and those areas with no fires. The time-since-fire areas are then calculated as a percentage of the total area and the percentages accumulated, starting with the most recent burn. Finally, the estimated truncated $A(t)$ distribution is plotted on the appropriate graph paper (Fig. 9C). Notice that the graph is truncated at 73 years, with only 63% of the area burned. Parameter estimation for truncated distributions is different from untruncated distributions. The Maximum Likelihood estimate for a Weibull can be found in Menon (1963). Here we will give the Maximum Likelihood estimator for the b parameter in a negative exponential distribution:

$$\hat{b} = \frac{(a_i x_i)^+ + \sum_{i=1}^{r} a_i x_i}{N} \tag{24}$$

where $(a_i x_i)^+$ is the time-since-fire age (x_i) in which the data is truncated (73 years in example above) multiplied by the area (a_i) in age x_i in the study which *did not* burn, $(a_i x_i)$ is the time-since-fire age multiplied by the area in each age

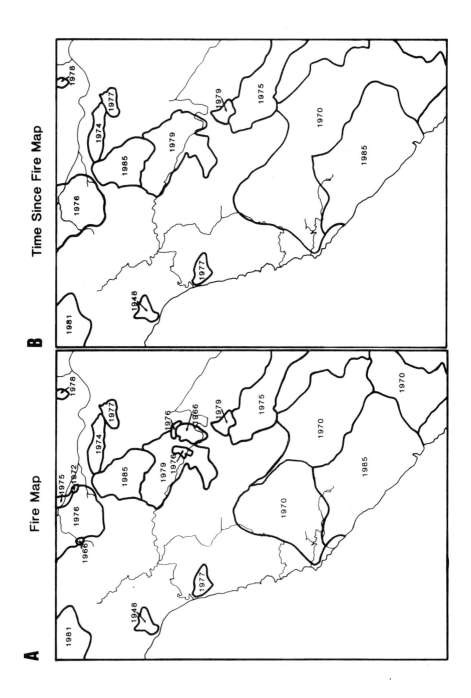

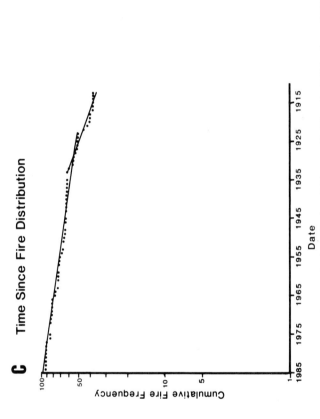

Fig. 9. Panel A is a small part of the fire maps for Los Padres National Forest, California, USA, from 1912 to 1985. It shows some fires reburning older fires and different fires occurring in the same year. Panel B shows the time-since-fire map constructed from the fire map in Panel A. Panel C is the time-since-fire distribution for the entire Los Padres National Forest. The distribution is truncated at 1912 (age 73 years). Notice the possible change in fire frequency at approximately 1930 (age 55 years).

which *did* burn, r is the total number of time-since-fire ages during the recorded period and N is the total area which burned during the study.

It is important to understand the difference between a truncated distribution and the individual components of the (temporally partitioned) mixed distribution. In a truncated distribution, the data beyond the truncation point exists but could not be obtained, i.e. there is some proportion of the whole study area which has not been included since the records do not go back far enough in time. The temporally partitioned components of mixed distributions are not truncated but the fire frequency has changed at certain times in the past. Consequently, the suvivorship has changed so that, for example, certain old stands may have been at risk under a longer average fire frequency during the Little Ice Age and are now at risk from a shorter average fire frequency.

C. Sampling for Time-since-fire

As we have just discussed, the time-since-fire map is not a sample of the study area, but rather a complete inventory. In many cases, a complete time-since-fire map may not be possible because the study area is too large, the area is difficult to access, there is a limited budget available or there is a limited time allowed for the study. In these cases, an unbiased sample of the area covered by different time-since-fire ages is required. In other words, the frequency at which different ages occur in a sample is an estimate of the area covered by these ages across the entire landscape. We will now propose a number of different sampling designs to collect statistically valid time-since-fire data. Unfortunately, most fire frequency studies have not used acceptable sampling designs (e.g. Houston, 1973; Arno, 1976; Zackrisson, 1977; Kilgore and Taylor, 1979; Talley and Griffin, 1980; Dieterich, 1983; Abrams, 1984; Jacobs *et al.*, 1985; Fisher *et al.*, 1986; Stein, 1988; Agee *et al.*, 1990; Baisan and Swetnam, 1990; Clark, 1990; Loope, 1991; Goldblum and Veblen, 1992). Cochran (1977) provides a more detailed discussion of sampling designs.

1. Simple Random Sampling

The essence of random sampling is that each unit in the study area must have an equal chance of being chosen. To accomplish this, the total study area must be divided into non-overlapping units using a sampling frame. The sampling frame is a list of all of the non-overlapping units in the study area. For example, let's assume that the sample area consists of 10 000 units of which a simple random sample of 500 units is to be drawn. These 500 units must be drawn in such a manner that every possible sample of 500 units has the same chance of being chosen. In order to do this, every unit in the sampling frame must have a number associated with it. The first unit has the number 0000 and the last unit the number 9999. Finally, 500 four-digit random numbers are chosen from a table of random numbers and then used to locate the appropriate unit in the sampling frame.

To determine the proportion of the study area in different time-since-fire ages from the simple random sample, first determine the number of units which are in a particular time-since-fire age. Next, express this number as a proportion of the total number of units sampled to give an estimate of the proportion of the study area covered by that particular age. Repeat this procedure for each age class. This gives an estimate of the values like those in column 3, Table 1. The remainder of the construction of the time-since-fire distribution follows what was given in Table 1. Remember that, as in any form of sampling, each time-since-fire frequency has sampling error associated with it.

2. Stratified Random Sampling

Simple random sampling will often require a very large number of samples to get fire frequency estimates with small standard errors. Stratified random sampling may reduce the number of samples by dividing the study area into smaller areas or strata, then randomly sampling within each of these strata. An easy way to define strata is to have a good set of air photos upon which fire boundaries can be seen. Strata can then be defined by drawing along different fire boundaries. Strata must completely cover the study area and are chosen so that they are as homogeneous as possible with respect to time-since-fire ages.

Within each stratum the proportion of area covered by each age class is estimated by simple random sampling. Each of the time-since-fire ages in the stratum is weighted by the proportion of the study area covered by that stratum. Finally, the sum of each of the weighted strata gives an estimate like the values in column 3, Table 1. Each stratum may not have all the same time-since-fire ages. Details of the calculation can be found in Cochran (1977).

3. Systematic Sampling and Other Means of Collecting Data

Systematic sampling collects data at prescribed intervals in the sample area. It does not imply subjective choice of sample units. It suffers, however, from one major limitation, that is, it is not necessarily a valid statistical design and therefore neither the precision nor accuracy of the results can be evaluated. One exception is the case where the occurrences of the time-since-fire dates on the map are already in random order so that a systematic sample is still a random sample. Of course, this is probably never true as a quick glance at any time-since-fire map reveals spatial autocorrelation between adjacent areas, i.e. fires burn contiguous areas. Consequently, systematic sampling is not generally a useful sampling technique. If it is used, justification must be given that all areas have an equal chance of being chosen, since this is the basis of statistical sampling designs.

Another technique widely used in calculating fire frequency is one proposed by Arno and Sneck (1977). Their technique involves subjective choices of areas using criteria of vegetation composition, topography, aspect, etc. Transects through the area are then used to locate *fire-scarred* trees which are then sampled. As Arno and Sneck (1977) make quite clear, this is not a statistical sampling design. Fire frequencies are usually calculated using a simple average of the fire intervals (see next section). Since this fire frequency value was not arrived at by a statistically valid sampling design, it is impossible to know how accurate and precise the calculations are. This kind of informal sampling is undesirable because it gives the impression of a valid quantitative estimate of fire frequency.

D. Fire Interval Sampling and its Analysis

An alternative approach to time-since-fire maps is to take a *random sample* and in each sample determine as many past fires as possible. Each sample will then have a record as in Fig. 3. This approach is useful, particularly in forests in which multiple event records are common (e.g. multiple fire scars). In some kinds of forests, multiple records of fires are rare and consequently this approach will not be very useful. Notice in Fig. 3 that the first and last time intervals are not *fire* intervals. By pooling the fire interval data from *all samples* a fire interval histogram can be constructed, i.e. the $f(t)$ distribution. Remembering that the time-since-fire distribution is

$$A(t) = \int_{t}^{\infty} f(t)\, \mathrm{d}t,$$

and hence $A(t)$ can be derived from the fire interval histogram by cumulative summary. This estimate of the $A(t)$ distribution is clearly different from that estimated from the time-since-fire map (in the preceding sections) in that it does not preserve the dates of fires on the abscissa. Time-since-fire maps allow the $A(t)$ to give the proportion of the study area which has gone without burning to date t (e.g. 1886). The fire interval approach gives $A(t)$ as the proportion of the area which has gone without burning to age t (e.g. 110 years). Both are survival distributions but differ in date or age on the abscissa. This is a subtle, but important difference.

If the $A(t)$ is a mixed distribution, then clearly some of the samples do not belong together, having come from areas with different hazard rates and/or at some date in the past the hazard rate changed. The techniques of partitioning for these spatial or temporal changes are different. For spatial heterogeneity, the solution is easy: simply divide the samples into subgroups according to different spatial criteria thought to be responsible for the differences in hazard rates until a good fit and significantly different hazard rates are arrived at. The spatial variable must be defined carefully and explicitly and must be related to known fire behaviour. All three of these concerns are much more difficult than may first seem.

Temporal heterogeneity can be recognized by at least two approaches. The least useful approach creates the time-since-fire distribution from the interval data for *each* sample unit. There are then as many time-since-fire distributions as sample units. Heterogeneity appears as breaks in slope as we have seen before. Clearly this approach, besides generating a large number of individual distributions, is also rarely possible because few samples will have enough fire intervals (approximately 30) to allow reasonable estimates of the distribution.

The most useful approach to detecting temporal heterogeneity is to pool the fire *occurrences* from all sample units (see Fig. 10). Assume that the fire occurrences in each sample unit are created by (independent) renewal processes (Cox and Lewis, 1966) which, when pooled, approximate a Poisson process. This would certainly be correct if each sample unit's fire frequency was a negative exponential. The pooled *intervals* are then combined into a time-since-fire distribution by

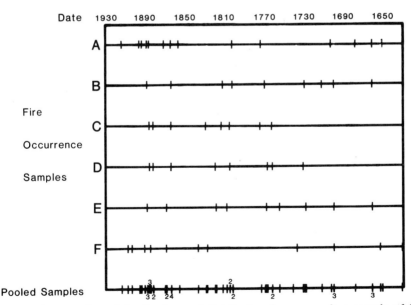

Fig. 10. Construction of the fire interval distribution from a random sample of fire occurrences from Fig. 3.

$$A(t) = \int_{t}^{\infty} f(t)\, dt,$$

graphed on appropriate graph paper and examined for evidence of a mixed distribution.

This approach is not very appealing because it requires several questionable assumptions, particularly the pooling of either the fire intervals or occurrences. It should be considered as a rather unsatisfactory approach and used with considerable care to make sure that the results can be supported by other independent data. Also, we have been unable to find a data set upon which to try these procedures.

E. Size of Landscape Required for Study

The study area of many fire frequency studies is often too small for time-since-fire maps or samples to have much meaning. As a rule of thumb, study areas in which one-third of the area has been burned by a single fire or fire year some time in the past make time-since-fire maps of limited use because of the small area covered by older fire dates. Figure 11 shows graphically the effects of a small study area and a large fire.

The deviations around the fire frequency may be the result of: (i) error in determining fire occurrence; (ii) areas that have mixed distributions; (iii) areas in which fires occur in temporal clumps; and (iv) areas that are small relative to the size of the largest fire. Only causes (iii) and (iv) will be discussed here. Cause (i) will be discussed in Section V and cause (ii) has already been discussed in Section III.F.

The area burned from year to year or even decade to decade is rarely uniform. For example, Marsden (1982) has shown that lightning fires follow a binomial distribution in time. Consistent with the assumptions of the binomial distribution, he showed that the fires occur in temporal clumps; that is, the clumps are distributed randomly in time and the number of fires in a clump is distributed logarithmically. Some years, for example, may have a large area burned while other years may have a much smaller area burned. The clumping of years with a large area burned (large fire years) is usually a result of the irregular occurrence of severe fire weather. The most common cause of severe fire weather seems to be persistent mid-tropospheric positive anomalies (blocking high pressure systems) which lead to lower precipitation and higher temperatures (Schroeder et al., 1964; Janz and Nimchuk, 1985; Knox and Hay, 1985; Flannigan and Harrington, 1988; Johnson and Wowchuk, 1993). Deviations in the fire frequency distribution may be associated with these large fire years. It must be remembered that the magnitude of these deviations will decrease over time as the large areas in large fire years are overburned by subsequent fires. The rate of decrease of the deviations is

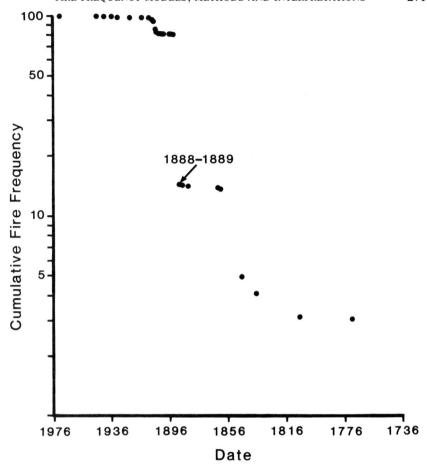

Fig. 11. A time-since-fire distribution for the Jasper townsite, Jasper National Park (Tande, 1979). The study area is 432 km^2; 78% of the area was burned by fires in 1888–89. This affect can be seen by the large drop in cumulative frequency in these years.

dependent upon the average fire frequency. With a long fire cycle, the damping of the deviation is slower than if the fire cycle is short.

The size of the study area relative to the size of the largest fire may also cause large deviations in the fire frequency or fire interval distribution. Clearly, a fire that is nearly as large as the study area will dominate the fire frequency distribution for several years. On the other hand, a study area that is very large relative to the size of the fire will show little deviation from this cause in its fire frequency distribution. As the number of different aged patches increases, the differences in age between these patches will increasingly cancel out any deviations.

V. DETERMINATION OF FIRE OCCURRENCE

Up to now we have discussed the mathematical models of fire frequency and the methods of sampling and analysing them. We will now deal with collecting the field data for fire occurrence. To do this we will divide this section into three parts: Recognizing Fire Evidence, Dating Fires, and Collecting and Preparing Tree-ring Data. Although this section may seem incidental compared to the problems of understanding the models and analysis, serious errors have resulted from incorrect determinations of fire occurrence.

A. Recognizing Fire Evidence

The most often used evidence of past fire is fire scars. Although fire scars may seem easy to recognize, they can be confused with scars resulting from other causes (e.g. Mitchell et al., 1983) such as those shown in Table 2. If the origin of a scar is in question, it should not be used. Fire scars always form at the base of the tree, usually in a triangular shape. They often form near the boundaries of a burn in active crown fires, but this is not always the case in ground fires or passive crown fires. Active crown fires have a fire front which is a wall of flame extending into the crowns, whereas passive crown fires occur when a ground fire extends into the crown and burns a single tree (Van Wagner, 1977). Trees with multiple fire scars often, but not always, have charring on the older scar faces and consequently could provide evidence that the most recent scar was caused by fire. Several investigations (e.g. McBride, 1983; McClaran, 1988) have found some evidence that a tree once scarred is more easily scarred subsequently. Consequently, unscarred and scarred trees may have different potential for scarring and hence have different fire records. Also, the speed at which healing allows closure of the scar and the time between fires interact with the potential for scarring. Clearly a good understanding of the physical processes involved in fire scar formation would be useful. Unfortunately, at the present time this understanding is incomplete (Spalt and Reifsnyder, 1962; Hare, 1965a,b; Gill and Ashton, 1968; Vines, 1968; Gill, 1974; Tunstall et al., 1976).

Without fire scars, there remain three kinds of evidence used to determine the dates of occurrence of fire. One type of evidence of fire involves the use of dendrochronology to date trees which are known to have been killed by a fire. It is important, in this case to ensure that the tree was, in fact, alive when the fire occurred; otherwise the date is clearly of no value. Trees that were alive at the time of the fire usually have little or no surface charring on the wood but the bark, if it has not sloughed off, may be heavily charred. Trees dead at the time of the fire, on the other hand, will have heavily charred wood. Trees killed by the fire, especially conifers, have their smaller branches consumed in the fire, leaving small branch stubs with burned tips. These burned tips can be

Table 2
Diagnostic features and appearance of some types of basal scars (modified from Molnar and McMinn, 1960)

Scar origin	Diagnostic features	Scar appearance
Fire	Charred bark, wood or twig ends Scar at base of tree	Triangular shape Open and charred face, or bark may gradually flake off
Mechanical (e.g. avalanche, tree fall)	Scored or gauged sapwood Broken branches around face Suitably positioned for tree fall or avalanche	Callused lower margin and sides Indication of impact
Armillaria infection (*Armillaria mellea* (Vahl:Fr.) P. Kumm)	Decayed root at scar base	Broadly triangular or rectangular in shape No healing at lower scar margin
Pole blight lesion	Scar spirals with wood grain	May extend well up bole but rarely below ground line Bark breaks as longitudinal crack with narrow lesions and peels off gradually
Mountain pine beetle (*Dendroctonus ponderosae* Hopkins)	Often with vertical "rope-like" ridges of callus tissue outside and above main scar	Usually longer than 1 m > 4 m Long scars with parallel sides, often spirally with the grain
Frost	Develops "frost-ring" on annual ring particularly in young stems (< 5 cm diameter) with thin bark	Abnormally rough and furrowed bark at tree base Accompanied by browning and death of leaves and shoots of current season, with characteristic curling of the shoot
Lightning	"Lightning" rings develop in annual ring, commonly characterized by an abnormal number of resin ducts	Long narrow strip of exposed wood and possibly a shallow layer of wood extending (sometimes spirally around the trunk) from the crown to the butt

differentiated from weathering or algal growth by rubbing the tip because on fire-caused branch stubs carbon will come off on the hand.

Another type of evidence of fire is in individual trees or groups of trees that survived the fire without fire scars but show growth release in their rings. A release is defined as an abrupt or sustained increase in radial growth which occurs in understorey trees after removal of the canopy (release from suppression) or in canopy trees when one or more neighbouring trees are removed (Lorimer and Frelich, 1989). To ensure that the ring pattern is not simply an anomaly, Lorimer and Frelich (1989) suggested the criteria of a slow growth before and rapid growth after a release be sustained for 15 years or more. Trees with dates of release from suppression can then be compared with trees which subsequently became established in open areas burned by the fire. By comparing the two, the date of the fire can then be established.

A third type of evidence used to date fire occurrence is obtained by ageing trees believed to have been recruited immediately after the fire. The major difficulty in dating a fire when no direct evidence of the fire exists is in finding the trees which recruited in the same or next year after a fire. In many cases, recruitment may not have occurred for several years and consequently, search of trees will be fruitless or will give fire dates several years after the fire. Sometimes comparison with nearby areas with good fire occurrence records (e.g. fire scars) can establish an accurate date. Probably the more realistic approach is to recognize that the dates are censored (Lawless, 1982). Censoring means that all one can say is that the date of a fire is at least as old as the "oldest" tree found but that the actual fire occurrence date is not known. Truncation, which we have previously discussed, is also a kind of censoring. The graphing of the time-since-fire distribution with this kind of censoring follows the rules given for truncated distributions. The censored dates are not used in the cumulative percentage and on the graph. Table 3 gives a simple example of how to calculate the $A(t)$ distribution for censored data. Compare the form of the calculation in Table 3 to that in Table 1 to see the difference.

B. Dating Fires

In the last section, we attempted to ensure that the data collected to determine fire occurrence were in fact a result of fire. In this section, we will be concerned with the reliability of the date of the fire. Clearly, the different evidences of fire vary in their reliability, with fire scars being the most reliable evidence.

Several methods have been proposed for evaluating how well fire evidence is replicated in and between samples. The most widely used is the method proposed by Wagener (1961), Arno and Sneck (1977), and Dieterich (1981). Dieterich calls his approach a fire master chronology. This is somewhat of an unfortunate choice of terms since it may be confused with a tree ring master chronology. A *tree ring* (width) master chronology is a long homogeneous ring-width chronology (cf. Fritts, 1976) which is constructed by dating and processing ring widths from many trees in a region. A *fire* master chronology, on the other hand, is used to evaluate the amount and type of evidence of fire occurrence *within* and sometimes between samples. Dieterich's fire master chronology portrays each tree in the sample as a time line on which is marked

Table 3

Example of how to construct a time-since-fire distribution when some dates are not time-since-fire ages but only age of the oldest tree, i.e. no exact date of fire occurrence is known. The dates shown with a cross (e.g. 1920+) are such censored dates. Compare to Table 1 for handling of uncensored dates.

Fire year*	Area in 1990 (ha)	Per cent of total area	Cumulative per cent of total area
1972	300	2·1	97·9
1965	801	5·5	92·4
1967	73	0·5	91·9
1933	2678	18·4	73·5
1920+	48	0·3	
1910	1616	11·1	62·4
1908	256	1·8	60·6
1888	32	0·2	60·4
1882	912	6·4	54·1
1853+	176	1·2	
1801	737	3·1	49·0
1732+	784	5·5	
1701+	6120	42·9	
1650+	23	0·2	
1641+	14	—	
	14 570	99·2	

* Years with + are not fire dates, but indicate that fire dates are at least as old as this date.

fire occurrence(s) and other pertinent information such as the pith. The fire master chronology uses the investigator's judgement of the reliability of certain fire events as revealed by the evidence of a sample.

Fire master chronologies cannot be used to adjust dates of individual fire occurrence which appear to be slightly misplaced from those given in other trees. Both Wagener (1961) and Arno and Sneck (1977) suggest a technique by which fire dates from years with few records are assigned the nearest year with large number of records. This confuses two sources of error: the amount and type of evidence and the accuracy of the date itself. The only accurate way to determine the correct date is to cross-date the tree against a ring-width master chronology (Madany et al., 1982).

C. Collecting and Preparing Tree-ring Data

A complete disk taken from the base of a tree is the best method for obtaining tree-ring data. Arno and Sneck (1977) proposed taking only a wedge from trees in situations where trees cannot be cut down. The limitation of this method is that the whole ring circuit cannot be examined. When fire has scarred a tree, the ring circuit near the scar is often distorted and consequently dating the scar from these rings will lead to incorrect results. The rings from the undistorted part of the ring circuit should instead be used. Barrett and Arno (1988) recognized this problem and proposed taking several cores at different angles (locations) around the tree. However, a large number of cores may seriously weaken the structural integrity of the tree. Further, increment cores do not allow the detection of locally absent rings which may lead to serious dating errors unless dendrochronological techniques are used (Madany *et al.*, 1982).

The careful sanding and cleaning of tree discs or cores is the most often ignored part of ring preparation. The techniques of sanding and surfacing discs and cores have been extensively discussed in dendrochronology texts (Stokes and Smiley, 1968; Fritts, 1976; Schweingruber, 1988). Briefly, first use a mechanical sander with coarse sandpaper and then in two or more steps work down to fine sandpaper (silicon carbide 240). The surface is finished when the whole circuit can be viewed in the case of discs and the individual cells are clearly visible in both discs and cores. Dead and decayed samples require that they be stabilized and then a clean surface prepared. An assortment of techniques have been proposed for stabilizing fragile wood. If the sample is wet, the easiest method is to simply freeze the sample in dry ice and then cut and sand the frozen surface. If the sample is dry, it must often be impregnated with a stabilizing material such as polyethylene glycol (molecular weight 600). The archaeological literature is often useful in providing techniques for preparing badly decayed samples (Cronyn, 1990). To identify dead trees, books by Core *et al.* (1979) and Hoadley (1990) will be useful.

After the disc has been properly surfaced, at least three radii on the disc should be drawn, and on each radius the rings counted and marked at each decade interval. In order to ensure that there are no locally missing or double rings, decade rings can be traced around the circuit from one marked radius to another. In situations where fire scars are present, the radii for age counting should be chosen where there is little or no distortion in the rings. From these undistorted rings, dates can then be traced back to the fire scars.

Ring counting is subject to several types of errors including counting error, missed rings, locally absent rings, and double rings in a single year. To overcome these errors, dendrochronology has developed an array of techniques which involve matching ring-width patterns (cross-dating) (see Spencer, 1964; Fritts, 1976; Swetnam *et al.*, 1985; Sheppard *et al.*, 1988). However, most fire frequency studies have used only ring counts despite graphic evidence of its possibilities for error (see Madany *et al.*, 1982). This is probably due to the time required to do cross-dating. Ring counting is not dendrochronology. In practice, most dendrochronologists quickly put together a list of ring-width patterns for a region which allows them to quickly and accurately establish the date of a particular ring. Recently, Yamaguchi (1991) has outlined this method which is useful in fire frequency studies in which there are large numbers of cores. The method is given here with very little change from Yamaguchi (1991):

(i) Choose a core that has relatively wide rings. Such cores are the least likely to contain missing rings. One does not need to worry about bias in selecting this core because in most applications, all cores will ultimately be cross-dated.

(ii) Identify the year of formation of the core's outermost ring according to its sampling date, relative to the growing season, and the degree of ring development. For example, cores collected in early summer will typically contain incompletely formed outer rings.

(iii) Starting with the core's outermost ring, count rings backward in time, labelling each decadal ring (e.g. 1990, 1980, etc.) with a single pencil dot, each half-century ring (e.g. 1950) with two pencil dots, and each century ring (e.g. 1900) with three pencil dots. Mechanical pencils with soft (HB) and fine (0·3 or 0·5 mm) lead are especially useful for counting and marking rings.

(iv) Double-check initial ring counts. (It is useful to put light "check marks" on the decadal rings of checked decadal intervals.) Also, label century rings on the mount adjacent to the core (e.g. 19 for 1900, 18 for 1800, etc.).

(v) List the years of formation of rings that are noticeably narrower than adjoining rings, as shown in Table 4 (under Wide-ring cores).

(vi) Repeat steps ii–vi with cores from four or more trees (Table 4, Wide-ring cores). If there is any uncertainty in assigning calendar years to rings in individual cores, (e.g. caused by breaks in cores, fine rings, apparent missing rings, etc.) postpone work on these cores until step ix.

(vii) Underline narrow rings shared by most (e.g. present in at least three out of five) or all cores. These rings are the local climatically controlled marker rings (Table 4, Marker rings).

(viii) Use the list of marker rings to assign calendar-year dates to rings in additional cores. Do this by repeating steps ii–v while continuing to work gradually from cores from fast-growing trees to cores from slow-growing trees. Check each core for the presence or absence of previously identified marker rings (Table 4, Cores containing fine or missing rings). Follow steps vi–viii to identify other tentative marker rings in the inner portions or older cores (Table 4, Cores containing fine or missing rings).

(ix) Date difficult cores against the extended narrow-ring record. For these cores, the presence of marker rings should be noted while ring counting because of the possible occurrence of missing rings. For cores where dating of rings is equivocal, it is helpful to compare two or more cores from the same tree and to make detailed notes on ring-width patterns

Table 4
Example of cross-dating using the list method (from Yamaguchi, 1991)

Tree	Core	Notes and tentative marker rings[a]
		Wide-ring cores
9	B	1989e, *84*, 79, *74*, *66*, *54vn*, *25*, 22, 17, 08;
		1899, 90, *87*, *80vn*, 76, *63*; (1843)
4	A	1989e, *84*, 71, *66*, 63, *54*, *44*, *35*, 32, *25*, 19;
		1899, *87*, *80vn*, 42; (1835)
2	A	1989e, 78, *74*, 70, *66*, 64vn, *54vn*, 49, 44, 40, *35*, 33, *25vn*, 22, 20, 18;
		1880mc, 76, 72, 68, 51, 42; (1827)
10	B	1989e, *84*, 77, *74*, 63, *54vn*, 46, *35*, *25vn*;
		1880vn, 68, *63*, 56, 51; (1826)
6	B	1989e, 77, *66vn*, *54mc*, 46, *35vn*, 32, *25vn*, 18;
		1887, *80mc*, 75, 67, *63*, 56, 44; (1824)
		Marker rings shared by most cores
		1989e, 84, 74, 66, 54vn, 35, 25vn;
		1887, 80vn, 63.
		Cores containing fine or missing rings
3	A	Rings clearly visible;
		1989e, *84*, 74, *35*, *25*;
		1880vn, *63*, 59, 55, 45, *42vn*, *34*, *21vn*, 13, 07, *04*;
		1798, 95, 91, 88, 85, 82, 77, 75, 72, *56f*, 48, 43vn, 41f; (1738)
3	B	Rings clearly visible; 1821M ring recognized from early 1-year offsets from core A;
		1989e, *84*, *25vn*;
		1880vn, *63*, 56, 51, *42*, *34*, 29, *21M*, 13, 11, 07;
		1798, 75, 70, 67, *60*, 55; (1745)
11	C	Rings narrow but clear; core contains two breaks in the 20th century but these appear to be properly joined;
		1984, *66*, *54vn*, *25*;
		1880p, *63*, 56, 48, *42vn*, *34*, 29, *21M*, 12, 04vn;
		1798, 91, 88, 70vn, 65vn, *60vn*, 58, 43vn; (1728)
11	B	Rings narrow but clear; "1863" and earlier marker rings offset by 1 year (decadal dots placed one ring early); 1881 ring is also mc, thus, cross-dated by assigning 1880 ring as M. This follows logically from it being p in pair core C;
		1984, *66*, *54mc*, *25*;
		1880M, *63*, *42vn*, *21p*, *04vn*;
		1798, 88, 70, 65, *60*, 58, *56f*, break, 46vn; (1743)
1	A	Rings narrow but clear; "1863" and earlier rings offset by 1 year (decadal dots placed one ring early); 1881 ring is also mc, thus cross-dated by assigning 1880 ring as M. This follows logically from it being vn or M in other cores;
		1989e, *84*, *66*, *54*;
		1887, *80M*, *63*, *42*, *34*, 31, 29, *21vn*, *04vn*;
		1798, 88, 85, 78, 70f, *60*, 58, *56f*, 43vn, 39, 31; (1730)

Table 4 (continued)

Tree	Core	Notes and tentative marker rings[a]
1	B	Rings narrow but clear; "1863" and earlier rings offset by 1 year (decadal dots placed one ring early); 1881 also vn; dated by assigning 1880 as M; *1984, 66vn, 54;* *1880M, 63, 42, 34,* 29, *21M, 04vn;* 1788, 71vn, 66, *60vn,* 45vn; (1734).
8	A	Rings narrow but clear; "1863" and earlier rings offset by 1 year; dated by assigning 1880 as M; 1989e, *84vn, 74, 54,* 25; *1887mc, 80M, 63, 42mc, 21,* 05mc; 1790, 83, 72p, 62vn, 49vn, 28, 16, 14, 06; 1691, 74vn; (1665)
		Revised marker rings[b] 1989e, 84, 74, 66, 54n, or vn; 35, 25n, or vn; 1880vn, mc, or M; 63, 42n or vn; 34, 21vn, p or M; 04vn; 1798, 60, 56f (occasionally)

[a] Listed rings are relatively narrow unless otherwise distinguished. To save space, only the last narrow ring of each century is listed in full. Italic rings are shared marker rings. Abbreviations: e, earlywood only present (an artefact of the growing-season sampling date that provides positive ring identification; n, narrow ring; vn, very narrow; mc, microring only a few cells wide; p, partial ring (not present along entire width of core); M, missing ring; f, false (double) ring. The year listed in parentheses is the year of formation of the innermost ring (latewood) in the core.
[b] Need to examine additional cores to establish earlier marker rings.

(Table 4, Cores containing fine or missing rings). The presence and approximate locations of missing rings, or more rarely false rings, are shown by the start of ring counts that are consistently offset, generally by 1 year, from marker rings. The direction of offset reveals whether missing or false rings are the cause. If a missing ring is detected, by convention it is assigned to the year of formation of the narrowest ring found in replicate cores or in other trees near the start of offset. By following these steps, one should be able to firmly establish which rings are reliable marker rings. Revise and extend the list of marker rings as needed (Table 4, Revised marker rings).

Another common problem with using increment cores from trees compared to discs is that the centre is sometimes missed. A large number of sometimes intricate approaches have been used (Ghent, 1955) to *guess* at the number of missed rings. It would seem best to examine each core in the field and, if the centre has been missed, to simply re-core the tree.

When the date of fire occurrence is, of necessity, based only on total tree age and not fire scars, investigators have often added corrections to the total tree age. Corrections have been added to take into account the fact that the tree was not sectioned or cored at the base. The correction is often taken as the intercept of the regression of "age at coring height" (independent variable) on "age at base" (dependent variable). Unfortunately, the variance around the regression line is often large and the slope is often not one, i.e. not parallel to the 45° line. In these cases, the correction (intercept) gives a false sense of accuracy. We feel it is better in situations where total tree age is required to core as close to the ground as possible using a rachet* instead of the normal increment borer handle. Also note that in areas in which no fire dates are certain (i.e. usually not fire scars) the date is censored (see Section IV.A).

VI. DISCUSSION

The principal difficulties in fire frequency studies have been: how to think about fire frequency, how to collect the data properly, how to locate the fire boundaries and recognize fire evidence, how to date fires carefully and what to do when exact dates of fire occurrence are not possible. In the remainder of this paper, we will consider a number of issues which have not yet been discussed but which appear in a number of fire frequency studies. Each issue discussed has one or more of the difficulties just mentioned.

Besides the approaches to fire frequency studies given in the preceding pages, there is another approach which we will call "fire chronologies". Fire chronology studies estimate the mean fire interval as the average of all fire intervals during a designated time and area (cf. Romme, 1980). The field procedures involve collecting fire scar data from trees encountered in a study area. The fire occurrence records are determined from individual fire-scarred trees (individual fire chronologies) and are then combined into a master fire chronology. This master fire chronology gives the complete record of fires that have occurred in the whole study area. The mean fire interval is then calculated from the master fire chronology by taking the sum of the intervals and dividing by the number of intervals. Notice that the interval "present to last fire" and "first fire to tree pith" are not necessarily fire intervals and consequently should not be included in the mean fire interval calculation. This elimination of the interval "present to last fire" has implications for interpretation when fire suppression effects are completely contained in this time period. The mean fire intervals for different but similarly defined study areas are often compared informally to see if they are different or by some statistical method (e.g. regression, ANOVA, or cluster analysis) to see if the mean fire intervals are related to some environmental or habitat variables.

The mean fire interval is the estimation method of the parameter of a fire

* An alternative is to cut off half of the handle of an increment borer so that the corer can be inserted very close to the ground level.

frequency distribution. The fire frequency model being estimated (e.g. negative exponential) is often not stated (e.g. Arno and Sneck, 1977; Romme, 1980) even though an explicit parameter estimation technique is being used. If the mean fire interval technique is a non-parametric estimator, no justification has ever been given.

In some fire chronology studies, the master fire chronology is assumed to be a $f(t)$ distribution of the negative exponential or Weibull model (Clark, 1990). In many cases, the number of intervals is small ($n < 20$) and consequently estimation errors are so large that the value of the estimates is very limited. Goodness of fit is often not given, only the parameters estimated without standard errors. When the $f(t)$ distribution is the negative exponential, the mean fire interval is the appropriate estimator of parameter b. However, if the $f(t)$ distribution is the Weibull, the mean fire interval is not the appropriate estimator of parameter b, rather the Maximum Likelihood estimates of b and c given in equations (9a) and (9b) should be used (or if the average fire interval is desired use equation 11).

Estimating only the mean fire interval and not plotting the $f(t)$ distribution often means that mixed distributions can go undetected. Sometimes mixed distributions have been assumed (e.g. Tande, 1979) and the intervals divided into temporal or spatial units without empirical basis in the distribution and without statistical (null) testing of the divisions (see Section III.F).

Finally, the most serious problem in fire chronology studies, which invalidates most of them, is that no sampling design is used in collecting individual fire chronologies. In most studies, sampling is subjective or systematic and all areas or fire-scarred trees do not have equal chance of being chosen. All estimation of statistical distributions or their parameters assume random sampling. If sampling is not random, the distribution is useless since bias is unknown and unknowable. Notice also that by not having a sampling design the estimation of the mean fire interval is often done incorrectly (see Section IV.D).

One of the most difficult aspects of fire frequency studies has been understanding what frequency means. This problem appears in several forms. Arno and Petersen (1983) point out that the mean fire interval (MFI) has a variance which may be as important in understanding fire frequency as the mean. The more general way of looking at this issue is to realize that any of the fire frequency distributions discussed ($A(t)$, $F(t)$ and $f(t)$) all graphically show the variation, i.e. distribution of frequencies. For example, the time-since-fire distribution $A(t)$ shows the variation in survivorship.

Although the scale and shape parameters are easy ways to compare the relative differences in fire occurrence, they can lead to misunderstandings and confusion. The scale parameter of a negative exponential, for example, might have a value of 100 years. This does not mean that a fire will recur every 100 years. Instead, by looking at a frequency distribution such as $A(t)$, the return time can be interpreted either on a study area basis (see Section III.E) or on a per element basis. Both interpretations will be equivalent. The study area interpretation says that if 1% of the area burned every year, then in 100 years an area equal to the study area would have burned. On a per element basis, the element is part of a cohort of elements all tracing the same survivorship (homogeneous distribution). An element which has survived to 100 years and which has a parameter $b = 100$ results in $A(t) = \exp(-100/100) = \exp(-1) = 36 \cdot 8\%$. In other words, 37% of the study area is equal to or older than 100 years.

The difference between the time axes in $A(t)$ and $f(t)$ distributions can cause confusion (e.g. Heinselman, 1981) since both are dimensioned in time. The $A(t)$, being a survivorship, follows a hypothetical cohort from $t = 0$ when it had 100% of its members (i.e. study area) and then follows the percentage reduction in area as burning occurs over time. The time t is the "running" time. On the other hand, $f(t)$ is the probability of interval between fires of length $t + \Delta t$. Consequently, the time t in $f(t)$ is an interval. Recognizing that the t in $f(t)$ are intervals, not the running time, clarifies most of the confusion. Also, the fire mortality $f(t)$ is mistaken for the hazard rate or the instantaneous rate of mortality. For example, Clark (1989) and others have interpreted the $t = 0$ in $f(t)$ as being unrealistic for a negative exponential because it suggests the highest probability of fire occurs at time zero. This appears to be both a confusion of the running time and intervals, and the meaning of $f(t)$ and $\lambda(t)$. In the negative exponential, the greater probability of fire occurring in shorter time intervals as opposed to longer time intervals is *not* due to a change in the hazard of burning (instantaneous rate of burning) which is constant. Instead, it is due to the annual *depletion* of the hypothetical cohorts. The Weibull $f(t)$ does not decrease (except when $c = 1$) because the hazard of burning increases from $t = 0$. Both the $A(t)$ and $f(t)$ are driven by $\lambda(t)$. If this is not clear, examine again the definition and equations in Section III.B.

Fire frequency investigations require an understanding of the heat transfer processes which operate in fires, and field experience with fire behaviour. Ecologists often have neither of these experiences because heat transfer is outside their normal training and large wildfires are infrequent. The recognition of fire evidence both for dating and determining fire age boundaries requires a good knowledge of how the fires behave in the study area. No amount of post-fire examination of evidence will give a picture of the actual fire intensity ($kW\,m^{-1}$), rate of spread, duff and larger fuel consumption, and pattern of vegetation survival as related to this fire behaviour.

Also, small fires that occur every few years are no substitution for the large fires. Years in which small areas burn have fires which burn with low intensities, slow rates of spread, and low duff consumption. They make up insignificant areas in most time-since-fire maps. Why? Because large fire years have fires with high intensities and high rates of spread which mean they quickly cover large areas. A good example of this is the decade of the 1980s in Canada. This period contains three of the largest area burned years since records were started in 1917. In this decade, 95% of the area burned was caused by 3% of the fires (personal communication, B. Stock). This rule appears in most fire records (Strauss *et al.*, 1989).

It is also important to understand the weather conditions before and during fires which determine the pattern on the time-since-fire map. This knowledge of the local and large-scale weather will often be useful in explaining the patterns in the fire frequency. For example, in the southern Canadian Rockies the negative exponential fire frequency model fits time-since-fire map data very well. This suggests a constant (age-independent) hazard rate. Although this does not at first seem reasonable, examination of the weather conditions seems to offer an explanation. Large fires are preceded by positive 50 kPa pressure height anomalies which cover a large (1000s km) area and which result in extreme fuel drying (Fryer and Johnson, 1988; Johnson and Wowchuk, 1993). The extreme fuel drying means that fuel differences play an insignificant role in fire behaviour. The consequence may explain time-since-fire map patterns which show no stand age, fuel type, or elevational pattern.

ACKNOWLEDGEMENTS

The following people read and offered helpful and critical comments: Y. Bergeron, A.H. Fitter, C. Dymond, K. Miyanishi, T.W. Swetnam, W.A. Romme, C.E. Van Wagner, J.M.H. Weir, and P. Woodard. We also thank F. Davis of the University of California—Santa Barbara and the staff of Los Padres National Forest for supplying unpublished data. Funding was provided by the Natural Sciences and Engineering Research Council of Canada to EAJ.

REFERENCES

Abrams, M.D. (1984). Fire history of oak gallery forests in northeast Kansas tallgrass prairie. *Am. Midl. Nat.* **114**, 188–191.

Agee, J.K., Finney, M. and De Gouvenain, R. (1990). Forest fire history of Desolation Peak, Washington. *Can. J. For. Res.* **20**, 350–356.

D'Agostino, R.B. and Stephens, M.A. (1986). *Goodness-of-fit Techniques.* Marcel Dekker, New York.

Arno, S.F. (1976). The historical role of fire on the Bitterroot National Forest. USDA Forest Service Research Paper INT–187.

Arno, S.F. and Petersen, T.D. (1983). Variation in estimates of fire intervals: a closer look at fire history on the Bitterroot National Forest. USDA Forest Service Research Paper INT–301.

Arno, S.F. and Sneck, K.M. (1977). A method of determining fire history in coniferous forests of the Mountain West. USDA Forest Service General Technical Report INT–42.

284 E.A. JOHNSON AND S.L. GUTSELL

Baisan, C.H. and Swetnam, T.W. (1990). Fire history on a desert mountain range: Rincon Mountain Wilderness, Arizona, USA. *Can. J. For. Res.* **20**, 1559–1569.

Barrett, S.W. and Arno, S.F. (1988). Increment-borer methods for determining fire history in coniferous forests. USDA Forest Service General Technical Report INT-244.

Begon, M. and Mortimer, M. (1986). *Population Ecology: A Unified Study of Animals and Plants.* Sinauer Assoc. Inc., Massachusetts.

Bormann, F.H. and Likens, G.E. (1979). Catastrophic disturbance and the steady state in northern hardwood forests. *Am. Sci.* **67**, 660–669.

Brown, J.K. (1975). Fire cycles and community dynamics in lodgepole pine forests. In: *Management of Lodgepole Pine Ecosystems: Symposium Proceedings*, pp. 429–456. Washington State University Cooperative Extension Service, Pullman, WA.

Clark, J.S. (1989). Ecological disturbances as a renewal process: theory and application to fire history. *Oikos* **56**, 17–30.

Clark, J.S. (1990). Fire and climate change during the last 750 years in northwestern Minnesota. *Ecol. Monogr.* **60**, 135–159.

Cochran, W.G. (1977). *Sampling Techniques.* Wiley, New York.

Cohen, A.C., Jr (1965). Maximum Likelihood estimation on the Weibull distribution based on complete and on censored samples. *Technometrics* **7**, 579–588.

Connell, J.H. (1978). Diversity in tropical rain forests and coral reefs. *Am. Assoc. Adv. Sci.* **199**, 1302–1310.

Cooper, C.F. (1960). Changes in vegetation, structure and growth of southwestern pine forests since white settlement. *Ecol. Monogr.* **30**, 129–163.

Core, H.A., Côté, W.A. and Day, A.C. (1979). *Wood Structure and Identification.* Syracuse University Press, NY.

Cox, D.R. (1953). Some simple tests for Poisson variates. *Biometrika* **40**, 354–360.

Cox, D.R. (1972). Regression models and life-tables. *R. Stat. Soc. J.* **34**, 187–220.

Cox, D.R. and Lewis, P.A. (1966). *The Statistical Analysis of a Series of Events.* Methuen, London.

Cronyn, J.M. (1990). *The Elements of Archeological Conservation.* Routledge, London.

Denslow, J.S. (1987). Tropical rainforest gaps and tree species diversity. *Ann. Rev. Ecol. Syst.* **18**, 431–451.

Dieterich, J.H. (1981). The composite fire interval—a tool for more accurate interpretation of fire history. USDA Forest Service General Technical Report RM-81.

Dieterich, J.H. (1983). Fire history of southwestern mixed conifer: a case study. *Ecol. Mgt.* **6**, 13–31.

Fisher, R.F., Jenkins, M.J. and Fisher, W.J. (1986). Fire and the prairie-forest mosaic of Devils Tower National Monument. *Am. Midl. Nat.* **117**, 250–257.

Flannigan, M.D. and Harrington, J.B. (1988). A study of the relation of meteorological variables to monthly provincial area burned by wildfire in Canada (1953–1980). *J. Appl. Meteorol.* **27**, 441–452.

Fritts, H.C. (1976). *Tree Rings and Climate.* Academic Press, NY.

Fryer, G.I. and Johnson, E.A. (1988). Reconstructing fire behaviour and fire effects in a subalpine forest. *J. Appl. Ecol.* **25**, 1063–1072.

Ghent, A.W. (1955). A guide for the re-alignment of off-center increment borings. *For. Chron.* **31**, 353–355.

Gill, A.M. (1974). Towards an understanding of fire-scar formation: field observations and laboratory simulation. *For. Sci.* **20**, 198–205.

Gill, A.M. and Ashton, D.H. (1968). The role of bark type in relative tolerance to fire of three central Victorian Eucalypts. *Aust. J. Bot.* **16**, 491–498.

Gilpin, M.E. (1987). Spatial structure and population viability. In: *Viable Populations for Conservation* (Ed. by M.E. Soulé), pp. 125–139. Cambridge University Press, Cambridge.

Goldblum, D. and Veblen, T.T. (1992). Fire history of a Ponderosa pine/Douglas fir forest in the Colorado Front Range. *Phys. Geog.* **13**, 133–148.

Hanski, I. (1985). Single-species spatial dynamics may contribute to long-term rarity and commonness. *Ecology* **66**, 335–343.

Hare, R.C. (1965a). Contribution of bark to fire resistance of southern trees. *J. For.* **63**, 248–251.

Hare, R.C. (1965b). Bark surface and cambium temperatures in simulated forest fires. *J. For.* **63**, 437–440.

Heinselman, M.L. (1973). Fire in the virgin forests of the Boundary Waters Canoe Area, Minnesota. *Q. Res.* **3**, 329–382.

Heinselman, M.L. (1981). Fire intensity and frequency as factors in the distribution and structure of northern ecosystems. In: *Fire Regimes and Ecosystem Properties*, pp. 7–57. USDA Forest Service (General Technical Report WO–26).

Hoadley, R.B. (1990). *Identifying Wood*. The Taunton Press. Newtown, CT.

Houston, D.B. (1973). Wildfires in Northern Yellowstone Park. *Ecology* **54**, 1111–1117.

Jackson, S.T., Futyma, R.T. and Wilcox, D.A. (1988). A paleoecological test of a classical hydrosere in the Lake Michigan sand dunes. *Ecology* **69**, 928–936.

Jacobs, D.F., Cole, D.W. and McBride, J.R. (1985). Fire history and perpetuation of natural coast redwood ecosystems. *J. For.* **83**, 494–497.

Janz, B. and Nimchuk, N. (1985). The 500-mb chart—a useful fire management tool. In: *Proceedings of the Eighth National Conference on Fire and Forest Meteorology*, pp. 233–238. American Meteorological Society, Detroit, Michigan.

Johnson, E.A. (1979). The relative importance of snow avalanche disturbance thinning on canopy plant populations. *Ecology* **68**, 43–53.

Johnson, E.A. (1992). *Fire and Vegetation Dynamics: Studies from the North American Boreal Forest*. Cambridge University Press, Cambridge.

Johnson, E.A. and Larsen, C.P.S. (1991). Climatically induced change in fire frequency in the southern Canadian Rockies. *Ecology* **72**, 194–201.

Johnson, E.A. and Van Wagner, C.E. (1985). The theory and use of two fire history models. *Can. J. For. Res.* **15**, 214–220.

Johnson, E.A. and Wowchuk, D.R. (1993). Wildfires in the southern Canadian Rocky Mountains and their relationship to mid-tropospheric anomalies. *Can. J. For. Res.* **23**, 1213–1222.

Kao, J.H.K. (1959). A graphical estimation of mixed Weibull parameters in life testing electron tubes. *Technometrics* **1**, 389–407.

Kilgore, B.M. and Taylor, D. (1979). Fire history of a sequoia-mixed conifer forest. *Ecology* **60**, 129–142.

Knox, J.L. and Hay, J.E. (1985). Blocking signatures in the Northern Hemisphere: frequency distribution and interpretation. *J. Climatol.* **5**, 1–16.

Lawless, J.F. (1982). *Statistical Models and Methods for Lifetime Data*. John Wiley & Sons, Toronto, Ontario.

Loope, W.L. (1991). Interrelationships of fire history, land use history, and landscape pattern within Pictured Rocks National Lakeshore, Michigan. *Can. Field Nat.* **105**, 18–28.

Lorimer, C.G. and Frelich, L.E. (1989). A methodology for estimating canopy disturbance frequency and intensity in dense temperate forests. *Can. J. For. Res.* **19**, 651–663.

Madany, M.H., Swetnam, T.W. and West, N.E. (1982). Comparison of two approaches for determining fire dates from tree scars. *For. Sci.* **28**, 856–861.

Marsden, M.A. (1982). A statistical analysis of the frequency of lightning-caused forest fires. *Environ. Mgt* **14**, 149–159.

Masters, A.M. (1990). Changes in forest fire frequency in Kootenay National Park, Canadian Rockies. *Can. J. Bot.* **68**, 1763–1767.

McBride, J.R. (1983). Analysis of tree rings and fire scars to establish fire history. *Tree Ring Bull.* **43**, 51–67.

McClaran, M.P. (1988). Comparison of fire history estimates between open-scarred and intact *Quercus douglasii*. *Am. Midl. Nat.* **120**, 432–435.

McCune, B. (1983). Fire frequency reduced two orders of magnitude in the Bitterroot Canyons, Montana. *Can. J. For. Res.* **13**, 212–218.

Menon, M.V. (1963). Estimation of the shape and scale parameters of the Weibull distribution. *Technometrics* **5**, 175–182.

Mitchell, R.G., Martin, R.E. and Stuart, J. (1983). Catfaces on lodgepole pine—fire scars or strip kills by the mountain pine beetle? *J. For.* **8**, 598–601.

Molnar, A.C. and McMinn, R.G. (1960). The origin of basal scars in the British Columbia interior white pine type. *For. Chron.* **36**, 50–60.

Muraro, S.J. (1971). The lodgepole pine fuel complex. Canadian Forestry Service Information Report BC–X–53.

Nelson, W. (1982). *Applied Life Data Analysis*. John Wiley & Sons, Toronto, Ontario.

Pickett, S.T.A. and White, P.S. (Eds) (1984). *The Ecology of Natural Disturbance and Patch Dynamics*. Academic Press, New York.

Prentice, R.L., Kalbfleisch, J.D., Peterson, A.V., Flournou, N. Jr, Farewell, V.T. and Breslow, N.E. (1978). The analysis of failure times in the presence of competing risks. *Biometrics* **34**, 541–554.

Romme, W.A. (1980). Fire history terminology: a report of the Ad Hoc committee. In: *Fire History Workshop: Symposium Proceedings*, pp. 135–137. USDA Forest Service General Technical Report RM-81, Tucson, AZ.

Schroeder, M.J., Glovinsky, M. and Hendricks, V.H. (1964). Synoptic weather types associated with critical fire weather. USDA, Forest Service, Berkeley, CA.

Schweingruber, F.H. (1988). *Tree Rings*. D. Reidel Publishing Company, Holland.

Sheppard, P.R., Means, J.E. and Lassoic, J.P. (1988). Cross-dating cores as a non-destructive method for dating living, scarred trees. *For. Sci.* **34**, 781–789.

Shugart, H.H. (1984). *A Theory of Forest Dynamics: The Ecological Implications of Forest Succession Models*. Springer-Verlag, New York.

Smith, W.L. and Leadbetter, M.R. (1963). On the renewal function for the Weibull distribution. *Technometrics* **5**, 393–395.

Sousa, W.P. (1984). The role of disturbance in natural communities. *Ann. Rev. Ecol. Syst.* **15**, 353–391.

Spalt, D.W. and Reifsnyder, W.E. (1962). Bark characteristics and fire resistance: a literature survey. USDA Forest Service Occasional Paper 193.

Spencer, D.A. (1964). Porcupine population fluctuations in past centuries revealed by dendrochronology. *J. Appl. Ecol.* **1**, 127–149.

Sprugel, D.G. (1976). Dynamic structure of wave-regenerated *Abies balsamea* forests in the northeastern United States. *J. Ecol.* **64**, 889–911.

Spurr, S.H. (1954). The forests of Itasca in the 19th century as related to fire. *Ecology* **35**, 21–25.

Stein, S.J. (1988). Fire history of the Paunsaugunt Plateau in southern Utah. *Great Basin Nat.* **48**, 58–63.

Stokes, M.A. and Smiley, T.L. (1968). *An Introduction to Tree Ring Dating.* University of Chicago Press, Chicago.

Strauss, D., Bednar, L. and Mees, R. (1989). Do one percent of forest fires cause ninety-nine percent of the damage? *For. Sci.* **35**, 319–328.

Swetnam, T.W., Thompson, M.A. and Sutherland, E.K. (1985). Using dendrochronology to measure radial growth of defoliated trees. USDA Forest Service Agriculture Handbook 639.

Talley, S.N., and Griffin, J.R. (1980). Fire ecology of a montane pine forest Junipero Sierra Peak, California. *Mondroño* **27**, 49–60.

Tande, G.F. (1979). Fire history and vegetation pattern of coniferous forests in Jasper National Park, Alberta. *Can. J. Bot.* **57**, 1912–1931.

Thoman, D.R. and Bain, L.J. (1969). Two sample tests in the Weibull distribution. *Technometrics* **11**, 805–815.

Tunstall, B.R., Walker, J. and Gill, A.M. (1976). Temperature distribution around synthetic trees during grass fires. *For. Sci.* **22**, 269–276.

Van Wagner, C.E. (1977). Conditions for the start and spread of crown fire. *Can. J. For. Res.* **7**, 23–34.

Van Wagner, C.E. (1978). Age class distribution and the forest fire cycle. *Can. J. For. Res.* **8**, 220–227.

Vines, R.G. (1968). Heat transfer through bark and the resistance of trees to fire. *Aust. J. Bot.* **16**, 499–514.

Wagener, W.W. (1961). Guidelines for estimating tree survival of fire-damaged trees in California. USDA Forest Service Miscellaneous Paper 60.

Weaver, H. (1974). The effects of fire on temperate forests: western U.S. In: *Fire and Ecosystems* (Ed. by T.T. Kozlowski and C.E. Ahlgren), pp. 279–319. Academic Press, NY.

Weins, J.A. (1976). Population responses to patchy environments. *Ann. Rev. Ecol. Syst.* **7**, 81–120.

White, P.S. (1979). Pattern, process, and natural disturbance in vegetation. *Bot. Rev.* **45**, 229–299.

Whittaker, R.H. and Levin, S.A. (1977). The role of mosaic phenomena in natural communities. *Theor. Pop. Biol.* **12**, 117–139.

Yamaguchi, D.K. (1991). A simple method for cross-dating increment cores from living trees. *Can. J. For. Res.* **21**, 414–416.

Yarie, J. (1981). Forest fire cycles and lifetables: a case study from interior Alaska. *Can. J. For. Res.* **11**, 554–562.

Zackrisson, O. (1977). Influence of forest fires on the North Swedish boreal forest. *Oikos* **29**, 22–32.

Index

Advances in Ecological Research
Volumes 1–24

Cumulative List of Titles